Electric Vehicles in Smart Grids: Integration, Optimization, and Sustainability

Electric Vehicles in Smart Grids: Integration, Optimization, and Sustainability

Guest Editors

Chun Sing Lai
Xi Chen

MDPI

Basel • Beijing • Wuhan • Barcelona • Belgrade • Novi Sad • Cluj • Manchester

Guest Editors
Chun Sing Lai
School of Electrical and
Information Engineering
Tianjin University
Tianjin
China

Xi Chen
GEIRI North America
San Jose
USA

Editorial Office
MDPI AG
Grosspeteranlage 5
4052 Basel, Switzerland

This is a reprint of the Special Issue, published open access by the journal *World Electric Vehicle Journal* (ISSN 2032-6653), freely accessible at: https://www.mdpi.com/journal/wevj/special_issues/8W0FGLE173.

For citation purposes, cite each article independently as indicated on the article page online and as indicated below:

Lastname, A.A.; Lastname, B.B. Article Title. *Journal Name* **Year**, *Volume Number*, Page Range.

ISBN 978-3-7258-7683-9 (Hbk)
ISBN 978-3-7258-7684-6 (PDF)
https://doi.org/10.3390/books978-3-7258-7684-6

Contents

About the Editors

Chun Sing Lai

Chun Sing Lai is currently a Full (Yingcai) Professor with the School of Electrical and Information Engineering at Tianjin University. His research interests are smart grid, energy system modeling, computational intelligence, and electric vehicle systems. He was Conference Chair for the 2024 5th International Conference on Intelligent Computing and Human–Computer Interaction, sponsored by the IEEE. He is a member of the International Advisory Committee of the 4th International Conference on Smart City and Green Energy, Sydney, Australia. He is a Senior Editor for *IEEE Transactions on Consumer Electronics* and an Associate Editor for *IEEE Transactions on Systems, Man, and Cybernetics (IEEE/SMCS): Systems* and IET *Energy Conversion and Economics*. He was an Associate Vice President for *Systems Science and Engineering* of the *IEEE/SMCS*. He is the Working Group Chair for IEEE P2814 and P3166 Standards and Chair of the IEEE/SMCS Intelligent Power and Energy Systems Technical Committee. Prof Lai was a recipient of the Best Paper award in the 2020 IEEE International Smart Cities Conference; the 2022 Meritorious Service Award from the IEEE SMC Society for "meritorious and significant service to IEEE SMC Society technical activities and standards development"; the IEEE Most Active SMC Technical Committee Award: Systems Science and Engineering in 2024. He was awarded first place in the Chester W. Sall Memorial Awards by the IEEE Consumer Technology Society in 2025. He was awarded a Best Associate Editor Award for *IEEE Transactions on Consumer Electronics*, and Excellent Reviewer Award for IET *Energy Conversion and Economics* in 2024. Prof Lai is recipient of IEEE SMC Early Career Award in 2025. He has been recognized as being in the top 2% of world active scientists by a survey conducted by Stanford University. He is a Senior Member of the IEEE, UK Chartered Engineer, and Fellow of the Higher Education Academy, UK.

Xi Chen

Xi Chen is the Operations Director of the Artificial Intelligence Department at the China Electric Power Research Institute, Beijing, China, and the Chief Information Officer at GEIRI North America, San Jose, California, USA. He received a BEng from Beijing Technology and Business University, Beijing, China, an MSc from King's College London, University of London, London, U.K., and a PhD from the Hong Kong Polytechnic University, Hong Kong SAR, China, in 2003, 2005, and 2009, respectively. From 2011 to 2013, he was a Postdoctoral Research Fellow at the Institute of Software, Chinese Academy of Sciences, Beijing, China. He joined the State Grid Information and Telecommunication Branch, State Grid Corporation of China (SGCC), Beijing, China, in 2009, as a Research Fellow and Senior Manager. He was the Research Director at the State Grid U.S. Representative Office, SGCC, New York, NY, USA. He currently serves as the Chair of the IEEE Technical Committee on Power and Energy Circuits and Systems (2023–2025), the Chair of the IEEE Computational Intelligence Society (CIS) Standards Committee (2025), an Editor of *IEEE Transactions on Wireless Communications*, and an Associate Editor of the *IEEE Systems Journal* and the *IEEE Open Journal of the Industrial Electronics Society*. He served as the Co-Chair of Santa Clara Valley Chapter, IEEE CASS and a Guest Editor of *IEEE Transactions on Intelligent Transportation Systems, IEEE Journal on Emerging and Selected Topics in Circuits and Systems*, and *IEEE Transactions on Consumer Electronics*. His current research interests include artificial intelligence, industrial large language models (LLMs), the Internet of Things, smart grids, electric vehicle charging networks, and nonlinear systems.

World Electric Vehicle Journal

MDPI

Editorial

Electric Vehicles in Smart Grids: Integration, Optimization, and Sustainability

Chun Sing Lai [1,*] and Xi Chen [2,3]

1 School of Electrical and Information Engineering, Tianjin University, Tianjin 300072, China
2 Department of Artificial Intelligence, China Electric Power Research Institute, Beijing 100192, China; xc@ieee.org
3 GEIRI North America, 250 W Tasman Dr., San Jose, CA 95134, USA
* Correspondence: cslai@tju.edu.cn

1. Introduction

Electric vehicles (EVs) have become increasingly popular due to their environmental benefits and potential for cost savings. However, the integration of EVs into the smart grid presents new challenges, including the need for charging infrastructure, communication protocols, and strategies to mitigate the impact of EVs on the grid. This Special Issue, "Electric Vehicles in Smart Grids: Integration, Optimization, and Sustainability", aims to provide a platform for researchers and practitioners to present and discuss the latest developments, challenges, and opportunities related to the integration of electric vehicles (EVs) into smart grids. The scope of this Special Issue includes, but is not limited to, the following topics:

- Integration of EVs into the smart grid: This topic covers the challenges and opportunities involved in integrating EVs into the smart grid, including vehicle-to-grid (V2G) technology, charging infrastructure, and communication protocols.
- Optimization of EV charging: This topic focuses on the development of optimal charging strategies for EVs, considering factors such as user preferences, grid conditions, and renewable energy sources.
- Grid impact of EVs: This topic covers the impact of EVs on the grid, including their potential to increase peak demand and affect grid stability. It also includes the development of strategies to mitigate these impacts and ensure the reliability of the grid.
- Sustainability of EVs and smart grids: This topic covers the environmental and economic benefits of EVs and smart grids, including reduction in greenhouse gas emissions and the potential for cost savings.
- Policy and regulation: This topic covers the policy and regulatory frameworks that support the integration of EVs into smart grids, including charging infrastructure requirements, incentives for EV adoption, and grid interconnection standards.

Overall, this Special Issue seeks to provide a comprehensive understanding of the integration of EVs into smart grids and the potential for this technology to promote sustainability and energy efficiency.

2. Short Presentation of the Papers

Guerreo-Silva et al. [1] presented a systematic review which examined recent research on EV charging network planning, with a particular focus on optimization techniques, machine learning applications, and sustainability integration. Using bibliometric methods

 https://doi.org/10.3390/wevj17030130

and Principal Component Analysis (PCA), the authors identified key thematic clusters, including smart grid integration, strategic station placement, renewable energy integration, and public policy impacts. This study revealed a growing trend toward hybrid models that combine artificial intelligence and optimization methods to address challenges such as grid constraints, range anxiety, and economic feasibility. The authors provided a taxonomy of computational approaches—ranging from classical optimization to deep reinforcement learning—and synthesized practical insights for researchers, policymakers, and urban planners. The findings highlight the critical role of coordinated strategies and data-driven tools in designing scalable and resilient EV charging infrastructures, and point to future research directions involving intelligent, adaptive, and sustainable charging solutions.

Qazi et al. [2] presented a survey paper to explore the key research challenges in smart grid operation from an engineering design perspective, such as the potential for voltage instability due to the integration of numerous EVs and distributed microgrids with fluctuating generation capacities and load demands. The authors investigated the need for a synergistic balance to optimize the energy supply and demand equation. Additionally, the authors discussed policies and incentives that may be enforced by national electricity carriers to maintain grid reliability and manage the influx of EVs. Furthermore, the authors addressed emerging issues regarding smart grid technology providing primary charging infrastructure for EVs, such as incentivizing green energy, the technical difficulties in integrating diverse hetero-microgrids based on High Voltage Alternating Current (HVAC) and High Voltage Direct Current (HVDC) technologies, challenges related to the speed of energy transaction processing during fluctuating prices, and vulnerabilities concerning cyber-attacks on blockchain-based smart grids architectures. Finally, future trends are discussed, including the impact of increased EV penetration on smart grids, advancements in vehicle-to-grid technologies, load-shaping techniques, dynamic pricing mechanisms, and artificial intelligence-based stability enhancement measures in the context of widespread smart grid adoption.

Zou et al. [3] provided a comprehensive literature survey on relocation optimization for shared electric vehicles. The literature is reviewed and categorized based on two types of relocation: static and dynamic relocation. Static relocation is analyzed in terms of operator-based relocation and user-based relocation, while dynamic relocation is analyzed in terms of four methodologies: optimization models, simulations, multi-stage methods, and deep reinforcement learning. The paper finally provided some interesting future research topics, such as considering the nonlinear charging process of electric vehicles in the process of constructing relocation optimization models and designing algorithms for shared electric vehicles.

Caminiti et al. [4] employed real-world datasets to propose a comprehensive spatial–temporal energy model that integrates a traffic model and geo-referenced data to realistically evaluate the flexibility potential embedded in the light-duty transportation sector for a given study region. The methodology involved assessing traffic patterns, evaluating the grid impact of EV charging processes, and extending the analysis to flexibility services, particularly in providing primary and tertiary reserves. The analysis is geographically confined to the Lombardy region in Italy, relying on a national survey of 8.2 million trips on a typical day. Given a target EV penetration equal to 2.5%, corresponding to approximately 200,000 EVs in the region, flexibility bands for both services are calculated and economically evaluated. Within the modeled framework, power-intensive services demonstrated significant economic value, constituting over 80% of the entire potential revenues. Considering European markets, the average marginal benefit for each EV owner is in the order of EUR 10 per year, but revenues could be higher for sub-classes of users better fitting the network needs.

https://doi.org/10.3390/wevj17030130

Duan et al. [5] proposed a novel bi-level optimization model for integrating solar, hydrogen, and battery storage systems with charging stations (SHS-EVCSs) to maximize social welfare. The first level employed a non-cooperative game theory model for each individual EV charging station to minimize capital and operational costs. The second level used a cooperative game framework with an internal management system to optimize energy transactions among multiple EV charging stations while considering EV owners' economic interests. A Markov decision process modeled uncertainties in EV charging times, and Monte Carlo simulations predicted charging demand. Real-time electricity pricing based on the dual theory enabled demand-side management strategies like peak shaving and valley filling. Case studies demonstrated the model's effectiveness in reducing peak loads, balancing energy utilization, and enhancing overall system efficiency and sustainability through optimized renewable integration, energy storage, EV charging coordination, social welfare maximization, and cost minimization. The proposed approach offers a promising pathway toward sustainable energy infrastructure by harmonizing renewable sources, storage technologies, EV charging demands, and societal benefits.

Xia et al. [6] proposed an optimization strategy for EV charging behavior based on deep reinforcement learning (DRL). The strategy aimed to minimize user charging costs while achieving load balancing across distribution networks. Specifically, the strategy divided the charging process into two stages: charging station selection and in-station charging scheduling. In the first stage, a Load Balancing Matching Strategy (LBMS) is employed to assist users in selecting a charging station. In the second stage, the authors used the DRL algorithm. In the DRL algorithm, the authors designed a novel reward function that enables charging stations to meet user charging demands while minimizing user charging costs and reducing the load gap among distribution networks. Case study results demonstrated the effectiveness of the proposed strategy in a multi-distribution network environment. Moreover, even when faced with varying levels of EV user participation, the strategy continued to demonstrate strong performance.

Vásquez-Cardona et al. [7] examined the effects of EV penetration on the frequency stability of Curaçao's power network, an aspect not previously studied for the island. As a key contribution, the authors presented a representative model of Curaçao's power network, adjusting the dynamic models of the speed governors of synchronous machines using data available to the academic community. Additionally, the authors analyzed the impacts of EVs on the grid's frequency stability under different EV participation scenarios. To achieve this, simulations were conducted considering various EV participation scenarios and different types of chargers to assess their impact on grid stability. The study evaluated key frequency stability metrics, including the rate of change of frequency (RoCoF) and the highest and lowest frequency values during the transient period. The results indicated that higher EV penetration can significantly impact frequency stability. The observed increase in the RoCoF and frequency zenith values suggests a weakening of the grid's ability to withstand frequency disturbances, particularly in high-EV-penetration scenarios.

Kovačević et al. [8] analyzed the impact of electromobility on distribution grids and voltage stability. In line with current legislation and the European Commission's plans for the future of electromobility, the aim is to increase the share of electric vehicles to 50% by 2050. However, achieving this target can be challenging due to the characteristics and features of the electric vehicle charging stations and the associated charging methods, which can lead to constraints within the network. The analysis included the integration of single-phase and three-phase chargers on a radial feeder, as well as the determination of the maximum number of vehicles that can be accommodated on a given feeder without compromising voltage stability. Five scenarios are evaluated to gain a better understanding of the impact of electromobility on the distribution grid.

https://doi.org/10.3390/wevj17030130

Saeseiw et al. [9] investigated a multi-port conversion system that connects Photovoltaic (PV) arrays, Battery Energy Storage (BES), and an EV to a single-phase grid, offering a flexible solution for smart homes. By integrating Vehicle-to-Grid (V2G) and Vehicle-to-Home (V2H) technologies, the system supports bidirectional energy flow, optimizing usage, improving grid stability, and supplying backup power. The proposed four-port converter consists of an interleaved bidirectional Direct Current (DC)-DC converter for high-voltage BES, a bidirectional buck–boost DC-DC converter for EV charging and discharging, a DC-DC boost converter with maximum power point tracking for PV, and a grid-tied inverter. Its non-isolated structure ensures high efficiency, compact design, and fewer switches, making it suitable for residential applications. A State-of-Charge (SoC)-based power management strategy coordinates operation among PV, BES, and EV in both on-grid and off-grid modes. It reduces reliance on EV energy when supporting V2G and V2H, while SoC balancing between BES and EV extends lifetime and lowers current stress. A 7.5 kVA system was simulated to validate feasibility. Two scenarios were studied: PV, BES, and EV with V2G supporting the grid and PV, BES, and EV with V2H providing backup power in off-grid mode. Tests under PV fluctuations and load variations confirmed the effectiveness of the proposed design. The system exhibited a fast transient response of 0.05 s during grid-support operation and maintained stable voltage and frequency in off-grid mode despite PV and load fluctuations. Its protection scheme disconnected overloads within 0.01 s, while harmonic distortions in both cases remained modest and complied with EN50610 standards.

Fantin et al. [10] analyzed the impact of Battery Electric Bus (BEB) charging on a Brazilian urban medium-voltage (MV) feeder using a novel methodology to convert utility geographic information system data into OpenDSS simulation models. The study utilized Geographic Database of the Distribution Company (BDGD) data from the Brazilian Electricity Regulatory Agency (ANEEL) and OpenDSS simulations. Motivated by Cuiabá's proposal to electrify its public bus fleet, four realistic scenarios were simulated, incorporating distributed PV generation and V2G operation. Results showed that up to 118 BEBs can be charged simultaneously without voltage violations. However, thermal overload occurs beyond 56 units, requiring conductor upgrades or load redistribution. PV systems can supply up to 64% of the daily energy demand but introduce reverse power flows and overvoltages, indicating the need for dynamic control. The findings suggest that while the current infrastructure partially supports fleet electrification, future scalability depends on integrating smart grid features and reinforcing the system. Although focused on Cuiabá, the methodology offered a replicable approach for low-carbon urban mobility planning in similar developing regions.

Author Contributions: Writing—original draft preparation, C.S.L. and X.C. All authors have read and agreed to the published version of the manuscript.

Funding: This research received no external funding.

Conflicts of Interest: The authors declare no conflicts of interest.

References

1. Guerrero-Silva, J.A.; Romeo-Gelvez, J.I.; Aristizábal, A.J.; Zapata, S. Optimization and Trends in EV Charging Infrastructure: A PCA-Based Systematic Review. *World Electr. Veh. J.* **2025**, *16*, 345. [CrossRef]
2. Qazi, S.; Khawaja, B.A.; Alamri, A.; AlKassem, A. Fair Energy Trading in Blockchain-Inspired Smart Grid: Technological Barriers and Future Trends in the Age of Electric Vehicles. *World Electr. Veh. J.* **2024**, *15*, 487. [CrossRef]
3. Zou, Y.; Yu, Q.; Jiang, D.; Deng, Y. Relocation Optimization for Shared Electric Vehicles: A Literature Review. *World Electr. Veh. J.* **2025**, *16*, 108. [CrossRef]

 https://doi.org/10.3390/wevj17030130

4. Caminiti, C.M.; Brigatti, L.G.; Spiller, M.; Rancilio, G.; Merlo, M. Unlocking Grid Flexibility: Leveraging Mobility Patterns for Electric Vehicle Integration in Ancillary Services. *World Electr. Veh. J.* **2024**, *15*, 413. [CrossRef]
5. Duan, L.; Taylor, G.; Lai, C.S. Solar–Hydrogen-Storage Integrated Electric Vehicle Charging Stations with Demand-Side Management and Social Welfare Maximization. *World Electr. Veh. J.* **2024**, *15*, 337. [CrossRef]
6. Xia, Y.; Cheng, Z.; Zhang, J.; Chen, X. User Cost Minimization and Load Balancing for Multiple Electric Vehicle Charging Stations Based on Deep Reinforcement Learning. *World Electr. Veh. J.* **2025**, *16*, 184. [CrossRef]
7. Vásquez-Cardona, D.; Saldarriaga-Zuluaga, S.D.; Bustamante-Mesa, S.; López-Lezama, J.M.; Muñoz-Galeano, M. Impacts of Electric Vehicle Penetration on the Frequency Stability of Curaçao's Power Network. *World Electr. Veh. J.* **2025**, *16*, 264. [CrossRef]
8. Kovačević, T.; Kljajić, R.; Glavaš, H.; Kljajin, M. Analysis of the Impact of Electromobility on the Distribution Grid. *World Electr. Veh. J.* **2025**, *16*, 358. [CrossRef]
9. Saeseiw, C.; Pongpri, K.; Kaewchum, T.; Somkun, S.; Pachanapan, P. Power Management for V2G and V2H Operation Modes in Single-Phase PV/BES/EV Hybrid Energy System. *World Electr. Veh. J.* **2025**, *16*, 580. [CrossRef]
10. Fantin, C.d.A.; Vasconcelos FMd Pardini, C.G.; Albuquerque FPd Abarca, M.E.R.; Bonaldo, J.P. Impact Assessment of Electric Bus Charging on a Real-Life Distribution Feeder Using GIS-Integrated Power Utility Data: A Case Study in Brazil. *World Electr. Veh. J.* **2025**, *16*, 621. [CrossRef]

https://doi.org/10.3390/wevj17030130

World Electric Vehicle Journal

MDPI

Article

Impact Assessment of Electric Bus Charging on a Real-Life Distribution Feeder Using GIS-Integrated Power Utility Data: A Case Study in Brazil

Camila dos Anjos Fantin [1,*], Fillipe Matos de Vasconcelos [1], Carolina Gonçalves Pardini [1], Felipe Proença de Albuquerque [1], Marco Esteban Rivera Abarca [2] and Jakson Paulo Bonaldo [1]

1 Department of Electrical Engineering, Federal University of Mato Grosso, Avenida Fernando Correa da Costa, 1367, Cuiaba 78060-900, Brazil; fillipe.vasconcelos@ufmt.br (F.M.d.V.); carolina.goncalves@sou.ufmt.br (C.G.P.); felipe.albuquerque@ufmt.br (F.P.d.A.); jakson.bonaldo@ufmt.br (J.P.B.)

2 Power Electronics, Machines and Control (PEMC) Research Institute, Department of Electrical and Electronic Engineering, Faculty of Engineering, University of Nottingham, 15 Triumph Rd., Lenton, Nottingham NG7 2GT, UK; marco.rivera@nottingham.ac.uk

* Correspondence: camila.fantin@ufmt.br

Abstract: The electrification of public transport with battery electric buses (BEBs) poses technical, regulatory, and environmental challenges. This paper analyzes the impact of BEB charging on a Brazilian urban medium-voltage (MV) feeder using a novel methodology to convert utility GIS data into OpenDSS simulation models. The study utilizes Geographic Database of the Distribution Company (BDGD) data from the Brazilian Electricity Regulatory Agency (ANEEL) and OpenDSS simulations. Motivated by Cuiabá's proposal to electrify its public bus fleet, four realistic scenarios were simulated, incorporating distributed photovoltaic (PV) generation and vehicle-to-grid (V2G) operation. Results show that up to 118 BEBs can be charged simultaneously without voltage violations. However, thermal overload occurs beyond 56 units, requiring conductor upgrades or load redistribution. PV systems can supply up to 64% of the daily energy demand but introduce reverse power flows and overvoltages, indicating the need for dynamic control. V2G operation enables peak shaving but also leads to overvoltages when more than 33 buses inject power concurrently. The findings suggest that while the current infrastructure partially supports fleet electrification, future scalability depends on integrating smart grid features and reinforcing the system. Although focused on Cuiabá, the methodology offers a replicable approach for low-carbon urban mobility planning in similar developing regions.

Keywords: distributed generation; electric bus; geographic information system; OpenDSS; smart grid; vehicle electrical

1. Introduction

Decarbonization of urban transport is essential to reduce greenhouse gas (GHG) emissions and mitigate the effects of climate change. Diesel-based public transportation systems are among the primary sources of emissions in urban areas. In Brazil, for example, the transport sector emitted a record 224 million tons of CO_2 in 2023, driven primarily by diesel consumption [1]. In this context, electrification of public fleets through battery electric buses (BEBs) has emerged as a strategic alternative to improve air quality and reduce dependence on fossil fuels.

While hydropower remains the main source of electricity in Brazil, the system is not entirely renewable in practice. During recurring droughts, when reservoirs reach critical levels, thermal plants running on oil, coal, and natural gas are called upon to keep the supply stable, increasing both costs and emissions. For this reason, photovoltaic (PV) generation plays an important complementary role, since it is not tied to hydrological conditions and can provide additional security to the system. Another advantage is that photovoltaic installations can be placed directly in urban areas, for example, bus depots or terminals, very close to the charging points of battery electric buses (BEBs). Locating generation near the demand reduces transmission losses, postpones the need for network reinforcements, and allows local energy management strategies that large hydro plants, concentrated in distant regions, cannot provide. Brazil's regulatory framework also encourages distributed generation, with policies such as ANEEL's normative Resolution 482/2012 and its subsequent updates, which have made photovoltaics financially attractive in urban projects. Together, these factors explain why photovoltaics should be considered alongside BEBs, even in a system still dominated by hydropower: they add diversity, increase resilience, and strengthen long-term sustainability.

Several international studies have demonstrated the technical feasibility and environmental benefits of this transition. The authors in [2] conducted a life cycle analysis of BEBs, indicating significant reductions in emissions and energy consumption compared to diesel vehicles. In [3], the operation of BEBs in Madrid has been analyzed, highlighting the importance of charging strategies for energy efficiency and battery life. The integration of photovoltaic generation into the terminals was investigated in [4], noting that although solar energy can reduce peak demand in the grid, it can also cause reverse flows and overvoltages during periods of low load.

To mitigate these effects, vehicle-to-grid (V2G) projects in Europe, China, India, and the United States are already exploring the use of fleets as a source of grid energy [5]. Recent studies have proposed intelligent charging algorithms and operating strategies [6–8]. Ref. [9] presents a comprehensive review of the integration of electric vehicles (EVs) into power grids, with a particular focus on V2G operation. A comparison of three different innovative change strategies, considering non-coordinated, bottom-up coordinated, and top-down coordinated control categories, has been presented in [7]. The study demonstrates that power quality issues on the grid side can be mitigated by selecting the appropriate smart charging strategy. Ref. [8] discusses the limitations of technical requirements related to battery degradation, communication protocols, and cybersecurity associated with grid connectivity and V2G technologies. Previous V2G cases have been analyzed considering the application of recent technologies.

The authors in [10] presented an optimization model based on dynamic pricing to minimize the impacts of charging on the distribution grid. In [11], a feasibility study considering the Manchester Science Park electric fleet with 50 vehicles for V2G as a case study concludes that the V2G strategies can lead to economic savings of GBP 400,000 in ten years.

In [12], the technical and economic feasibility of using V2G buses for ancillary services, such as frequency regulation and reduction of peak demand, was analyzed, but warned about the risks of instability if the operation is not properly coordinated. Similarly, the authors in [13] demonstrated that coordinated control of BEB fleets can increase grid reliability when real-time constraints are considered.

Despite these advances, many studies still rely on theoretical models or simplified representations of the power grid. Reference [14] points out that the use of real-world data, including power consumption and feeder constraints, is essential to understand the actual operational limits of the integration of the electric fleet, particularly in urban areas with weak infrastructure.

Feeder-level studies also highlight technical challenges that are often overlooked in higher-level planning. For example, ref. [15] has developed a feeder reconfiguration strategy to reduce losses in distribution systems with a high penetration of electric vehicles, considering the real topology of the grid.

Other researchers explore policy, economic, and spatial planning perspectives, but often neglect the physical constraints of actual distribution feeders [16]. For example, ref. [17] reviewed regulation and market integration strategies for electric vehicles but did not take into account the behavior of the real-time network.

As reported in [18–21], recent studies in the Latin American context remain limited. Although some initiatives model urban electrification in specific Brazilian cities [22–25], comprehensive technical analyses based on GIS-integrated feeder data are still lacking, particularly in scenarios involving large BEB fleets and bidirectional energy flows through V2G technology. To address these limitations, this study presents a novel comprehensive technical analysis based on GIS-integrated feed data from the real world in a Brazilian urban centre, evaluating large BEB fleets and bidirectional energy flows using V2G technology.

Given these gaps, in this paper, a detailed technical analysis of BEB charging and distributed generation integration is presented based on real data from a medium voltage feeder serving a medium-sized city in an emerging economy.

Using real GIS data from an urban feeder in Cuiabá, Brazil, OpenDSS simulations were conducted to evaluate the network's performance in various electrification scenarios, including those with and without photovoltaic generation and with the vehicle-to-grid (V2G) application.

This study makes the following key contributions to the existing body of knowledge:

- The simulations performed revealed that reverse flows can induce voltage violations and thermal overloads even in grid segments located far from the generation source, particularly in areas with ageing infrastructure and undersized conductors. Such spatial effects can only be identified through detailed real-network modeling, which has not been thoroughly investigated in previous works.
- The combined presence of BEB charging and photovoltaic generation creates complex load–generation interactions that vary over time and between locations, requiring fine-grained coordination to prevent unbalanced voltage profiles and power quality issues. Our results demonstrate how these dynamics manifest themselves in a realistic urban feeder.
- We evaluated reinforcement schemes considering different rated sections of the conductors, showing how strategic upgrades can mitigate voltage violations and thermal overloads while extending the feeder hosting capacity. This provides a practical pathway for distribution utilities to plan targeted investments in ageing networks.
- The integration of V2G strategies under coordinated dispatch is shown to stabilise voltage levels and reduce peak loads without exceeding conductor ampacity. This highlights the potential of electric bus fleets to provide adequate grid support when operational and network constraints are explicitly considered.

Although this study focusses on Cuiabá, its contributions are broadly applicable. It offers a scalable roadmap that can be adopted by cities around the world to accelerate low-cost, zero-emission urban mobility. More than a case study, this work provides a disruptive redprint for policymakers, transportation agencies, and stakeholders in the energy sector to accelerate electrification.

In addition, the findings are of special significance for regions like Mato Grosso, where electricity transmission and distribution systems face growing constraints due to increasing demand. By modeling real-world feeder behavior under electrification and distributed generation scenarios, this study provides a tool to identify localized reinforcement needs, promote the strategic use of local energy resources, and ultimately reduce the stress on the broader transmission infrastructure.

The remainder of this paper is structured as follows. Section 2 describes the proposed methodology for converting GIS-integrated distribution data into simulation models using {OpenDSS (Version 9.6.1.1, Electric Power Research Institute—EPRI, Palo Alto, CA, USA). Section 3 introduces the case study based on a real distribution feeder in Cuiabá, Brazil, detailing the modeling of electric bus charging, photovoltaic generation, and vehicle-to-grid operation. Section 4 presents the mathematical formulation adopted for the power flow simulations. Section 5 discusses the results obtained under different electrification scenarios, including sensitivity analyses of feeder capacity and reinforcement strategies. Finally, Section 6 summarizes the main findings and outlines directions for future research.

2. Methodology

The proposed methodology for feeder simulation is based on the conversion of the Geographic Database of the Distribution Company (BDGD, *i.e., Banco de Dados Geográfico da Distribuidora* in Portuguese) data using dataset "Energisa_MT_405_2023-12-31_V11_20250210-1825" (Version 11) provided by National Electric Energy Agency (ANEEL) [26], enabling the extraction of technical characteristics for various entities of the electrical system. Initially, these data are extracted using QGIS 3.34.16 Prizren software and exported to Excel spreadsheets, which must then be interpreted following the guidelines established in the BDGD manual. Subsequently, these spreadsheets are imported into the Google Colab environment, where relevant data are filtered and structured to generate the DSS files used for simulation in OpenDSS. Figure 1 presents the flow chart of the methodological process, which highlights the main steps from data extraction to the final simulation in OpenDSS.

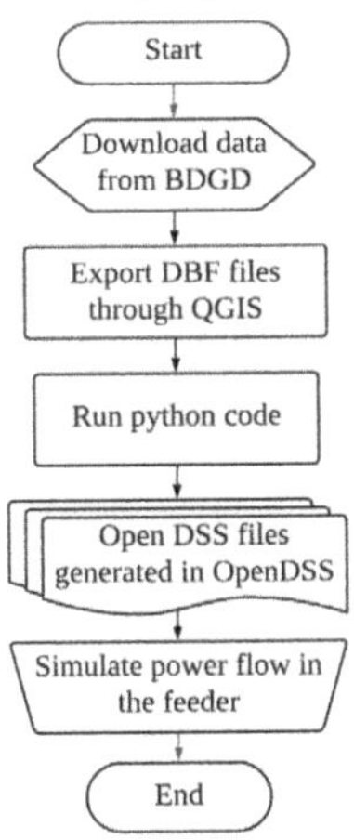

Figure 1. Methodology flowchart.

Converting BDGD Data to OpenDSS

The BDGD is a georeferenced model that consolidates information provided by distributors to the ANEEL of Brazil, representing the electrical system's configuration in a structured and standardized manner. Its structure organizes data into entities, each corresponding to a specific piece of equipment or structure within the electrical grid. The organization and standardization of this information follow the guidelines established in Module 10 of the Brazilian Procedures for Energy Distribution Standard (PRODIST).

The information contained in the BDGD is classified into two main groups:

- Geographic entities: represent the spatial location of electrical grid elements, such as feeders, transformers, and poles;
- Non-geographic entities: contain technical and operational attributes of equipment, such as electrical characteristics and identification codes.

The current version of the BDGD data model (V1.0) comprises 23 Geographic Entities and 20 Non-Geographic Entities, whose specifications are detailed in the ANEEL Data Dictionary (DDA) [27]. Geospatial data are publicly available on ANEEL's open data portal and are provided in Geodatabase (.gdb) format, organized by distributor and reference year. For this study, the ENERGISA MT-2022-12-31 database was used.

After downloading, the files must be unzipped and loaded into the QGIS software as vector layers. QGIS is a free and open source tool widely used for the visualization, analysis, and manipulation of geospatial data [28].

Once the layers have been imported into the QGIS, it is possible to view the elements of the electrical grid operating within the distributor's area, specifically in the state of Mato Grosso. To facilitate spatial interpretation, the QuickMapServices plugin was used, which allows the overlay of base maps, such as Google Maps.

Figure 2 illustrates the spatial visualization of the segments of the high, medium, and low voltage distribution system in the municipality of Cuiabá.

Geographic entities are stored in Shapefile format, a standard for representing spatial data. This format is composed of multiple files, two of the primary ones being:

- shp: contains the geometric characteristics of the elements (points, lines, or polygons).
- dbf: stores the attribute tables associated with each geographic element.

Attribute tables can be exported and accessed using spreadsheet software, such as Excel, facilitating data analysis and manipulation. To select the feeder of interest, its location can be identified on the map, and then the CTMT field, representing the medium voltage circuit, can be located in the attribute table. This parameter is essential for filtering other geographic entities, ensuring that only relevant data for the study is processed.

To carry out the simulations, it is necessary to previously export the geographic and non-geographic entities described in Table 1.

Table 1. Entities of the Distribution System.

Acronym	Name
UCBT	Low Voltage Consumer Unit
UCMT	Medium Voltage Consumer Unit
UGBT	Low Voltage Generation Unit
UGMT	Medium Voltage Generation Unit
SSDBT	Low Voltage Distribution System Segment
SSDMT	Medium Voltage Distribution System Segment
UNCRBT	Low Voltage Reactive Compensation Unit
UNCRMT	Medium Voltage Reactive Compensation Unit
UNSEBT	Low Voltage Switching Unit
UNSEMT	Medium Voltage Switching Unit
UNTRMT	Medium Voltage Transformer Unit
RAMIG	Service Line
EQTRMT	Medium Voltage Transformer Equipment
SEGCON	Conductor Segment
CRVCRG	Load Curve

Figure 2. Spatial visualization of the segments of the high-, medium-, and low-voltage distribution system in the municipality of Cuiabá. The colors represent different voltage levels across the feeder.

After exporting the data, the next step is to transfer this information to the Google Colab environment. Google Colab is a platform based on Jupyter Notebooks, which does not require prior configuration for its use. In addition, it offers free access to computing resources, including GPUs and TPUs, making it a highly suitable tool for applications in machine learning, data science, and educational activities [29].

A Python code was developed to automate the process of exporting data from BDGD spreadsheets. The code prompts the user for the feeder code (COD-ID), loads the necessary spreadsheets, and filters the information based on this identifier. For each spreadsheet, data is analyzed line by line; if the feeder matches the one specified, the relevant information is extracted, formatted, and written to text files in .dss format. At the end of the process, files are generated that are ready for use in OpenDSS simulations.

The implemented code is capable of generating the electrical circuit corresponding to any feeder in the BDGD database from 2022 onwards. However, any data entry errors or inconsistencies present in the database itself are not automatically identified by the code, requiring subsequent analysis of the generated files to ensure the integrity of the modeled circuit.

Figure 3 shows the flowchart of this export process, highlighting the main steps from data entry to the final generation of the files. In this study, OpenDSS was selected as the simulation tool due to its flexibility and suitability for distribution system analysis. OpenDSS is an open-source software widely used in both academia and industry for feeder-level studies, with native support for unbalanced three-phase systems, distributed generation, storage, and electric vehicle modeling [30]. Its scripting interface enables direct integration with Python 3.13.5, which was essential in this work for automating the conversion of BDGD data and running large-scale simulations in Google Colab. Compared to commercial alternatives such as DIgSILENT PowerFactory or CYME, OpenDSS offers full transparency of the modeling process and reproducibility of results, while avoiding licensing restrictions that could limit scalability and accessibility. These features make it particularly suitable for research applications in emerging economies, where open-source and flexible tools are critical for replicability and knowledge transfer.

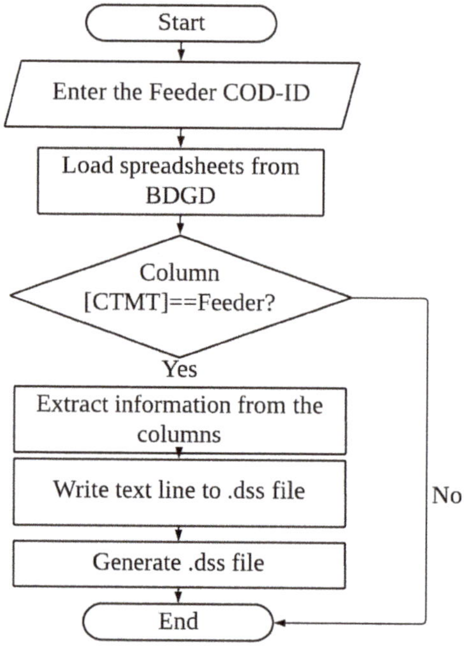

Figure 3. Flowchart export process.

3. Case Study

To carry out the simulations, it was necessary to define the location for the BEBs fleet garage initially. The chosen location was the former Light Rail Vehicle (VLT) yard, located near the Cuiabá-Marechal Rondon International Airport, in Várzea Grande-MT. This location was chosen because of its proximity to the routes of the operated line, favoring the logistics of operating and recharging the vehicles. Figure 4 shows a satellite image of the region in question.

Figure 4. Location for the electric bus fleet garage.

Figure 5 illustrates the electrical distribution network around the VLT yard, where it is possible to observe the arrival of a feeder in the area. This feeder, identified as 069083, will be the subject of study in this work.

Figure 5. Electrical network around the garage.

Taking into account the medium voltage circuit of the selected feeder, the network to be analyzed was delimited, as illustrated in Figure 6. This is the feeder of the Várzea Grande 138 kV substation, situated in the Vila Pirineu neighbourhood, in the municipality of Várzea Grande, MT. For simulation purposes, the entire circuit upstream of the substation was represented by a Thévenin equivalent, modeling the downstream elements based on data extracted from the BDGD. The main characteristics of the feeder and the elements that make up the system are summarized in Table 2.

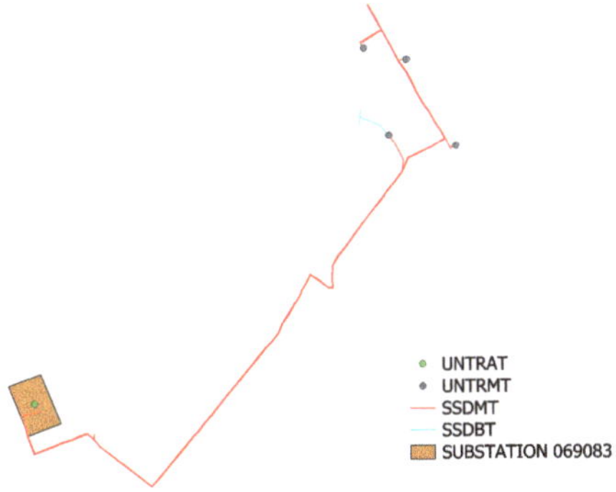

Figure 6. Medium voltage circuit.

Table 2. Feeder characteristics under study.

Characteristic	Value
Medium voltage line voltage	13.8 kV
Low voltage line voltage	0.22 kV
Number of low-voltage consumer units	4
Number of medium-voltage consumer units	2
Number of public lighting points	5
Maximum demand	662.11 kW
Low-voltage segments	7
Medium-voltage segments	170
Medium-voltage transformer units	4

The feeder in question primarily serves low-consumption residential loads, resulting in reduced demand for active power. As it is a small urban feeder, there are no voltage regulators or capacitor banks along the route. However, losses in the conductors remain insignificant, because of the short length of the lines and the moderate load profile.

3.1. Electric Bus Charging Modeling

In this work, in addition to the typical elements of a distribution system, the Storage element of OpenDSS was used to model the BRT charging points. This component represents a versatile electrical energy storage system, being treated as an energy conversion element.

The storage element can operate in two distinct modes: as a constant power load, when configured in charging mode, or as a generator, in discharge mode, supplying energy to the grid.

In this study, plug-in charging at night was considered, with only one daily charge for 50 articulated BEBs. The charging was modeled as uncoordinated, as it is a reality in Brazil that the power distribution infrastructure still lacks the advanced automation required for smart charging, and allows analysis of the worst-case scenario in terms of impact on the grid. For this study, simultaneous charging of all BEBs will be considered between 00:00 and 03:00. The bus selected for analysis is the articulated model D11A, developed by BYD for urban applications, especially for BRT projects [31]. Table 3 presents the main electrical characteristics of these vehicles. In China, it is often seen that 30-kW or 60-kW wireless are charged. However, in this pilot project, a 3-h charging time was defined as a requirement. Thus, a 160 kW charging power was considered for a 516 kWh battery capacity.

Table 3. Specifications of the BYD D11A Bus.

Technical Specification	Description
Battery Model	Byd LiFePO4
Battery Capacity	516 kWh
Maximum Charging Power	160 kW
Charging Time	3 h
Voltage Magnitude	380 V
Travel Distance per Charge	250 km

Although wireless charging is not in the scope of this study, it is well-known that the advantages of wireless over conventional (i.e., it eliminates mid-day downtime and supports smaller batteries). This technology is an emerging alternative for future Brazilian BEBs integration. Although wireless charging presents challenges, such as the detection of foreign objects, advanced solutions are already applicable in real-time [32].

3.2. Photovoltaic System Power Calculation Methodology

To assess the feasibility of meeting the energy demand to charge the BEB fleet through distributed generation, the calculations proposed in the study in [33] were used as a basis. This study presents a methodology for sizing photovoltaic systems for electric mobility applications.

The total daily energy consumption of the fleet, in kilowatt-hours (kWh), is determined by multiplying the power of the charger (P_{charger}, in kW) by the number of charging hours (H_{charge}) and the quantity of available chargers (N_{chargers}), as represented in Equation (1).

$$E_{\text{daily}}\ [\text{kWh}] = P_{\text{charger}} \times H_{\text{charge}} \times N_{\text{chargers}} \tag{1}$$

The power output of the photovoltaic system (in kW) is determined by dividing the daily energy production of the system (P_{DS}) by the number of hours of full sun (H_{SP}), as shown in Equation (2).

$$P_{PV}\,[\mathrm{kW}] = \frac{P_{DS}\,[\mathrm{kWh}]}{H_{SP}\,[\mathrm{h}]} \tag{2}$$

Based on the total daily consumption, the required power of the photovoltaic system (P_{SF}) in kWp is calculated considering the system efficiency (η) and H_{SP}, as indicated in Equation (3). Solar panel efficiency, also known as yield, represents the ratio between the incident solar energy on the panels and the electricity they generate, typically expressed as a percentage.

$$P_{SF}\,[\mathrm{kWp}] = \frac{E_{\mathrm{daily}}\,[\mathrm{kWh}]}{\eta \times H_{SP}} \tag{3}$$

To characterize the solar resource in the case study, official data were obtained using the Sundata software version 3.0, Instituto Nacional de Meteorologia—INMET, Brasília, Brazil [34]. For the city of Cuiabá, the Direct Normal Irradiance (DNI) corresponds to an average of 5.13 h of peak sun per day. Given the daily energy consumption of the fleet of up to 24,000 kWh, and assuming an average system efficiency of 75%, the required photovoltaic capacity was estimated at 6237 kWp using the Equation (3).

Based on this model and with the aim of complying with the legal limits for micro and mini distributed generation established by Brazilian Law No. 14,300/2022, a distributed generation (DG) installed capacity of 3 MW was adopted [35]. This limit applies to intermittent sources such as photovoltaic solar energy and follows ANEEL regulations for nondispatchable systems. The complete technical specifications of the PV system are provided in Appendix A.

According to the referenced study, a photovoltaic plant with an installed capacity of 3 MW requires the installation of a system with a total power of 4000 kWp, resulting in an estimated daily energy production of 15,390 kWh, considering performance factors and local solar irradiation. This size allows for the supply of approximately 64.12% of the daily energy required to charge a fleet of 50 articulated BEBs fully [33].

To enable the system's operation, the three-phase WEG inverter model SIW500G T100 W0 was considered. These inverters, with a unit power of 100 kW, are suitable for commercial and industrial applications and offer features such as intelligent string monitoring, natural convection cooling, and protection against surges and grounding faults [36]. The system requires the installation of 40 photovoltaic inverters to meet the proposed generation capacity, taking into account modularity, efficiency, and compatibility with the vehicle battery charging voltage. In the plant modeling in OpenDSS, each inverter was represented as a PVSystem element, allowing for individualized simulation of solar generation throughout the day.

A cost analysis was also conducted based on the sizing of the plant using the reference cost parameters for solar photovoltaics published by EPE in the Decennial Energy Expansion Plan 2034 [37], which establishes a CAPEX of USD 728/kWp. Considering the average exchange rate of BRL 5.22/USD from the base date, the total investment for the 4000 kWp system would be on the order of approximately USD 2.91 million. Although additional expenses related to controllers and power quality mitigation are expected, these costs can be offset by long-term benefits, such as reducing purchased energy from the grid, lower dependence on thermoelectric power plants during drought periods, and alignment with the national regulatory framework for distributed generation. Thus, the integration of PV within the current legal limits demonstrates not only technical feasibility but also an economic perspective that strengthens its adoption in the electrification of urban transportation.

4. Mathematical Description of the Problem

The system was simulated using a three-phase power flow formulation for unbalanced networks, since distribution lines are generally untransposed. This condition requires solving the complete set of equations directly, avoiding the use of symmetrical components [38]. Given the three-phase power injections $P_{i,\phi}$ and $Q_{i,\phi}$ for $\phi \in \{a,b,c\}$, an initial guess $\mathbf{V}^{(0)}$, and the admittance elements $\mathbf{Y}_{ij}^{\text{series}}$ and $\mathbf{Y}_{ij}^{\text{sh}}$, the problem of evaluating node voltages $\mathbf{V}_i$, branch currents $\mathbf{I}_{ij}$, and power flows $\mathbf{S}_{ij}$ is formulated as follows:

$$\mathbf{Y}_{\text{bus}}\,\mathbf{V} = \mathbf{I}^{\text{inj}}(\mathbf{V}). \tag{4}$$

$$I_{i,\phi}(\mathbf{V}) = \left(\frac{S_{i,\phi}}{V_{i,\phi}}\right)^{*} = \frac{P_{i,\phi} - jQ_{i,\phi}}{V_{i,\phi}^{*}}, \qquad \phi \in \{a,b,c\}. \tag{5}$$

$$\mathbf{I}_{ij} = \mathbf{Y}_{ij}^{\text{series}}\,(\mathbf{V}_i - \mathbf{V}_j) + \mathbf{Y}_{ij}^{\text{sh}}\,\mathbf{V}_i, \qquad \mathbf{I}_{ji} = \mathbf{Y}_{ij}^{\text{series}}\,(\mathbf{V}_j - \mathbf{V}_i) + \mathbf{Y}_{ij}^{\text{sh}}\,\mathbf{V}_j. \tag{6}$$

$$\mathbf{I}^{\text{bus}} = \mathbf{Y}_{\text{bus}}\,\mathbf{V}, \qquad \mathbf{I}_{i,\text{bus}} \text{ is the } i\text{-th } (3\times 1) \text{ block of } \mathbf{I}^{\text{bus}}. \tag{7}$$

$$\mathbf{S}_{ij} = \text{diag}(\mathbf{V}_i)\,\mathbf{I}_{ij}^{*}, \qquad \mathbf{S}_{ji} = \text{diag}(\mathbf{V}_j)\,\mathbf{I}_{ji}^{*}, \qquad \mathbf{S}_{i,\text{bus}} = \text{diag}(\mathbf{V}_i)\,\mathbf{I}_{i,\text{bus}}^{*}. \tag{8}$$

The loads are modeled as constant power and follow a 24-h daily profile discretized into time intervals of 15 min. At each sampling instant k, the active and reactive powers are specified and allocated among the three phases as follows:

$$\alpha_{i,a} + \alpha_{i,b} + \alpha_{i,c} = 1, \qquad \alpha_{i,\phi} \geq 0, \tag{9}$$

where $\alpha_{i,\phi}$ represents the fraction of the total demand at bus i assigned to phase $\phi \in \{a,b,c\}$.

$$S_i[k] = P_i[k] + j\,Q_i[k], \qquad S_{i,\phi}[k] = \alpha_{i,\phi}\,S_i[k], \quad \phi \in \{a,b,c\}. \tag{10}$$

$$Q_{i,\phi}[k] = P_{i,\phi}[k]\,\tan\big(\arccos(\text{pf}_{i,\phi})\big), \tag{11}$$

where $\text{pf}_{i,\phi}$ denotes the constant power factor of phase ϕ at bus i. For the load, the complex power is considered negative in the power flow solution.

A similar approach was used for PV generation, which follows a daily production curve. As the generation is equally distributed among the phases, the model can be written as:

$$\beta_{i,a} = \beta_{i,b} = \beta_{i,c} = \tfrac{1}{3}, \tag{12}$$

where $\beta_{i,\phi}$ denotes the fraction of the total photovoltaic generation at bus i assigned to phase $\phi \in \{a,b,c\}$.

$$S_i^{\text{PV}}[k] = P_i^{\text{PV}}[k] + j\,Q_i^{\text{PV}}[k], \qquad S_{i,\phi}^{\text{PV}}[k] = \beta_{i,\phi}\,S_i^{\text{PV}}[k], \quad \phi \in \{a,b,c\}. \tag{13}$$

$$Q_{i,\phi}^{\mathrm{PV}}[k] = P_{i,\phi}^{\mathrm{PV}}[k] \tan\big(\arccos(\mathrm{pf}_{i,\phi}^{\mathrm{PV}})\big), \tag{14}$$

where the power factor of PV generation depends on the selected inverter control strategy.

In addition to the loads and generation, electric buses are modeled as storage elements corresponding to batteries characterized by rated charging/discharging power, maximum energy storage capacity, and dynamic state of charge (SOC). These devices can operate in both charging and discharging modes following a specific curve. Mathematically, the model is given by:

$$0 \le P_i^{\mathrm{ch}}[k] \le P_i^{\mathrm{ch,max}}, \qquad 0 \le P_i^{\mathrm{dis}}[k] \le P_i^{\mathrm{dis,max}}, \tag{15}$$

$$E_i[k] = E_i[k-1] + \left(\eta_i^{\mathrm{ch}} P_i^{\mathrm{ch}}[k] - \frac{1}{\eta_i^{\mathrm{dis}}} P_i^{\mathrm{dis}}[k]\right) \Delta t, \tag{16}$$

$$\mathrm{SOC}_i[k] = \frac{E_i[k]}{E_i^{\mathrm{max}}}, \qquad 0 \le \mathrm{SOC}_i[k] \le 1, \tag{17}$$

where E_i^{max} is the maximum energy storage capacity (kWh), $P_i^{\mathrm{ch,max}}$ and $P_i^{\mathrm{dis,max}}$ are the rated charging and discharging powers (kW), η_i^{ch} and η_i^{dis} are the charging and discharging efficiencies, and Δt is the simulation time step (here, $\Delta t = 15$ min).

In the power flow formulation, positive power injection at the bus is considered. Thus, at discrete time k and bus i, the net active power is given by:

$$P_i^{\mathrm{bus}}[k] = P_i^{\mathrm{dis}}[k] - P_i^{\mathrm{ch}}[k]. \tag{18}$$

Therefore, the presented formulation can be fully described as follows:

- If $P_i^{\mathrm{dis}}[k] > 0$ and $P_i^{\mathrm{ch}}[k] = 0$, then $P_i^{\mathrm{bus}}[k] > 0$, meaning the storage injects power into the network (generator behavior).
- If $P_i^{\mathrm{ch}}[k] > 0$ and $P_i^{\mathrm{dis}}[k] = 0$, then $P_i^{\mathrm{bus}}[k] < 0$, meaning the storage absorbs power from the network (load behavior).
- If $P_i^{\mathrm{ch}}[k] = P_i^{\mathrm{dis}}[k] = 0$, then $P_i^{\mathrm{bus}}[k] = 0$, meaning no power exchange with the network.

Thus, a single variable $P_i^{\mathrm{bus}}[k]$ can represent both charging and discharging modes, and the corresponding complex power injection is written as:

$$S_i^{\mathrm{bus}}[k] = P_i^{\mathrm{bus}}[k] + j\, Q_i^{\mathrm{bus}}[k]. \tag{19}$$

5. Results

This section presents and analyzes the results obtained from simulations carried out using the OpenDSS software. Different scenarios of electric fleet charging and distributed generation integration were considered in order to assess the impact on the electrical behavior of the feeder studied. For all scenarios, power flow simulations were performed over a one-day horizon at 15-min intervals, totaling 96 simulations.

5.1. Base Case: Real Primary Feeder

Figure 7 shows the three-phase active power demand curve in a real primary feeder, i.e., without the presence of BEBs or distributed generation. The active power of the transformer reaches 662.11 kW at 9 p.m. at the time of highest demand, and 605.78 kW at the time of lowest demand.

The behavior of the voltage at the bus station where the future bus charging station will be installed can be seen in Figure 8. The voltage at the bus station assumes its lowest value, 1.0261 pu at 2:00 p.m, still presenting stable voltage levels within the limits established in the Brazilian grid code PRODIST Module 8 [39].

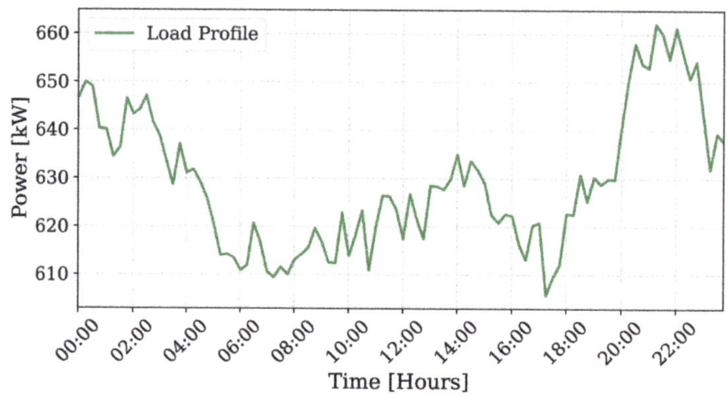

Figure 7. Three-phase power of the feeder throughout the day.

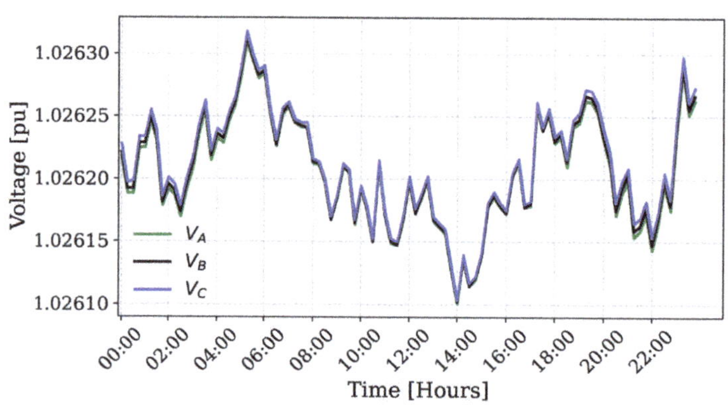

Figure 8. Voltage at the charging station throughout the day.

5.2. Insertion of 50 Electric Buses with Distributed Generation

In this scenario, simultaneous overnight charging of 50 articulated BEBs is considered, scheduled between 12:00 a.m. and 3:00 a.m., using four dedicated 2000 kVA transformers, as shown in Figure 9.

The charging stations were modeled in OpenDSS using the storage element configured as a constant power load during the charging mode. Distributed photovoltaic generation with a total installed capacity of 3 MW was modelled using 40 WEG SIW500G three-phase inverters (100 kW each), represented in the simulation by PVSystem elements operating at a unity power factor.

To account for variability in solar resource availability, three characteristic irradiance profiles were simulated: sunny, partly cloudy, and rainy, corresponding to high, moderate, and low solar generation days, respectively. These profiles were developed from typical meteorological patterns for Cuiabá, representing the range of daily PV output conditions that can occur throughout the year. Figure 10 shows the generation curves for a single inverter under each condition, highlighting the substantial differences in total daily energy production and the mid-day peak output. While the sunny profile achieves full rated power between 11:00 a.m. and 2:00 p.m., the partly cloudy and rainy days exhibit a significant reduction in generation, reaching approximately 70% and 40% of the clear-sky peak, respectively.

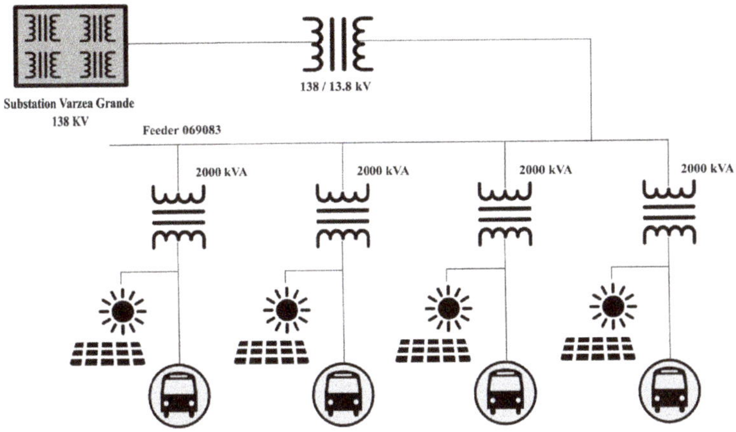

Figure 9. System one-line diagram.

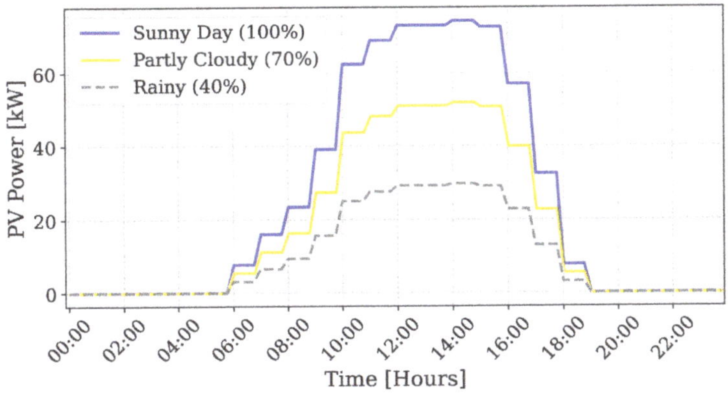

Figure 10. Three-phase power generated by a single inverter over the course of a day.

Figure 11 shows the three-phase active power profile required in the feeder for the three cases. During daylight hours, distributed generation is dominant, resulting in reverse power flow between 7:00 a.m. and 6:00 p.m. in the sunny case, as local PV output exceeds feeder demand. Under partly cloudy and rainy conditions, the magnitude and duration of reverse flow are substantially reduced, mitigating potential issues such as transformer overloading and voltage rise at the feeder ends. During the night-time charging period, the active power in the feeder's main bus reaches a peak of 8855.1 kW, representing a 1331.6% increase compared to the base case.

Figure 12 illustrates the voltage profile on the busbar connected to the chargers. During peak charging, the observed minimum voltage was 0.987 pu, reflecting a 3.8% drop compared to the nominal operating point (1.026 pu). At solar generation peak, the voltage increased to 1.037 pu in the sunny case and remained below 1.02 pu in the rainy case, all within the operational limits established by PRODIST Module 8 (0.93–1.05 pu) [39]. These results demonstrate that even under high irradiance conditions, the feeder remains within acceptable voltage and thermal limits, though reverse power flow is more pronounced during clear-sky days.

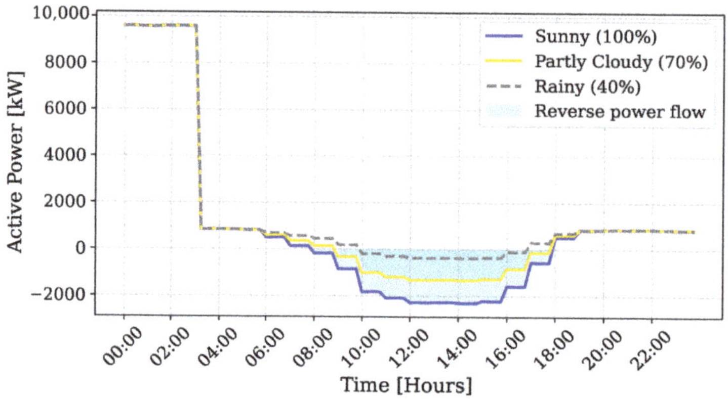

Figure 11. Three-phase power on the feeder throughout the day, with charging of 50 buses and distributed generation.

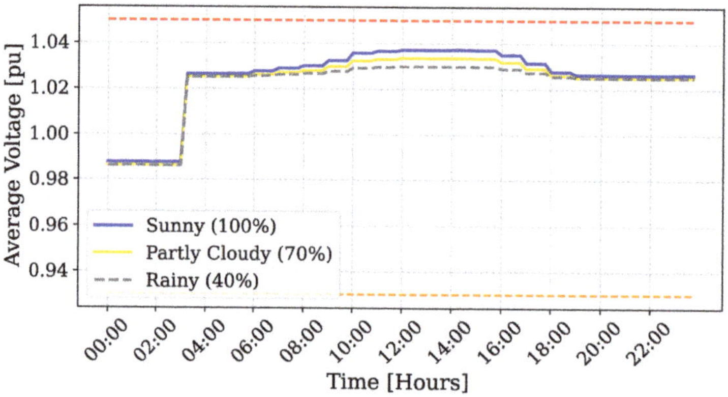

Figure 12. Voltage at the charging station throughout the day, with charging of 50 buses and distributed generation.

It is important to emphasize that, although modern inverters are already equipped with voltage regulation algorithms and the capability to absorb reactive power to mitigate overvoltage, these functionalities are not yet widely adopted in Brazil due to existing regulatory and market constraints. Since end-consumers are primarily incentivized to maximize active energy production (kWh), operating in kVAR control mode typically offers no direct financial benefit and, in some cases, may reduce active energy output when inverters operate near their apparent power limit. For this reason, in this study, PV systems were modeled at unity power factor, reflecting the prevailing operational reality. Nonetheless, future research should investigate scenarios where advanced inverter functionalities are both actively implemented and adequately incentivized, as these could play a decisive role in improving feeder voltage stability and mitigating operational challenges. For instance, inverter-based reactive power control strategies, such as Q–V mode, could be applied to alleviate reverse power flow effects. In addition, daytime Battery Energy Storage Systems (BESS) could be deployed to absorb surplus solar generation and shift it to the nighttime charging window, further enhancing system flexibility and reliability.

5.3. Assessment of the Load Limit on the Feeder in Relation to the Voltage

Figure 13 shows the voltage curve on the charging cable for different scenarios of expansion of the electric fleet: 100, 250, and 500 buses charging simultaneously over 3 h. The objective is to investigate the feasibility of electrifying the entire municipal fleet, which currently comprises approximately 500 vehicles [40]. The simulation shows that the maximum number of buses that can be charged simultaneously without affecting the voltage profile of the feeder is 118. From this value, violations of the minimum voltage limits begin to occur, characterizing undervoltage during the charging period. This shows that total electrification of the fleet would not be feasible through a single feeder, therefore, requiring a multifeeder or a distributed charging infrastructure.

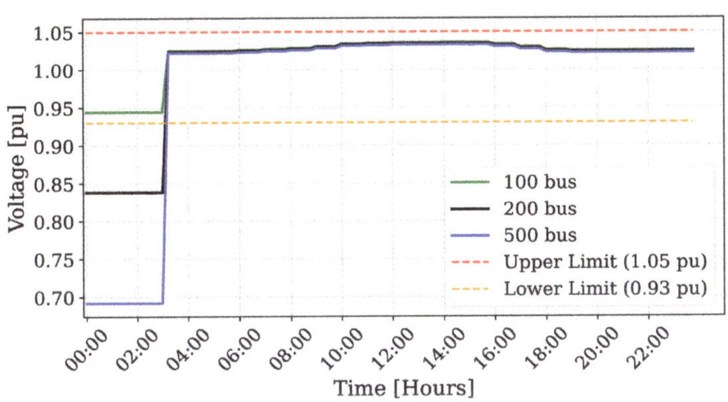

Figure 13. Comparison of voltage at charging station for expansion of electric fleet.

5.4. Evaluation of the Use of V2G Technology

This section analyzes the technical implications of implementing V2G technology, whereby BEBs discharge stored energy back into the distribution network during peak demand hours from 5:30 p.m. to 8:30 p.m. Note that the timing of this peak demand is a characteristic of the Brazilian sector and may vary depending on the industrial and economic configurations of other regions around the world. Several charging station scenarios were simulated, changing the number of buses that inject power into the grid. Each vehicle was allowed a maximum depth of discharge of 70% of its nominal battery capacity, with at least 30% for operational mobility and battery longevity. These limits follow current best practices to minimize battery degradation over time [41].

Figure 14 illustrates the voltage profile on the busbar where the chargers are connected. It is observed that when more than 33 buses simultaneously inject energy into the system, voltage levels exceed the regulatory upper limit of 1.05 pu, resulting in overvoltage conditions. This phenomenon is especially prominent during the overlap between residual solar generation in the late afternoon and V2G activity, which reinforces the upward pressure on the local voltage.

These results confirm that while V2G presents a promising solution for grid support during peak hours—potentially contributing to demand response strategies and ancillary services—its deployment must be carefully controlled. To avoid voltage violations and ensure grid stability, it is recommended to implement local voltage-dependent reactive power controls, such as volt-var or volt-watt functions, integrated into the vehicle inverter interface.

In addition, real-time coordination between V2G assets and the distribution system operator is crucial. Time-based injection windows, load-following algorithms, and communication protocols should be established to adjust discharge profiles according to grid conditions dynamically. V2G operation should ideally be restricted or curtailed when combined with significant local distributed generation, as demonstrated in the simulations.

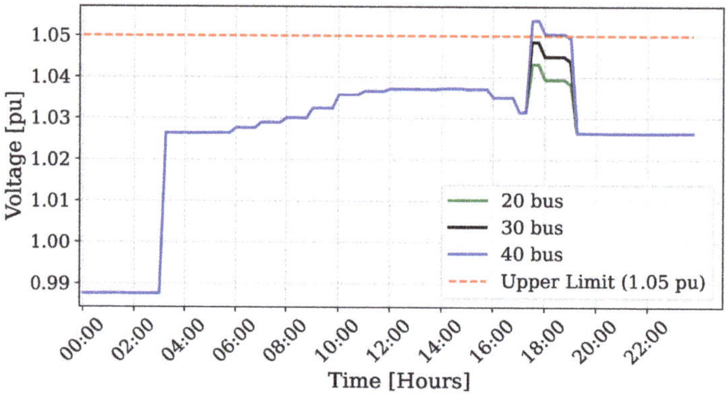

Figure 14. Comparison of voltage at charging station using V2G technology.

The effective deployment of the proposed smart charging and V2G strategies depends on a supporting infrastructure of sensors and communication systems. A robust automatic control system would require the installation of measurement sensors (e.g., Phasor Measurement Units (PMUs) or advanced smart meters) at strategic points in the feeder to provide real-time voltage and current data. Furthermore, a reliable low-latency communication network (e.g., fibre optic, cellular, or power line communication) is essential to transmit these sensor data to a centralized Energy Management System (EMS) and to issue control actions back to the inverters (both PV and V2G) to adjust active and reactive power setpoints. Although this represents an additional investment, the benefits of avoided infrastructure upgrades and improved grid stability can provide a strong economic justification for this necessary enabling infrastructure. The design and cost-benefit analysis of this sensor and communication network, although beyond the scope of this study, are critical next steps for practical implementation.

In a broader planning context, EMS and Advanced Metering Infrastructure (AMI) can enhance V2G feasibility by enabling dynamic tariffs and flexible dispatch of stored energy. Further research should investigate optimized scheduling algorithms and the integration of V2G within innovative grid architectures to maximize technical benefits while maintaining regulatory compliance.

Although the use of V2G in the system can play a relevant role in meeting energy demand, its implementation should be carefully considered due to practical barriers. First and foremost, the primary function of BEBs is to transport people in the city. Therefore, any V2G strategy should only be implemented during non-operational periods, such as overnight parking or extended stops, to ensure optimal performance. This approach must also ensure that vehicles are fully recharged for the following scheduled route, guaranteeing operational energy availability. V2G participation should never prevent buses from reaching the required charge levels by morning, which imposes strict limits on the volume and timing of discharge. Our simulations assume that exported energy originates only from excess capacity within safe operational margins.

Second, the implementation of V2G requires bidirectional inverters and associated upgrades to switchgear and communication systems. These adjustments imply additional costs, making the solution more viable in newly built charging depots rather than in existing infrastructure, thus, highlighting the need for policy incentives and innovative business models to justify investment.

Third, lithium-ion battery degradation is a significant concern. Deep discharges and frequent cycling can accelerate aging and reduce the useful life of the equipment. Advanced Battery Management Systems (BMS) and warranty frameworks for V2G use are essential to mitigate these effects, for example, through optimized SoC windows and thermal management. This issue has been emphasized in recent studies such as [25], which demonstrated that coordinated charging strategies considering battery degradation are key to enabling sustainable V2G applications. Similarly, ref. [42] highlighted the importance of optimizing bus operations together with V2G integration to balance technical feasibility with service reliability in urban transit.

Beyond technical aspects, regulatory, market, and social barriers also constrain V2G adoption. Tariff structures, interconnection standards, and compensation mechanisms remain under development in Brazil and elsewhere. Likewise, public perception of using buses as grid assets requires transparent communication to build trust among operators, regulators, and passengers.

In summary, while the simulations highlight the technical potential of V2G in supporting the distribution grid, practical deployment will depend on addressing operational, economic, and regulatory challenges. Further research should combine grid impact analysis with techno-economic feasibility, battery degradation modeling, and policy frameworks to assess the true scalability of V2G for electric bus fleets.

5.5. Feeder's Current Carrying Capacity During Charging

This section evaluates the current-carrying capacity of the main feeder under different scenarios involving bus fleet charging and energy injection into the grid. The analysed conductor segment corresponds to a compact overhead aluminium cable with a cross-sectional area of 150 mm^2 and PVC insulation, as described in Table 4. According to technical specifications, this conductor has a continuous current rating of 291 A and a maximum allowable current of 415 A.

For this evaluation, the following cases were simulated:

- Case 1: Original feeder, without connecting BEBs.
- Case 2: Nighttime charging of 50 BEBs simultaneously, with insertion of distributed generation at day.
- Case 3: Nighttime charging of 118 BEBs, which has been shown as a threshold for adequate voltage levels according to the methodology discussed in Section 5.3.
- Case 4: Power injection through V2G by 32 BEBs, during peak hours.

Figure 15 presents the peak current values recorded in the main feeder segment for each scenario. Case 1 showed a current well below the thermal limit, peaks at 29.2 A, reflecting only local residential and commercial loads. In Case 2, the maximum current rose to 382.3 A, still within the absolute limit but close enough to warrant attention under real operating conditions.

Furthermore, our analysis revealed a critical thermal overload threshold at 56 simultaneously charged buses, where the current exceeded the 415 A limit. This finding indicates that, while 118 buses might be feasible from a voltage perspective, the thermal capacity of the feeder is saturated much earlier, with only 56 buses. This result was not shown in the figure because of its similar behavior to Case 2.

In Case 3, with 118 buses charging simultaneously, the maximum current reached 890 A, exceeding the permissible limit by more than twice. This result characterizes a severe thermal overload condition, exposing the conductor to accelerated degradation or failure. In a real system, this condition would trigger protection devices and would not be operationally viable without the reinforcement of the infrastructure.

Case 4, referring to V2G injection, showed the same maximum current as Case 2 (382.3 A), confirming that controlled discharge during peak hours does not compromise the thermal capacity of the conductor under the simulated conditions. However, it is essential to note that prolonged injection or exposure to high ambient temperatures can reduce the effective current-carrying capacity.

Table 4. Technical Specifications of the Conductor Segment.

Technical Specification	Description
Geometry	Compact Overhead
Configuration Type	Single
Conductor Size	150 mm^2
Conductor Material	Aluminum
Insulation Type	Protected-PVC
Rated Current	291 A
Maximum Current	415 A

To evaluate the sensitivity of feeder performance to conductor sizing, alternative cross-sections of 185 mm^2, 240 mm^2, and 300 mm^2 aluminium conductors (multifilar, class 2, IEC 60228/EN 60228) were considered in addition to the existing 150 mm^2 cables. The nominal current ratings were obtained from IEC 60364-5-52 [43]. Table 5 summarizes the maximum current supported and the corresponding number of BEBs that could be simultaneously charged without exceeding thermal limits.

Table 5. Sensitivity of feeder performance to conductor sizing.

Conductor Size	Maximum Current (A)	Number of BEBs
150 mm^2	414	56
185 mm^2	510	71
240 mm^2	607	82
300 mm^2	823	114

The sensitivity analysis confirms that increasing the size of the conductor can significantly expand the feeder hosting capacity, reducing both voltage deviations and thermal overloads. Specifically, replacing the 150 mm^2 conductors with 240 mm^2 or 300 mm^2 alternatives increases the feasible number of buses. These findings highlight the urgent need for infrastructure reinforcement in large-scale electrification scenarios. Some viable strategies include:

- Infrastructure reinforcement: Upgrading conductors to larger cross sections (e.g., 240 mm^2 or 300 mm^2), as demonstrated in the sensitivity analysis.

- Load segmentation: Distributing charging points across different feeders to reduce current density per circuit. This strategy is well established in the literature [44,45]. Commonly referred to as *charging infrastructure placement*, it represents a methodology aimed at minimizing the total charging cost of electric buses while reducing the cross-sectional area required to accommodate the bus fleet. Although this approach yields promising results [46], it requires the presence of pre-existing infrastructure at the installation site, or at least the possibility of expanding the city's electrical capacity.
- Staggered charging: Implementing charging strategies at alternate times helps to avoid simultaneous peak loads. This type of strategy is particularly useful in the studied case, given the existing infrastructure in Cuiabá city. Therefore, proposing a smart charging schedule for the bus fleet may not only reduce the required feeder ampacity but also optimize the operational routes of the different buses. In the literature, this type of solution is known as charging scheduling [45], in which the main goal is to minimize charging costs and manage the power demand at charging stations. For instance, [47] proposed an approach to handle the charging schedule of large-scale bus depots, aiming to minimize peak loads. For the problem under consideration, this solution should also address the lack of multiple depots, the requirement for nighttime charging, and the differing routes assigned to each bus.
- Real-time thermal monitoring: Installing current sensors and smart protection devices enables dynamic assessment of thermal capacity and proactive overload management. This methodology is a current practice in transmission systems, known as Dynamic Line Rating (DLR), in which environmental conditions and system characteristics are considered in real time to evaluate the dynamic ampacity of conductors [48]. In this context, such an approach can also be adopted for critical feeders, as in the case of the studied bus feeder. However, its implementation depends on a robust sensor infrastructure and on a reliable methodology to integrate these measurements into a complex state estimation framework [49].

Although this study provides a static view of current peaks, future work should consider conductor thermal time constants and ambient temperature effects for a more accurate assessment of ageing and allowable operation time under overload conditions and staggered charging algorithms.

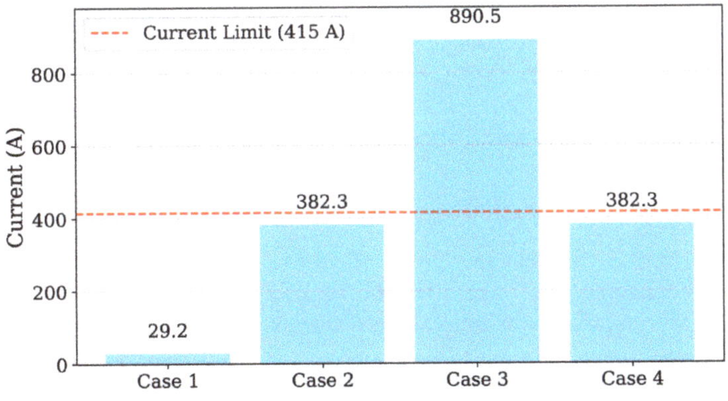

Figure 15. Comparison of maximum current in the feeder.

5.6. Discussion

The simulation results presented in this study provide important insights into the technical challenges and opportunities associated with the large-scale integration of the BEBs related with distributed photovoltaic generation in a medium-voltage distribution feeder. In the initial case, the system operates within the voltage limits, and the presented infrastructure can deal with the thermal stress caused by the operative current. However, as the number of buses increases, the feeder becomes more loaded and presents operational constraints, especially considering the ampacity, requiring a reinforcement scheme.

Numerically, without the presence of the buses, the lowest registered voltage level was 1.0261 pu, whereas after the insertion of 50 buses, the minimum observed voltage dropped to 0.987 pu. It is also relevant to emphasize that the peak load was shifted from the afternoon to the evening. However, simulation of photovoltaic generation revealed the occurrence of reverse power flow during certain hours of the day, highlighting the importance of inverter-based voltage regulation strategies. When considering the electrification of the entire fleet, the results indicated that while voltage limits were respected up to 118 buses, the current carrying capacity of the feeder was already exceeded with only 56 buses, exposing the system to critical thermal overload. Therefore, considering the primary role of buses, that is, transporting people along predefined routes, these results reveal the necessity of a dedicated charging scheme for BEVs. This approach would enable scheduling the charging process according to the minimum energy required for daily operation on each route, aligning technical feasibility with the operational goals of the fleet.

The evaluation of V2G technology confirmed that this strategy can provide adequate support through power injection during peak hours, thereby enhancing the system's robustness in meeting energy demand. Nevertheless, several concerns arise when this process is considered in practice. The simulations revealed overvoltage conditions when V2G coincided with PV generation, underscoring the need for real-time monitoring, inverter-based functionalities such as volt-var or volt-watt control, and reliable communication between operators and charging infrastructure. Although the analysis demonstrated the potential of V2G as an auxiliary support mechanism for the distribution network, further studies are required to consolidate it as a viable solution, particularly under contingency scenarios where system resilience is critical.

It should be noted that the V2G strategy evaluated in this study represents a technical feasibility assessment rather than an operational implementation. In practice, BEBs must prioritize their primary function of passenger transportation and, thus, cannot regularly discharge energy to the grid without affecting service availability. Nevertheless, as electric bus fleets expand, it is foreseeable that some vehicles will remain idle or off-duty for extended periods, such as during maintenance or non-peak hours. In such cases, these buses could be strategically deployed to provide grid support services (e.g., voltage regulation or emergency backup) in specific or contingency scenarios, offering additional flexibility to the distribution network.

The performed studies demonstrated the technical feasibility of integrating BEBs into the system, including the potential reinforcement of conductors to address ampacity constraints, which proved to be more critical than voltage limitations. However, since the analysis was carried out in a computational environment, additional challenges are expected in practical field implementation. Furthermore, the loads were modeled deterministically, without explicitly capturing stochastic variability or operational uncertainties. In this context, future work could include experimental validation, non-linear stochastic loads, and optimization techniques to define charging load shapes that take into account route requirements and state-of-charge conditions.

6. Conclusions

The new proposal evaluated the technical impacts of BEB implementation in a real medium-voltage distribution feeder in Cuiabá, Brazil. This study considered PV generation, V2G operation, and the potential for conductor reinforcement. The findings provided clear insights into the system's limitations, especially in the context of bus fleet electrification.

Considering only voltage constraints, up to 118 BEBs can be charged without violating the regulatory minimum voltage requirement. The system registered a minimum of 0.987 pu, compared to 1.026 pu in the base case. Nevertheless, the thermal limits related to cable ampacity proved to be more restrictive, since only 56 buses can be charged without exceeding the conductor's capacity. This confirms a common observation in power systems: ampacity is usually more critical than voltage stability.

PV generation covered about 64.12% of the fleet's daily energy demand; however, reverse power flow and rising voltage levels were observed, confirming the need for inverter-based voltage regulation. This analysis introduced a quantitative contribution by establishing a relationship between PV penetration levels, voltage deviations, and reverse flow in the feeder in Cuiabá. Regarding V2G operation, although its development in field applications raises some concerns, the study showed that BEB operation under such conditions can support the grid during peak hours.

Moreover, the sensitivity analysis of conductor reinforcement showed that increasing the nominal cross-section from 150 mm^2 to 300 mm^2 increased the number of buses that can be charged simultaneously from 56 to 114, nearly doubling the initial value. This result demonstrates that such infrastructure modification provides an immediate engineering solution to allow more BEBs to be integrated into the charging process at the same time.

Finally, it is important to highlight that a thorough economic feasibility study must precede any decision on large-scale implementation. This study should consider not only initial investment costs, but also operational costs, revenue potential from V2G services, infrastructure reinforcement costs, and investments in monitoring and control systems.

Author Contributions: Conceptualization, C.G.P., C.d.A.F. and F.M.d.V.; methodology, C.G.P., C.d.A.F. and F.P.d.A.; software, C.G.P., C.d.A.F. and J.P.B.; validation, C.G.P., C.d.A.F., F.P.d.A., F.M.d.V., J.P.B. and M.E.R.A.; formal analysis, C.G.P., C.d.A.F. and F.P.d.A.; investigation, C.G.P., C.d.A.F., F.P.d.A. and M.E.R.A.; resources, F.M.d.V., F.P.d.A. and M.E.R.A.; data curation, C.G.P., C.d.A.F., J.P.B. and M.E.R.A.; writing—original draft preparation, C.G.P., C.d.A.F., F.P.d.A. and F.M.d.V.; writing—review and editing, C.d.A.F., F.P.d.A., F.M.d.V., J.P.B. and M.E.R.A.; visualization, C.G.P. and C.d.A.F.; supervision, C.d.A.F., J.P.B. and M.E.R.A.; project administration, C.d.A.F. and J.P.B.; funding acquisition, C.d.A.F. and M.E.R.A. All authors have read and agreed to the published version of the manuscript.

Funding: This research was funded by the Agencia Nacional de Investigación y Desarrollo (ANID), through the FONDECYT Regular Grant No. 1220556 and the FONDAP SERC Grant No. 1523A0006; the National Council of Science and Technology (CONACYT), through Grant No. PINV01-743; the British Council, through the IND/CONT/G/23-24/14 – Going Global Partnerships Industry-Academia Collaborative Grant; the Programa de Redução de Assimetrias na Pós-Graduação (PRAPG)—Call No. 14/2023—DRI—CAPES (ID No. 046.821.818-15); the National Council for Scientific and Technological Development—Brazil (CNPq), through Grant No. 441715/2023-0; the Mato Grosso Research Foundation (FAPEMAT), through Grant No. 0001047/2022; and the Faculty of Architecture, Engineering and Technology of the Federal University of Mato Grosso (FAET/UFMT).

Data Availability Statement: The raw data supporting the conclusions of this article will be made available by the authors on request.

Conflicts of Interest: The authors declare no conflicts of interest.

Abbreviations

The following abbreviations are used in this manuscript:

BEBs	Battery Electric Buses
GIS	Geographic Information System
MV	Medium Voltage
BDGD	Geographic Database of the Distribution Company
ANEEL	Brazilian Electricity Regulatory Agency
PV	Photovoltaic
EMS	Energy Management System
AMI	Advanced Metering Infrastructure
V2G	Vehicle-to-Grid
OpenDSS	Open Distribution System Simulator
GHG	Greenhouse Gas
PRODIST	Procedures for Energy Distribution Standard
BYD	Build Your Dreams (vehicle manufacturer)
LiFePO4	Lithium Iron Phosphate (battery type)
PVC	Polyvinyl Chloride (insulation material)
kW	Kilowatt
kWh	Kilowatt-hour
HSP	Full Sun Hours
PMUs	Phasor Measurement Units
BMS	Advanced Battery Management Systems
DLR	Dynamic Line Rating

Appendix A. PV System Technical Specifications

Appendix A.1. Characteristics and Parameters of the PV System

This appendix provides the complete technical specifications and modeling parameters of the photovoltaic system used in the simulations.

Appendix A.1.1. System Configuration Overview

The distributed generation system consists of a 4000 kWp photovoltaic plant designed to operate within the 3 MW limit established by Brazilian Law No. 14,300/2022 for mini distributed generation.

Table A1. PV System Component Specifications.

Component	Parameter	Value
PV Panel	Model	Titan RDM120-8-600BMDG
	Rated Power	600 Wp
	Panel Area	2.83 m^2
	Panel Efficiency	21.2%
Inverter	Model	WEG SIW500G T100 W0
	Unit Power	100 kW
	Power Factor	1
	Efficiency	98%
System Configuration	Number of Inverters	40 units
	Total Installed Power	4000 kWp
	Connection Voltage	0.38 kV
	Total Estimated Generation	15,390 kWh/day
	Overall System Efficiency	75%

Appendix A.1.2. OpenDSS Modeling Parameters

The photovoltaic generation systems were modeled in OpenDSS using the PVSystem class, configured to represent realistic operational behavior under varying temperature and irradiance conditions. The power and efficiency variations were defined through the `XYCurve` function, which allows modeling nonlinear relationships between temperature, irradiance, and output power. In this study, the temperature–power curve accounted for power derating with increasing temperature, while the efficiency curve reflected inverter conversion efficiency across different loading levels. These curves were applied to all `PVSystem` instances to ensure realistic dynamic responses to environmental variations.

Table A2 summarizes the main operational parameters adopted for each PVSystem element in the simulations.

Table A2. Operational parameters for each PVSystem instance.

Description	Parameter (Value)
Inverter capacity	100 KVA
Maximum power point tracking (MPPT)	100 kW
Reference irradiance	0.98 kW/m^2
Reference temperature	40 °C
Minimum irradiance for operation	0.1%
Minimum irradiance to remain connected	0.1%
Power factor	1.0

Appendix A.1.3. Performance Calculation Basis

The system performance estimates are based on:

- Local Irradiation: 5.13 HSP daily average for Cuiabá
- System Efficiency: 75% overall system efficiency factor
- Performance Ratio: Includes losses from temperature, soiling, inverters, and wiring

This comprehensive technical specification provides the complete dataset used for the PV system modeling in the OpenDSS simulations, ensuring the reproducibility and transparency of the analysis presented in the main paper.

References

1. SEEG—Sistema de Estimativas de Emissões de Gases de Efeito Estufa [SEEG —Greenhouse Gas Emissions Estimation System]. Emissões de Gases de Efeito Estufa no Setor de Transportes-2023 [Greenhouse Gas Emissions in the Transportation Sector—2023]. 2024. Available online: https://seeg.eco.br/ (accessed on 2 July 2025).
2. Lajunen, A. Lifecycle costs and charging requirements of electric buses with different charging methods. *J. Clean. Prod.* **2018**, *172*, 56–67. [CrossRef]
3. López, I.; Calvo, P.L.; Fernández-Sánchez, G.; Sierra, C.; Corchero, R.; Chacón, C.O.; Juan, C.d.; Rosas, D.; Burgos, F. Different Approaches for a Goal: The Electrical Bus-EMT Madrid as a Successful Case Study. *Energies* **2022**, *15*, 6107. [CrossRef]
4. Zhang, Y.; He, Y.; Song, Z. Optimal planning and scheduling for fast-charging electric bus system with distributed photovoltaics. *Transp. Res. Part D Transp. Environ.* **2025**, *139*, 104584. [CrossRef]
5. Manzolli, J.A.; Trovão, J.P.F.; Henggeler Antunes, C. Chapter 3-Forecasting of vehicle electrification in modern power grids. In *Vehicle Electrification in Modern Power Grids*; Monteiro, V., Afonso, J.L., Williamson, S., Eds.; Elsevier: Amsterdam, The Netherlands, 2024; pp. 47–73. [CrossRef]
6. Li, B.; Li, Z.; He, D. Research and optimization of energy management system for photovoltaic vehicles. *Energy* **2024**, *289*, 129776. [CrossRef]
7. Huang, P.; Munkhammar, J.; Fachrizal, R.; Lovati, M.; Zhang, X.; Sun, Y. Comparative studies of EV fleet smart charging approaches for demand response in solar-powered building communities. *Sustain. Cities Soc.* **2022**, *85*, 104094. [CrossRef]
8. Rao, S.P.; Olusegun, T.S.; Ranganathan, P.; Kose, U.; Goveas, N. Vehicle-to-Grid technology: Opportunities, challenges, and future prospects for sustainable transportation. *J. Energy Storage* **2025**, *110*, 114927. [CrossRef]

9. Jia, H.; Ma, Q.; Li, Y.; Liu, M.; Liu, D. Integrating Electric Vehicles to Power Grids: A Review on Modeling, Regulation, and Market Operation. *Energies* **2023**, *16*, 6151. [CrossRef]
10. Qu, Z.; Song, J.; Liu, Y.; Yang, S. Optimization Model of EV Charging and Discharging Price Considering Vehicle Owner Response and Power Grid Cost. *J. Electr. Eng. Technol.* **2019**, *14*, 2251–2261. [CrossRef]
11. Gough, R.; Dickerson, C.; Rowley, P.; Walsh, C. Vehicle-to-grid feasibility: A techno-economic analysis of EV-based energy storage. *Appl. Energy* **2017**, *192*, 12–23. [CrossRef]
12. Mattos, M.M.; Archetti, J.A.G.; Bitencourt, L.d.A.; Wallberg, A.; Castellucci, V.; Dias, B.H.; de Oliveira, J.G. Analysis of voltage control using V2G technology to support low voltage distribution networks. *Iet Gener. Transm. Distrib.* **2024**, *18*, 1133–1157. [CrossRef]
13. Hungbo, M.; Gu, M.; Meegahapola, L.; Littler, T.; Bu, S. Impact of electric vehicles on low-voltage residential distribution networks: A probabilistic analysis. *IET Smart Grid* **2023**, *6*, 536–548. [CrossRef]
14. Li, S.; Jenn, A. The impact of electric vehicle charging on electricity distribution networks: Evidence from real-world data. *Proc. Natl. Acad. Sci. USA* **2024**, *121*, e2312345119. [CrossRef]
15. Singh, B.; Verma, K.; Tripathi, R. Feeder reconfiguration for loss minimization in distribution networks with electric vehicles. *J. Electr. Syst. Inf. Technol.* **2019**, *6*, 1–11. [CrossRef]
16. Manzolli, J.A.; Trovão, J.P.F.; Henggeler Antunes, C. Electric bus fleet charging management: A robust optimisation framework addressing battery ageing, time-of-use tariffs, and energy consumption uncertainty. *Appl. Energy* **2025**, *381*, 125137. [CrossRef]
17. Jia, H.; Zhang, X.; Li, Z. A comprehensive review on electric vehicles integration into power systems: From market and policy to technology perspectives. *Renew. Sustain. Energy Rev.* **2023**, *176*, 113177. [CrossRef]
18. Basso, F.; Feijoo, F.; Pezoa, R.; Varas, M.; Vidal, B. The impact of electromobility in public transport: An estimation of energy consumption using disaggregated data in Santiago, Chile. *Energy* **2024**, *286*, 129550. [CrossRef]
19. Basnak, P.; Giesen, R.; Muñoz, J.C. Estimation of crowding factors for public transport during the COVID-19 pandemic in Santiago, Chile. *Transp. Res. Part A Policy Pract.* **2022**, *159*, 140–156. [CrossRef]
20. Proboste, F.; Muñoz, J.C.; Gschwender, A. Comparing social costs of public transport networks structured around an Open and Closed BRT corridor in medium sized cities. *Transp. Res. Part A Policy Pract.* **2020**, *138*, 187–212. [CrossRef]
21. Hidalgo, D.; Giesen, R.; Muñoz, J.C. Bus Rapid Transit: End of trend in Latin America? *Data Policy* **2024**, *6*, e2. [CrossRef]
22. Camargo, R.S.; De Melo, V.S.; Brunoro, M.; Nunes, W.T.; Krebel, F.; Fernandes, M.C.; Nunes, R.B. Impact of Vehicle-to-Grid (V2G) Technology on the Brazilian Southeast/Midwest Subsystem. In Proceedings of the 2023 15th IEEE International Conference on Industry Applications (INDUSCON), São Bernardo do Campo, Brazil, 22–24 November 2023; pp. 270–277. [CrossRef]
23. Drude, L.; Pereira Junior, L.C.; Rüther, R. Photovoltaics (PV) and electric vehicle-to-grid (V2G) strategies for peak demand reduction in urban regions in Brazil in a smart grid environment. *Renew. Energy* **2014**, *68*, 443–451. [CrossRef]
24. Soares, L.O.; Sodré, J.R.; Hernández-Callejo, L.; Boloy, R.A.M. Electric vehicle adoption in Brazil: Economical analysis and roadmap. *Transp. Res. Part D Transp. Environ.* **2024**, *137*, 267–310.
25. Manzolli, J.A.; Trovão, J.P.F.; Henggeler Antunes, C. Electric bus coordinated charging strategy considering V2G and battery degradation. *Energy* **2022**, *254*, 124252. [CrossRef]
26. Agência Nacional de Energia Elétrica (ANEEL) [Brazilian National Electric Energy Agency]. Base de Dados Geográfica da Distribuidora (BDGD) [Distributor Geographic Database]. 2023. Available online: https://dadosabertos.aneel.gov.br/dataset/base-de-dados-geografica-da-distribuidora-bdgd (accessed on 10 September 2025).
27. Agência Nacional de Energia Elétrica [Brazilian National Electric Energy Agency]. 2022. Dados Abertos [Open Data]. Available online: https://dadosabertos.aneel.gov.br/ (accessed on 7 July 2025).
28. QGIS Development Team. QGIS Geographic Information System. Available online: https://www.qgis.org/ (accessed on 12 July 2025).
29. Google LLC. Google Colaboratory. Available online: https://colab.google/ (accessed on 12 July 2025).
30. Electric Power Research Institute (EPRI). *OpenDSS Manual*; EPRI: Palo Alto, CA, USA, 2010; Version 7.6, Revision 2.
31. BYD. The BYD D11A Datasheet. 2020. Available online: https://bydbrasil.com.br/wp-content/uploads/2024/07/D11A.pdf (accessed on 2 July 2025).
32. Niu, S.; Zhang, C.; Shi, Y.; Niu, S.; Jian, L. Foreign object detection considering misalignment effect for wireless EV charging system. *ISA Trans.* **2022**, *130*, 655–666. [CrossRef]
33. Farah, B.M.C.; Bonaldo, J.P.; Fantin, C.A. Análise da viabilidade econômica para implantação de ônibus elétrico em Cuiabá com base em dados e experiências similares em outras cidades [Analysis of the economic feasibility of implementing an electric bus in Cuiabá based on data and similar experiences in other cities]. In Proceedings of the Congresso Brasileiro de Automática 2024 [Brazilian Congress of Automation], Rio de Janeiro, Brazil, 15–18 October 2024. Available online: https://www.sba.org.br/cba2024/papers/paper_1934.pdf (accessed on 10 September 2025).

34. Centro de Referência para a Energia Solar e Eólica Sérgio de Salvo Brito (CRESESB) [Sérgio de Salvo Brito Reference Center for Solar and Wind Energy]. Potencial Solar-SunData v 3.0 [Solar Potential-SunData v 3.0]. Available online: http://www.cresesb.cepel.br/sundata/ (accessed on 8 September 2025).
35. BRASIL. Lei Nº 14.300, de 6 de Janeiro de 2022 [Brazilian Law 14.300, January 6th of 2022]. Institui o Marco Legal da Microgeração e Minigeração Distribuída. Brasília, DF: Diário Oficial da União [Establishes the Legal Framework for Distributed Microgeneration and Minigeneration. Brasília, Official Gazette of the Union]. Available online: https://www.planalto.gov.br/ccivil_03/_ato2019-2022/2022/lei/l14300.htm. (accessed on 10 June 2025).
36. WEG. SIW—Solar Inverter Weg Datasheet. 2023. Available online: https://static.weg.net/medias/downloadcenter/h09/h17/WEG-SIW-inversores-string-50105878-pt.pdf (accessed on 10 June 2025).
37. Empresa de Pesquisa Energética (EPE) [Brazilian Energy Research Company]. Caderno de Parâmetros de Custos-Geração e Transmissão: Estudos do Plano Decenal de Expansão de Energia 2034 [Cost Parameters Notebook - Generation and Transmission: Studies of the 2034 Ten-Year Energy Expansion Plan]. 2024. Available online: https://www.epe.gov.br/sites-pt/publicacoes-dados-abertos/publicacoes/PublicacoesArquivos/publicacao-804/topico-709/Caderno%20de%20Custos%20Gera%C3%A7%C3%A3o%20e%20Transmiss%C3%A3o_PDE2034_2024.09.06.pdf#search=custo%20de%20gera%C3%A7%C3%A3o (accessed on 8 September 2025).
38. Montoya, O.D.; Giraldo, J.S.; Grisales-Noreña, L.F.; Chamorro, H.R.; Alvarado-Barrios, L. Accurate and efficient derivative-free three-phase power flow method for unbalanced distribution networks. *Computation* **2021**, *9*, 61. [CrossRef]
39. Agência Nacional de Energia Elétrica [Brazilian National Electric Energy Agency]. Procedimento de Distribuição. Prodist: Módulo 8-Qualidade do Fornecimento de Energia Elétrica [Distribution Procedures, Prodist: Module 8-Quality of Electricity Supply]. 2021. Available online: https://www2.aneel.gov.br/cedoc/aren2021956_2_7.pdf (accessed on 1 July 2025).
40. Associação Matogrossense dos Transportadores Urbanos (AMTU) [Mato Grosso Association of Urban Transporters]. Conheça a Frota [Get to Know the Fleet]. 2025. Available online: https://amtu.com.br/conheca-a-frota/ (accessed on 20 July 2025).
41. Tan, K.M.; Ramachandaramurthy, V.K.; Yong, J.Y. Integration of electric vehicles in smart grid: A review on vehicle to grid technologies and optimization techniques. *Renew. Sustain. Energy Rev.* **2016**, *53*, 720–732. [CrossRef]
42. Son, J.; Im, J.; Kim, D. Urban transit optimization: Efficient electric bus operations and vehicle-to-grid integration. *Comput. Ind. Eng.* **2025**, *205*, 111169. [CrossRef]
43. *IEC 60364-5-52*; Low-Voltage Electrical Installations—Part 5-52: Selection and Erection of Electrical Equipment—Wiring Systems. International Electrotechnical Commission: Geneva, Switzerland, 2023.
44. Yang, S.; An, K.; Rich, J. Electric Bus Battery Swapping and Charging Management across Multiple Depots. *Omega* **2025**, *138*, 103434. [CrossRef]
45. Perumal, S.S.; Lusby, R.M.; Larsen, J. Electric bus planning & scheduling: A review of related problems and methodologies. *Eur. J. Oper. Res.* **2022**, *301*, 395–413. [CrossRef]
46. Iliopoulou, C.; Kepaptsoglou, K. Integrated transit route network design and infrastructure planning for on-line electric vehicles. *Transp. Res. Part D Transp. Environ.* **2019**, *77*, 178–197. [CrossRef]
47. Jahic, A.; Eskander, M.; Schulz, D. Charging schedule for load peak minimization on large-scale electric bus depots. *Appl. Sci.* **2019**, *9*, 1748. [CrossRef]
48. Lawal, O.A.; Teh, J. Assessment of dynamic line rating forecasting methods. *Electr. Power Syst. Res.* **2023**, *214*, 108807. [CrossRef]
49. Foss, S.D.; Maraio, R.A. Dynamic line rating in the operating environment. *IEEE Trans. Power Deliv.* **2002**, *5*, 1095–1105. [CrossRef]

World Electric Vehicle Journal

MDPI

Article

Power Management for V2G and V2H Operation Modes in Single-Phase PV/BES/EV Hybrid Energy System

Chayakarn Saeseiw [1], Kosit Pongpri [1], Tanakorn Kaewchum [2], Sakda Somkun [2] and Piyadanai Pachanapan [1,3,*]

[1] Department of Electrical and Computer Engineering, Faculty of Engineering, Naresuan University, Phitsanulok 65000, Thailand; chayakarns66@nu.ac.th (C.S.); kositp65@nu.ac.th (K.P.)

[2] School of Renewable Energy and Smart Grid Technology, Naresuan University, Phitsanulok 65000, Thailand; tanakornka65@nu.ac.th (T.K.); sakdaso@nu.ac.th (S.S.)

[3] Centre of Excellence on Energy Technology and Environment (CeTe), Faculty of Engineering, Naresuan University, Phitsanulok 65000, Thailand

* Correspondence: piyadanip@nu.ac.th

Abstract: A multi-port conversion system that connects photovoltaic (PV) arrays, battery energy storage (BES), and an electric vehicle (EV) to a single-phase grid offers a flexible solution for smart homes. By integrating Vehicle-to-Grid (V2G) and Vehicle-to-Home (V2H) technologies, the system supports bidirectional energy flow, optimizing usage, improving grid stability, and supplying backup power. The proposed four-port converter consists of an interleaved bidirectional DC-DC converter for high-voltage BES, a bidirectional buck–boost DC-DC converter for EV charging and discharging, a DC-DC boost converter with MPPT for PV, and a grid-tied inverter. Its non-isolated structure ensures high efficiency, compact design, and fewer switches, making it suitable for residential applications. A state-of-charge (SoC)-based power management strategy coordinates operation among PV, BES, and EV in both on-grid and off-grid modes. It reduces reliance on EV energy when supporting V2G and V2H, while SoC balancing between BES and EV extends lifetime and lowers current stress. A 7.5 kVA system was simulated in MATLAB/Simulink to validate feasibility. Two scenarios were studied: PV, BES, and EV with V2G supporting the grid and PV, BES, and EV with V2H providing backup power in off-grid mode. Tests under PV fluctuations and load variations confirmed the effectiveness of the proposed design. The system exhibited a fast transient response of 0.05 s during grid-support operation and maintained stable voltage and frequency in off-grid mode despite PV and load fluctuations. Its protection scheme disconnected overloads within 0.01 s, while harmonic distortions in both cases remained modest and complied with EN50610 standards.

Keywords: electric vehicles; high-voltage battery; multi-port converter; photovoltaic; vehicle-to-grid; vehicle-to-home

1. Introduction

Many countries worldwide are seeing considerable growth in residential solar photovoltaic (PV), battery energy storage (BES), and electric vehicles (EVs) [1]. The progress highlights a shift toward sustainable energy solutions, with significant investments in PV, BES, and EVs aimed at achieving energy security and reducing carbon emissions. According to the International Energy Agency (IEA) in [2], the global EV fleet is projected to exceed 250 million units by 2030, representing an aggregated energy storage capacity of more than

10 TWh. This enormous potential positions EVs not only as a means of transportation but also as distributed energy resources capable of supporting the grid and end-users.

The integration of Vehicle-to-Grid (V2G) and Vehicle-to-Home (V2H) technologies can transform EVs into active energy resources, rather than passive consumers; therefore, enhancing both grid stability and residential resilience [3]. V2G offers bidirectional energy flow, allowing EVs to participate in demand response, peak shaving, ancillary services such as frequency and voltage regulation, as well as facilitating renewable energy integration. These capabilities help to balance supply and demand effectively, thereby improving overall grid performance [4]. Meanwhile, V2H enables EVs to improve household/business resilience by supplying backup power during grid outages and enabling optimized energy management through smart home systems [5]. Despite these advantages, challenges such as battery degradation and user participation remain barriers to large-scale deployment.

To realize these benefits, efficient and flexible power electronic interfaces are required. Multi-port converters present an effective solution to combine those multi energy sources and the grid [6,7] which integrating sources at multiple ports can achieve weight and size reduction and efficiency improvement. Research on V2G and V2H has gained considerable attention over the last decade. Early works mainly focused on dual-port converters connecting EVs and the grid [8,9]. However, dual-port solutions lack the flexibility to integrate additional energy resources.

Three-port converters were then introduced to integrate PV or BES with EVs and the grid. A PV/EV/grid interface with V2G capabilities is proposed in [10,11], and offers significant benefits in terms of sustainable EVs and robust grid support, particularly in enhancing grid stability and integrating renewable energy. Similarly, a three-port converter that combines fuel cell/BES, EV, and grid can enhance the backup power capacity to support V2H applications [12,13]. Nonetheless, these benefits come with potential downsides such as system complexity, cost, and dependence on intermittent renewable sources.

Recently, several studies have focused on four-port converters for grid-integrated PV/BES/EV hybrid energy systems [14–16]. These multi-port configurations are more advantageous for integrating an increasing number of energy sources into a modern power system, which typically includes a unidirectional port for PV and three bidirectional ports for BES, EV (with V2G and V2H functionalities), and the electricity grid. Compared to dual-port or three-port designs, four-port converters offer enhanced functionality by supporting multiple bidirectional interfaces, extending backup power during outages, and improving overall efficiency. However, the existing works on four-port architectures frequently lack detailed consideration of V2G/V2H operational scenarios, as well as limited bidirectional power flow capabilities across all ports, often operating in fixed configurations with only one power flow path between ports.

Upcoming topology trends and challenges for bidirectional DC-DC and DC-AC converters are discussed in [17]. It was discovered that non-isolated DC-DC converters connect BES or EVs directly to the grid or residential loads without a galvanic isolation barrier, making them simpler and often more efficient [18]. The non-isolated configuration offers compact, higher efficiency by eliminating transformer losses, and is generally cost-effective. These benefits make them attractive for household energy applications and V2G/V2H implementations. In this work, a non-isolated four-port converter is selected to leverage these advantages, providing direct power sharing among PV, BES, EVs, and the grid with fewer conversion stages and faster dynamic response. Although the absence of isolation raises concerns regarding back-feeding and grounding, these risks are mitigated through advanced fault detection [19], the use of higher-rated components [20], and proper grounding schemes [21], ensuring compliance with safety standards such as IEEE 1547. Thus, the non-isolated architecture represents a practical trade-off, offering superior efficiency, compactness, and cost-effectiveness while maintaining safety through protective measures rather than galvanic separation.

One of the most serious concerns about V2G/V2H is the risk of increased battery degradation. As battery degradation is highly influenced by the depth of discharge (DoD) and current stress, frequent deep cycling or high rate charging/discharging can significantly decrease the number of lifetime cycles. For example, recent works in [22,23] demonstrated that increased current stress and deeper DoD are critical factors influencing lithium-ion battery aging, both fresh and second-life batteries. These stress factors, along with state of charge, cycle count, and temperature, all have an impact on battery operational efficiency and economic viability. Hence, it is important to manage the DoD of BES and EV batteries alongside managing the power flow. Power sharing, depending on the battery's state of charge (SoC), can be an effective approach to improve battery life and reduce current stress [24]. EVs should be encouraged to provide more power to the grid or house higher than BES, due to an EV's battery capacity (40–70 kWh) being typically much larger than that of a residential-scale BES (5–10 kWh).

Contribution: Motivated by these challenges, a single-phase four-port converter with power flow management based on SoC is implemented to support grid and home during on- and off-grid operations. The proposed non-isolated multi-port converter consists of (1) interleaved bidirectional DC-DC converter for high-voltage BES, (2) buck–boost bidirectional DC-DC converter for EV charger, (3) DC-DC boost converter with the maximum power point tracking (MPPT) for PV arrays, and (4) grid-tied DC-AC inverter to integrate with the single-phase electrical grid. The DC link voltage is controlled by the BES system which is essential for the power-balancing mechanism. The DC-AC inverter operates in grid-forming or grid-following mode through a phase lock loop (PLL) synchronization. Additionally, the SoC-based power-sharing between BES and EVs is used to avoid the high DoD during V2G and V2H operations.

The main contributions of this work are the following:

1. Design of a novel four-port topology tailored for PV/BES/EV hybrid systems with V2G and V2H functionalities, to support both on-grid and off-grid operations.
2. Development of a coordinated control strategy that ensures stable bidirectional power flow and fast dynamic response. Also, SoC-based power management can help to reduce battery degradation by reducing DoD and charge/discharge current stress.
3. Comprehensive performance evaluation through small time-step simulations, demonstrating the performance benefits of the proposed four-port converter in various operating modes.

2. Multi-Port Conversion System

The configuration of a single-phase multi-port converter is illustrated in Figure 1. Note that the direction of arrow means the positive power flow. The details of each port are as follows.

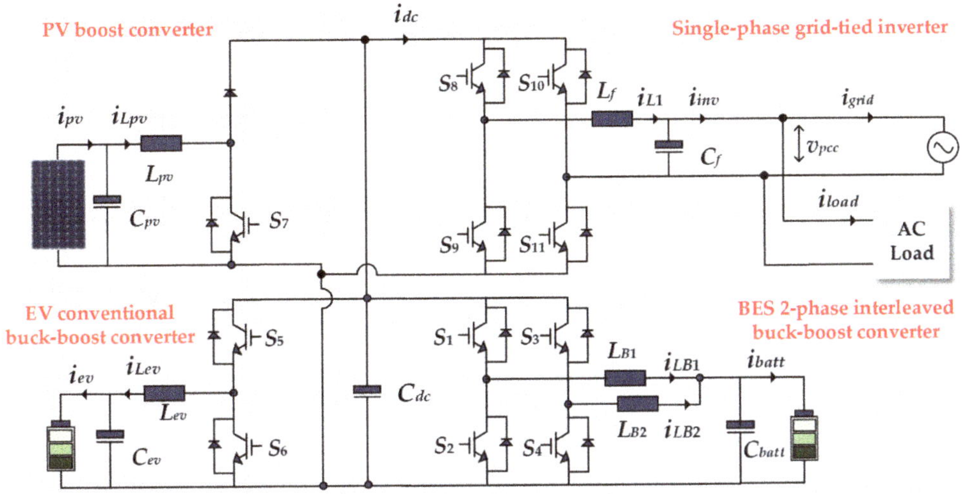

Figure 1. A proposed single-phase four-port conversion system.

2.1. Interleaved Bidirectional DC-DC Converter for High-Voltage BES

The choice of DC-DC converter topology in a BES system depends on factors such as voltage levels, efficiency needs, cost constraints, and the desired complexity. The N-phase interleaved topologies are particularly favored in high-power applications because they distribute current evenly across multiple phases, reducing ripples and heat dissipation [25]. This leads to enhanced system efficiency, better scalability, and reduced passive component requirements, making them ideal for energy storage implementations. The basic operation of two-phase interleaved bidirectional DC-DC converter, both in charge and discharge mode, is illustrated in Figure 2.

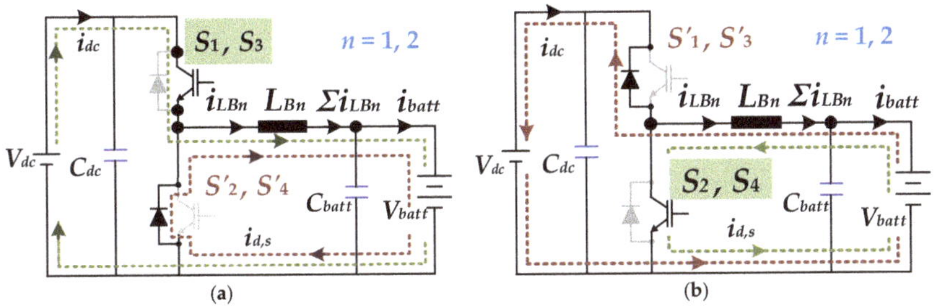

Figure 2. Basic operation of two-phase interleaved bidirectional DC-DC converter: (**a**) charging mode and (**b**) discharging mode.

Switches S_1, S'_2 and S_3, S'_4 are operated with a 180° phase shift and the switches on each phase (S_1-S_2 and S_3-S_4) are complementary. The change in each inductor current, i_{LBn}, can be written as follows:

$$L_{Bn}\frac{di_{LBn}}{dt} = \begin{cases} V_{dc}(t) - V_{batt}(t) & : S_1\,, S'_2 \text{ and } S_3\,, S'_4 \\ -V_{batt}(t) & : S'_1\,, S_2 \text{ and } S'_3\,, S_4 \end{cases} \tag{1}$$

where n is 1, 2, ..., N, and the symbol (′) denotes the complementary switches.

The control structure is divided into two systems: the voltage loop, which is the outer control system, and the inner loop, which is the average inductor current control system. The voltage loop regulator compares the reference DC link voltage ($V_{dc,ref}$) with the sensed voltage ($V_{dc,sensed}$), the voltage error value is eliminated by the PI controller and generates the reference battery current value (i_{batt}) for the current loop. The block diagrams of their control are shown in Figure 3.

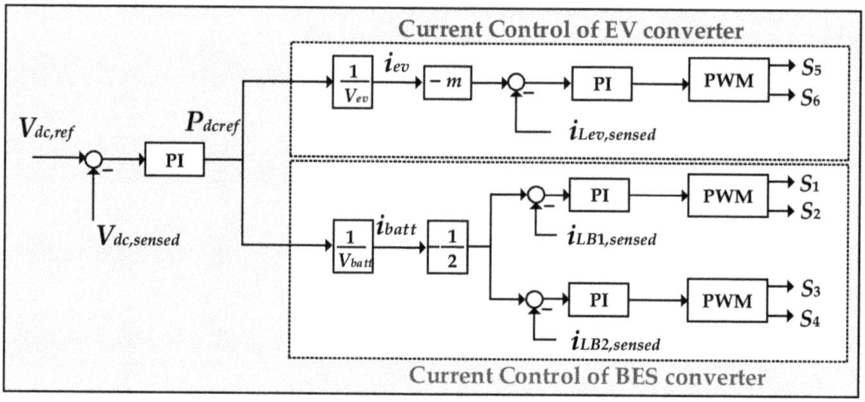

Figure 3. BES and EV control systems with power sharing.

After i_{batt} is sent to the current control loop, it is split in half into the parallel circuits (i_{batt}/N) which $N = 2$ for a two-phase interleaved converter. This divided value is equal to the reference current of the inductor in each phase, i_{LB1} and i_{LB2}. Then, the system sends a pulse width modulation (PWM) signal to control the operation of switches S_1 to S_4. In addition, those switches can be IGBTs, or power MOSFETs.

2.2. Buck–Boost Bidirectional DC-DC Converter for EV

The buck–boost DC-DC converter architecture was chosen to provide DC charge to EVs and give V2G/V2H capability to the grid and residential houses. This converter can step down (buck) voltage level for EV charging or step up (boost) voltage for supporting V2G/V2H function. The structure is similar to the interleave topology, but there is only one phase, making it generally simpler to design and implement when only two switches, S_5 and S_6, are controlled. However, the current ripple is higher which requires a larger inductor. The heat and power losses are consequently increased.

The inductor current, i_{Lev}, is regulated by PI controller as shown in Figure 3. Since the power sharing is applied to BES and EV, the DC powers contributed by BES and EV and then delivered to the DC link (P_{dcref}) are written as follows:

$$P_{dcref} = \underbrace{P_{BESS} + m \times P_{EV}}_{power\ sharing} \text{ where } m \in \{0, 1, 2, 3\} \quad (2)$$

where P_{BESS} is the power contribution from the BES; P_{EV} is the power contribution from the EV; and m is a proportional gain that varies according to the SoC level of the EV battery, adjusting the power-sharing ratio.

This allows the EV to assist in supplying power based on the parameter m, which determines the power-sharing ratio, as presented in Table 1. The value of m ranges from 0 to 3, depending on the SoC level of the EV battery. In addition, $m = 0$ indicates that the SoC is too low to support power, while higher values (i.e., 1, 2, or 3) offer more power-sharing contributions from the EV.

Table 1. Power-sharing ratio of EV and BES batteries.

SoC of EV	Status	Power-Sharing Ratio (EV:BES)	m
50% > SoC_EV	No discharge	0:1	0
50% ≤ SoC_EV < 70%	Low	1:1	1
70% ≤ SoC_EV < 90%	Medium	2:1	2
90% ≤ SoC_EV ≤ 100%	High	3:1	3

Although the primary power-sharing ratio is based on EV's SoC, the BES's SoC is monitored in parallel. The power management strategy includes conditions where the BES stops providing power support if its SoC falls below a certain threshold (e.g., 50%). In such cases, the BES only provides voltage control at the DC link, and only the PV system can deliver power to the grid and load. This indicates that while the EV's SoC dictates the ratio of power sharing, the BES's SoC acts as a 'hard limit' to prevent over-discharge and guarantee its operational reliability.

2.3. DC-DC Boost Converter with MPPT for PV

A DC-DC boost converter with MPPT algorithm is widely used in PV systems to optimize energy harvesting by raising the panel's low voltage to a higher, stable output voltage. This design improves efficiency by adjusting the duty cycle of the boost converter to adapt to variable sunlight conditions.

The incremental conduct (IC) approach is suitable for this application since it employs the slope of the PV array power curve to determine the maximum power point (MPP), as seen in Figure 4. When the power derivative with respect to voltage (dP/dV) is zero, the system has reached MPP. This algorithm outperforms traditional Perturb and Observe (P&O) approaches by eliminating oscillations around MPP [26].

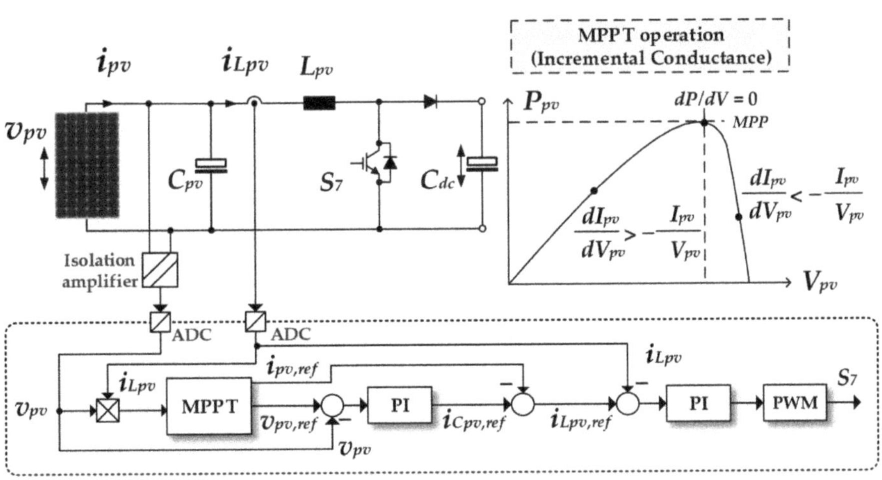

Figure 4. The control structure of DC-DC boost converter with MPPT for PV.

The algorithm for the MPPT defines the voltage and current references, $v_{pv,ref}$ and $i_{pv,ref}$ that can extract the highest power. The v_{pv} and i_{Lpv} are regulated using PI controllers, resulting in a change in duty ratio (D), of PMW signal to control switch S_7.

2.4. Grid-Tied Full Bridge DC-AC Inverter

A typical single-phase grid-tied inverter consists of several key components: a DC power source, a full-bridge inverter stage for DC-AC conversion, a filter for waveform shaping, and a grid synchronization unit. The grid synchronization ensures that the inverter output matches the grid's frequency, phase, and voltage, while the filter reduces high frequency switching noise and smooths the output waveform.

The DC-AC conversion is achieved using a full-bridge inverter circuit, which consists of four switches, S_8 to S_{11}. The PWM technique is often used to refine the waveform and reduce harmonic distortion. Moreover, the LC filter, which is of low cost, easy to implement, and has a simpler design, was chosen.

The current control approach is based on the synchronous reference frame (dq), which transforms AC components in the stationary frame ($\alpha\beta$) to the DC signals. Steady-state errors can be eliminated by standard PI controllers. In consequence, the PI controllers output components in the dq-axes are transformed back to the $\alpha\beta$-axes. Only the modulation signal in the real axis m_α is provided to the PWM.

Unlike the three-phase system, the single-phase system requires an extra imaginary orthogonal signal for the axis transformation. This signal is typically provided by an orthogonal signal generator (OSG), by 90° phase shifting of the measured real component. However, the OSG methods rely on filters that introduce significant signal delay, especially in tracking the fundamental component, leading to poor dynamic response.

The unbalanced synchronous reference frame technique [27] which is famous for the single-phase grid-tied inverter applications is employed as demonstrated in Figure 5. This control technique transforms control signals from the $\alpha\beta$ axes to the dq axes without requiring the artificial quadrature signal from the OSG (extra block). In addition, the orthogonality is achieved through signal decomposition using specific transformation matrices involving the grid angle, θ, generated from the PLL.

The reference and feedback currents can be expressed as follows:

$$\begin{bmatrix} i'_{dref} \\ i'_{qref} \end{bmatrix} = \begin{bmatrix} \cos\theta & \sin\theta \\ -\sin\theta & \cos\theta \end{bmatrix} \begin{bmatrix} i_{\alpha ref} \\ i^*_\beta \end{bmatrix} \tag{3}$$

and

$$\begin{bmatrix} i'_{d} \\ i'_{q} \end{bmatrix} = \begin{bmatrix} \cos\theta & \sin\theta \\ -\sin\theta & \cos\theta \end{bmatrix} \begin{bmatrix} i_{\alpha} \\ i^*_\beta \end{bmatrix} \tag{4}$$

where $i_{\alpha ref}$ is the reference inductor current ($i_{L,ref}$); i_α represents the feedback inductor current (i_{L1}); and $i^*{}_\beta$ is the arbitrary signal, which can be assigned as zero. The symbol ($'$) denotes the signals in the unbalanced synchronous reference frame.

The current error signals in dq axes, $e'{}_{id}$ and $e'{}_{iq}$, are determined from (3) subtracted by (4), resulting in the following:

$$\begin{bmatrix} e'_{id} \\ e'_{iq} \end{bmatrix} = \left(i_{\alpha ref} - i_\alpha\right) \begin{bmatrix} \cos\theta \\ -\sin\theta \end{bmatrix} = e_{i\alpha} \begin{bmatrix} \cos\theta \\ -\sin\theta \end{bmatrix} \tag{5}$$

It was found that the present of i^*_β resulted in a zero orthogonal current error, which $e_{i\beta} = 0$. Applying the inverse Park transformation, the line inductor current error, $e_{i\alpha}$, can be written as follows:

$$e_{i\alpha} = e_{id}\cos\theta - e_{iq}\sin\theta \tag{6}$$

where e_{id} and e_{iq}, are the current error components in the balanced synchronous reference frame. Substitution of (6) into (5) results in the following:

$$\begin{bmatrix} e'_{id} \\ e'_{iq} \end{bmatrix} = \frac{1}{2}\begin{bmatrix} (1+\cos 2\theta) & -\sin 2\theta \\ -\sin 2\theta & (1-\cos 2\theta) \end{bmatrix}\begin{bmatrix} e_{id} \\ e_{iq} \end{bmatrix} \tag{7}$$

Assuming $\theta \approx \omega t$, where ω represents the grid's fundamental frequency. It is observed that (7) contains both DC and double-line frequency AC components. This imbalance is temporary. If the DC terms are forced to zero, the AC terms, which are oscillation components, and the physical current error will be simultaneously eliminated.

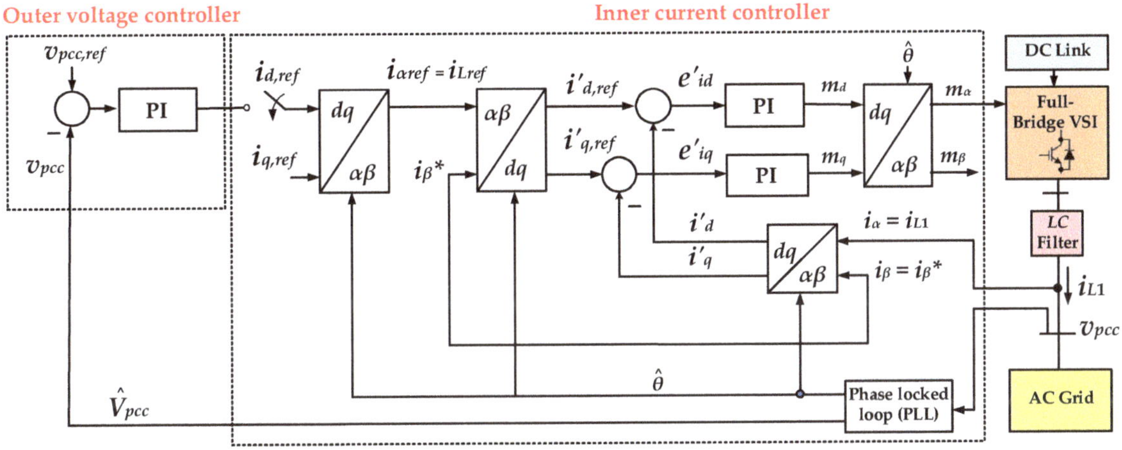

Figure 5. The unbalanced synchronous reference frame applied in full-bridge DC-AC inverter.

The full-bridge DC-AC inverter control is made up of an outer voltage control loop that cascades with an inner current control loop as illustrated in Figure 6a. The inductor current, $i_{L1} \approx i_{inv}$, is controlled using a PI controller. During grid-support operations, only the inner current loop is engaged. The value of $i_{L,ref}$ is adjusted according to the amount of power required from the grid. Additionally, the feed-forward compensation is also added to improve the control performance, ensuring that the inverter's current references remain accurate despite changes in the DC link voltage.

On the other hand, the outer voltage loop is only active when the grid is lost. The voltage at the point of common coupling, v_{pcc}, is then regulated using the PI controller to keep the inverter's output voltage at the reference value, $v_{pcc,ref}$.

Figure 6b depicts a phase lock loop (PLL) system that can operate in both on-grid and off-grid modes. A mode-selection block enables switching between these two modes using an S_o, signal. When $S_o = 0$, it means the grid is lost and the reference frequency, ω_n, remains constant at the fundamental (i.e., 50 Hz), and there is no frequency tracking error ($\Delta\omega = 0$). However, when $S_o = 1$, the PLL performs the grid frequency tracking. The PI controller adjusts the frequency difference between the DC-AC inverter and the grid, $\Delta\omega$, based on the phase difference, ensuring that the DC-AC inverter's frequency and phase angle always synchronize with the main grid.

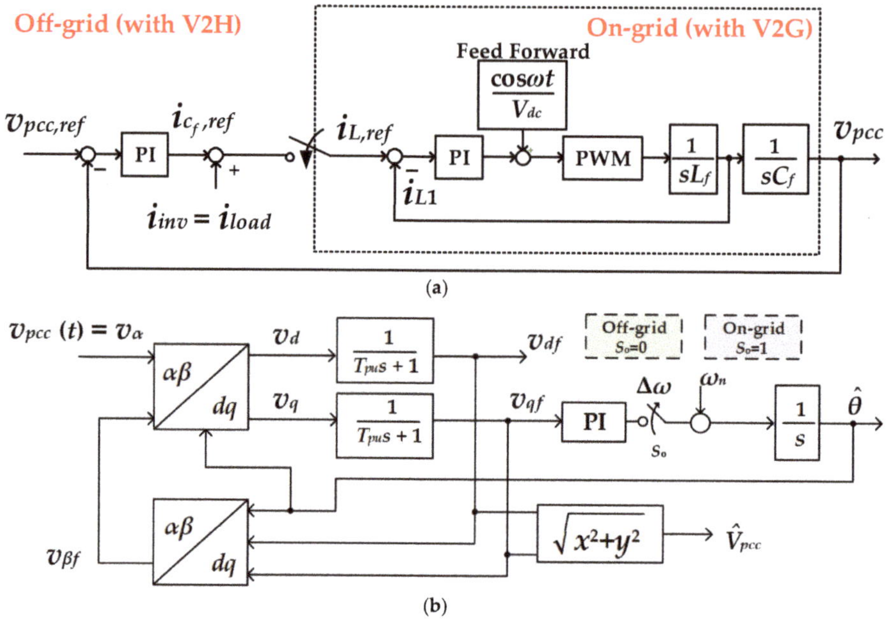

Figure 6. The control structure of grid-tied DC-AC full bridge inverter: (**a**) voltage and current controls; (**b**) phase lock loop.

3. Power Management for V2G and V2H Operation Modes

The design of the multi-port converter with V2G and V2H is focused on three primary features: (1) effective power balancing, (2) intelligent power management, and (3) rigid safety protocols. These ensure that EVs can safely and efficiently contribute power back to the grid or home, enhancing energy flexibility and stability. Coordination of energy storage based on the SoC levels is required to maximize power supply to the grid or residence during both on- and off-grid operations.

Power management begins by checking the availability of each key component: EN_BES, EN_EV, EN_PV, EN_Load, and EN_Grid. Each of these symbols represents the state of an individual subsystem, indicating whether it is online and operational. When all components are enabled, the PV/BES/EV hybrid energy system proceeds to manage power flows based on load demand, PV generation, and the SoC levels of BES and EV, as illustrated in Figure 7.

When EN_EV is enabled and V2G or V2H is active, the EV is ready to engage in energy management and can either charge or discharge energy according to its SoC. The power management system assesses the EV's SoC to determine its power contribution:

- 50% > SoC_EV: The system does not allow the EV to discharge power, while the V2G or V2H function is disabled, preserving the EV's battery for other uses.
- 50% ≤ SoC_EV < 70%: The EV provides a baseline discharge current, $i_{Lev} = -P_{dcref} / V_{ev}$ The EV contributes a minimal amount of energy back to the system, helping stabilize the grid or meet load demands without draining its battery too quickly.
- 70% ≤ SoC_EV < 90%: The EV discharges at a higher rate with $i_{Lev} = -2 \times P_{dcref} / V_{ev}$, offering more support to the grid and load.

- 90% $\leq$ SoC_EV $\leq$ 100%: The EV provides maximum support with $i_{Lev} = -3 \times P_{dcref} / V_{ev}$. This mode aims to utilize the high SoC for supporting grid and load efficiently.

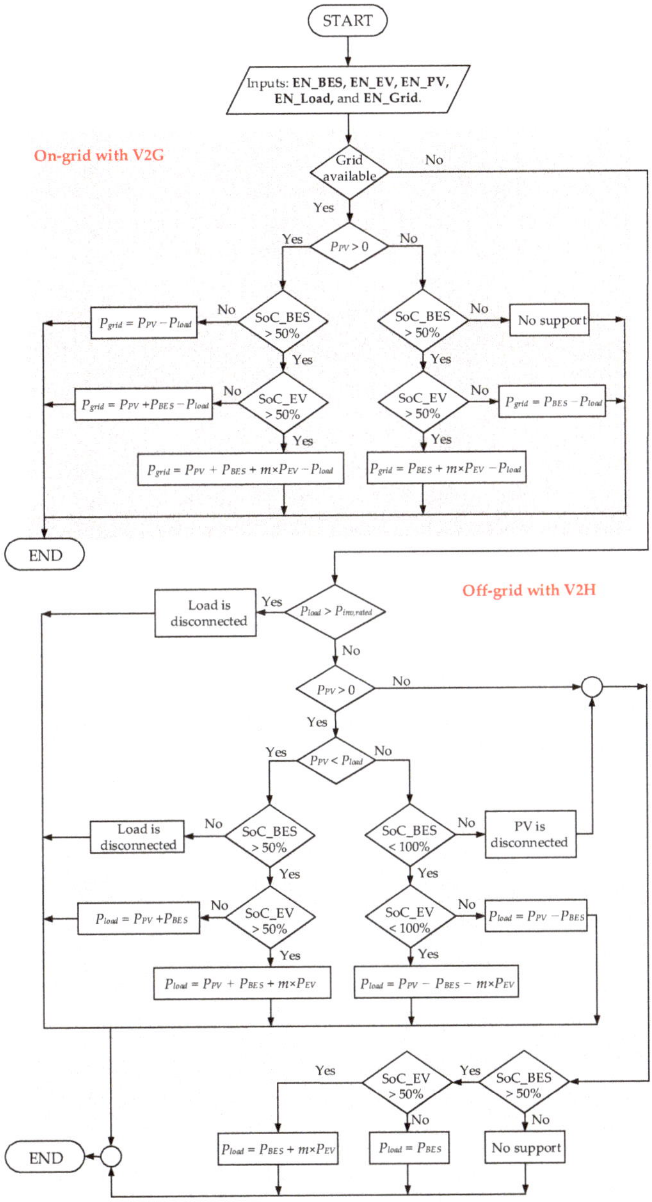

Figure 7. Flow chart of power management system with V2G and V2H.

3.1. On-Grid Operation with V2G

If the EN_PV, EN_Load, and EN_Grid are enabled, the PV system operates under MPPT to maximize solar generation. The DC-AC invertor's output power, P_{inv}, delivered to grid and load, is the combination of power from PV, BES, and EV, respectively ($P_{inv} = P_{PV} + P_{BES} + m \times P_{EV}$). The power sharing between BES and EV is determined using Table 1. In addition, the power output of the inverter is controlled by the grid-tied DC-AC inverter, which the i_{Lref} can be adjusted till it reaches the inverter's rated capacity, $P_{inv,rated}$ ($i_{L,ref} \approx 1.0$).

If the load demand is lower than the inverter's rated power ($P_{load} \leq P_{inv,rated}$), the excess power will flow back to the grid ($P_{grid} = P_{inv,rated} - P_{load}$). However, when the power generation from the PV/BES/EV hybrid energy system is not enough to meet the load demand ($P_{load} > P_{inv,rated}$), the remaining power will be drawn from the grid as a supplement, ensuring a consistent energy supply.

The V2G is turned off when the SoC of EVs falls below 50%. Similarly, when the BES's SoC falls below a certain threshold, such as 50%, it stops providing power support and only provides voltage control at the DC link. Meanwhile, when both the BES and EV hit their lower SoC limits, the system's power management strategy dictates that they stop providing power support. In this scenario, only the PV system can deliver power to the grid and load, which $i_{L,ref} \approx i_{PV}$.

3.2. Off-Grid Operation with V2H

When the main grid is unavailable, the PV/BES/EV hybrid energy system serves as the backup generation source to the local load or home. The DC-AC inverter switches to the grid-forming mode to keep the voltage and frequency at nominal values. The challenge of this off-grid operation is to balance power generation and demand. In addition, the BES and EV act as a power balancing tool to deal with fluctuations in PV generation and load usage.

The PV system operates at full capacity to meet demand via MPPT function. If the available power from the PV/BES/EV hybrid energy system equals or exceeds the load demand, then

- If the $P_{PV} > P_{load}$, the BES and EV assist the power injection to the local load, based on the power-sharing ratio in Table 1. In contrast, BES and EV absorb the surplus power from the PV generation, if $P_{PV} > P_{load}$ and BES are not yet fully charged.
- When $P_{PV} > P_{load}$ and BES are fully charged, the PV is then disconnected to prevent excessive power generation which may increase the system voltage and frequency. Hence, BES and EV assist energy supply with power sharing among them. The PV will resume generating power after the BES's SoC decreases below the specific value, such as 50%.
- When the BES and EV hit their lower SoC limits, only the PV system supplies power to load.

If the power supplied by the PV/BES/EV hybrid energy system is inadequate to fulfill load demand, or if the load demand exceeds the inverter's rated capacity ($P_{load} > P_{inv,rated}$), the local load is disconnected by the built-in protection system, potentially causing a residence outage.

The PV disconnection can be achieved by suppressing the MPPT duty cycle (D), which effectively reduces or stops power extraction from the PV array. Setting $D = 0$ disables the main switch of the PV boost converter, causing the PV current to drop instantly to zero. This method would be a softer disconnection rather than an abrupt hardware relay opening. In addition, the PV is disconnected when (1) the BES battery is fully charged (e.g., SoC reaches 100%) and (2) the PV power generation is greater than the load demand

($P_{PV} > P_{load}$). Once the BES's SoC falls below a certain threshold, such as 50%, the PV will resume power generation.

4. Test System

The model of four-port conversion system integrated with PV, BES, EV, grid, and residential load, implemented in MATLAB/Simulink version 2023a software, is illustrated in Figure 8. Although power electronic converters can be simulated using specialized software such as PSCAD/EMTDC or PSIM, these platforms are costly and require extensive modeling expertise. In contrast, MATLAB/Simulink provides a cost-effective and flexible environment where the switched-circuit model of each port can be directly implemented using basic blocks and easily integrated with available control libraries [28]. This allows simultaneous modeling of converter dynamics, PWM, and control systems within a single platform, making MATLAB/Simulink particularly suitable for multi-port converter studies.

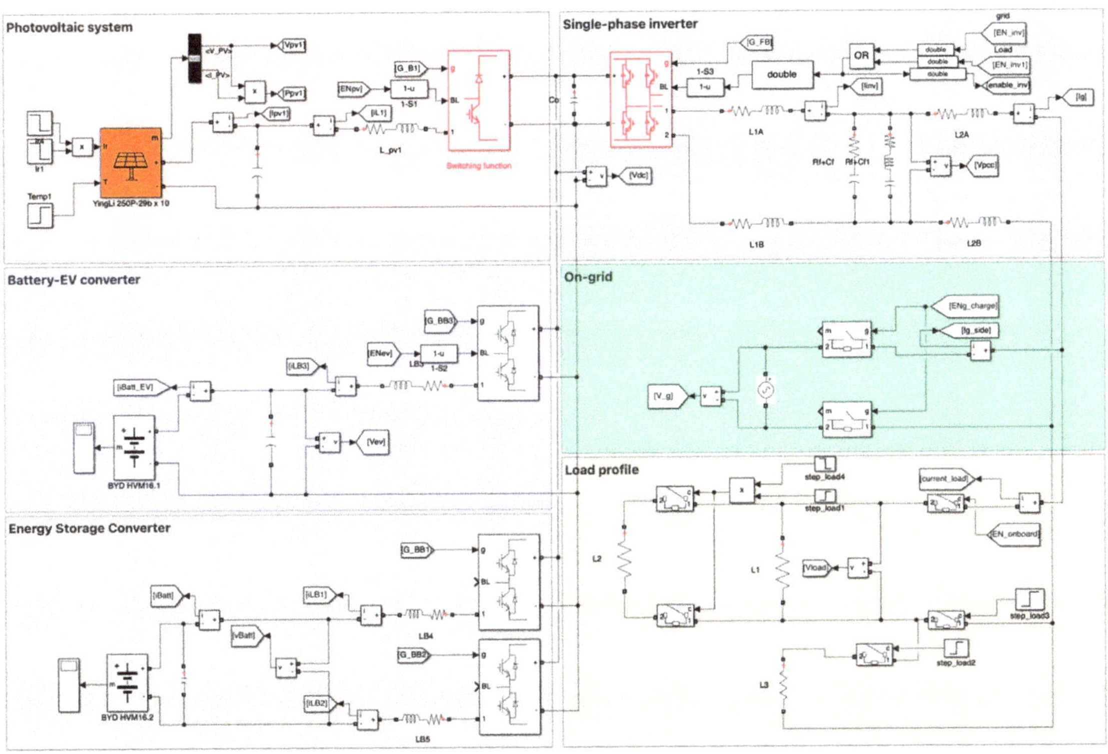

Figure 8. The proposed four-port conversion system in MATLAB/Simulink.

The AC grid is a single-phase 220 Vrms, 50 Hz system. The resistive load can be stepped up from 6.4 kW to 12.9 kW. The PV power varies in response to solar irradiance at the maximum power point. This proposed multi-port converter has a capacity of 7.5 kVA, which is comparable to the size of a household EV charger, making it compact and suited for residential use. Assuming BES and EV batteries have SoC levels between 90% and 100%, the EV:BES power-sharing ratio becomes 3:1.

The PV system consists of ten series-connected modules, each having a maximum power of 600 W with voltage and current at the maximum power point of 35.20 V and

17.05 A, respectively. BES and EV systems use lithium-ion batteries with nominal voltages of 300 V and 400 V correspondingly. For simulation purposes, the BES and EV batteries are both assumed to have a nominal capacity of 1 Ah to align with the specified simulation timeframe.

The controller gains are designed using extended symmetrical optimization (ESO) and pole-placement techniques [29,30]. The ESO method is often used for inner current control loop. It is recommended that to have an acceptable overshoot, the PI controller should have the phase margin in the range of 30°–65° to assure the stability of the control system. The parameter β is introduced allowing designers to balance performance characteristics such as overshoot and settling time, with a recommended range of 4 to 16 to avoid excessively small or unnecessarily large phase margins. Once β is selected, the PI controller parameters, K_c and T_c, can be determined using the following relations:

$$K_c = 1/\left(\beta\beta^{1/2}T_2K_p\right) \tag{8}$$

$$T_c = \beta T_2 \tag{9}$$

where T_2 is a small time constant of the plant, often representing parasitic effects, while K_p is the plant's gain.

For the outer voltage control loop, the pole-placement technique allows for precise placement of the closed-loop poles to achieve desired transient response characteristic, such as damping ratio, ζ, and natural frequency, ω_o. After ζ and ω_o are chosen, the PI controller parameters, K_c and T_c, can be determined by the following:

$$K_c = \frac{2\zeta\omega_o}{K_p} \tag{10}$$

$$T_c = \frac{2\zeta}{\omega_o} \tag{11}$$

The information for each port is as follows:

1. The interleaved DC-DC bidirectional converter has a capacity of 7 kW. The nominal battery voltage is 300 V, and the DC bus voltage is 500 V. The capacitance of BES side is 150 μF and the capacitance of DC bus is 2000 μF. Each inductance is 2 mH. The PI parameters of voltage control loop are K_{pv} = 13.6435 and K_{iv} = 0.0888, while the current control loop consists of K_{pi} = 0.3760 and K_{ii} = 0.0313. The switching frequency is 20 kHz.
2. The power rating of the buck–boost DC-DC bidirectional converter is 7 kW. The nominal EV voltage is 400 V. The capacitance of EV side is 390 μF and the inductance is 3 mH. The PI parameters of the current control loop are K_{pi} = 0.9219 and K_{ii} = 0.0768. The switching frequency is 20 kHz.
3. The capacity of the DC-DC boost converter is 6 kW. The nominal PV voltage is 400 V. At the PV side, the capacitance is 50 μF and the inductance is 2 mH. The PI parameters of voltage control loop are K_{pv} = 1.3329 and K_{iv} = 0.0178, while the current control loop consists of K_{pi} = 0.1633 and K_{ii} = 0.0136. The switching frequency is 20 kHz.
4. The grid-tied DC-AC full-bridge inverter has a rated capacity of 7.5 kW. The *LC* filter consists of 1 mH and 20 μF, the PI parameters of voltage control loop are K_{pv} = 0.3732 and K_{iv} = 0.0223, while the current control loop consists of K_{pi} = 0.6732 and K_{ii} = 0.0235. The switching frequency is 20 kHz. Moreover, the PI parameters of

the PLL are K_p = 0.4082 and K_i = 0.0011, and the low pass filter time constant, T_{pu}, is 0.0032 s.

5. Case Studies

The performance of the proposed multi-port converter with power management system is evaluated using computer simulations on MATLAB/Simulink in two scenarios: on-grid operation with V2G and off-grid operation with V2H.

Case 1: The PV/BES/EV hybrid energy system connects to the grid and local load, trying to generate the inverter's output power at the rated capacity (about 7.5 kW). The list of changes in PV generation and load consumption is presented in Table 2. The planned islanding transition is also added in this case by disconnecting the multi-port converter from the main grid, which BES and EV must perform the power balancing mechanism quickly to ensure the consistent energy supply for the local load. Moreover, the DC-AC inverter switches to grid-forming mode to maintain stable voltage and frequency.

Table 2. List of events in Case 1 (On-grid operation).

Time	Events
0.1 s	EV's battery is connected to the system and prepared for operating in power sharing with BES's battery.
0.105 s	The grid is connected to the system, and the DC-AC inverter performs in grid-following mode. Since there is no load, the inverter operates at maximum capability to offer grid support.
0.35 s	The load demand steps up to 6.4 kW, which is less than the inverter's rated capacity, resulting in a surplus of power available for export to the grid.
0.4 s	PV begins to gradually provide power until it reaches 3.5 kW.
0.6 s	The load demand is increased to 12.9 kW, which exceeds the inverter's rated capacity, requiring assistance from the grid in delivering the remaining power to the loads.
0.8 s	PV power drops by 50%, and the load demand is stepped down to 6.4 kW, resulting in surplus power export to the grid.
1 s	Apply the planned islanding by isolating the converter from the main grid. Therefore, the DC-AC inverter switches to grid-forming mode to fulfill the load demand of about 6.4 kW.

The planned islanding is assigned by manually separating the main grid from the multi-port converter, without using any islanding detection techniques such as under/over voltages, frequency shifts, and rate of change in frequency. Meanwhile, the standardized thresholds for islanding detection, such as in IEEE 1547, are unnecessary and not included in this work. However, those established standards are required in the case of unintended islanding, for which automatic islanding detection is required.

Case 2: After the grid is lost, the changes in PV and load demand are listed in Table 3. To avoid residential outage, the coordinated power management between PV, BES, and EV must respond quickly while preserving the voltage and frequency required by the local load. Three conditions are investigated:

1. The BES and EV absorb the surplus PV power in the system when the load demand is less than the PV generation.
2. When the load fluctuates but the inverter continues to supply power, the system must maintain reliability and stability.

3. If the inverter's power supply capacity is insufficient to supply the load demand, the multi-port converter will instantly disconnect.

To protect the converter's power circuit from over-currents, the built-in protection system disconnects the local load when the inverter current exceeds a predefined threshold, which is often a percentage higher than the rated current. Typically, the fault current contribution from the voltage source converter is its maximum current rating. In this work, the threshold value for overcurrent protection is therefore set to the rated current/power. However, voltage and frequency protections, which are defined thresholds by established standards, are not yet included.

Table 3. List of events in Case 2 (Off-grid operation).

Time	Events
0.35 s	The load demand of about 2.16 kW is connected to the system, but PV has yet to provide power. As a result, BES and EV share power to supply load demand and regulate DC bus voltage.
0.4 s	The PV power gradually increases to 4.8 kW causing BES and EV to lower the amount of power supplied to the load.
0.42 s	PV generates more power than the load demand, allowing BES and EV to begin absorbing this excessive power in order to maintain power balancing.
0.6 s	The demand rises to 6.4 kW, which exceeds PV power. Hence, BES and EV share the additional power to meet the remaining demand.
0.9 s	PV generation drops rapidly to 2.5 kW, while the load demand is stepped up to exceed the inverter's rated capacity (>7.5 kW). The load is immediately disconnected, and PV power is then used to charge EV and BES batteries.

6. Simulation Results

6.1. Case 1: On-Grid and Islanding Operations with V2G

The results in Figure 9 show that the PV/BES/EV hybrid energy system provides grid support aiming to supply power into the grid and local load at the rated capacity of 7.5 kW. When the local load is not yet connected, EV and BES provide power into the grid with power ratio of 3:1. After the resistive load of 6.4 kW is connected at t = 0.35 s, the power transferred to the grid is reduced to approximately 1.1 kW while the power sharing between EV and BES still maintains at the ratio of 3:1. It is discovered that the step response of the control system is approximately 0.05 s.

The PV starts to inject power at t = 0.4 s and reaches 3.5 kW at t = 0.6 s. When PV generation increases, the system reduces the power contribution from BES and EV in order to maintain the overall power balance and ensure that the inverter's output power matches the grid and load requirements. However, the inverter 's output power cannot exceed the rated capacity of the grid-tied DC-AC inverter, which is 7.5 kVA. It was observed that when the PV power hits approximately 3.5 kW, the aggregated power supplied by BES and EV is then decreased to around 4 kW.

After that, at t > 0.6 s, the load is stepped up to 12.9 kW which exceeds the DC-AC inverter's rated capacity, the power from grid is suddenly brought in to support this load increase while the power from each port continues to generate as much as possible. On the other hand, at t = 0.8 s, the multi-port converter can re-inject excess power into the grid once the load has been reduced below the converter rated capacity of 7.5 kW.

After the grid is disconnected at t = 1.0 s, the DC-AC inverter switches to grid-forming mode to offer consistent power supply to the load. The BES and EV serve as power balancing controllers, ensuring that power generation always matches load demand. As a result, the BES and EV lower power output, making the total power output to 6.4 kW which is equivalent to the load requirement.

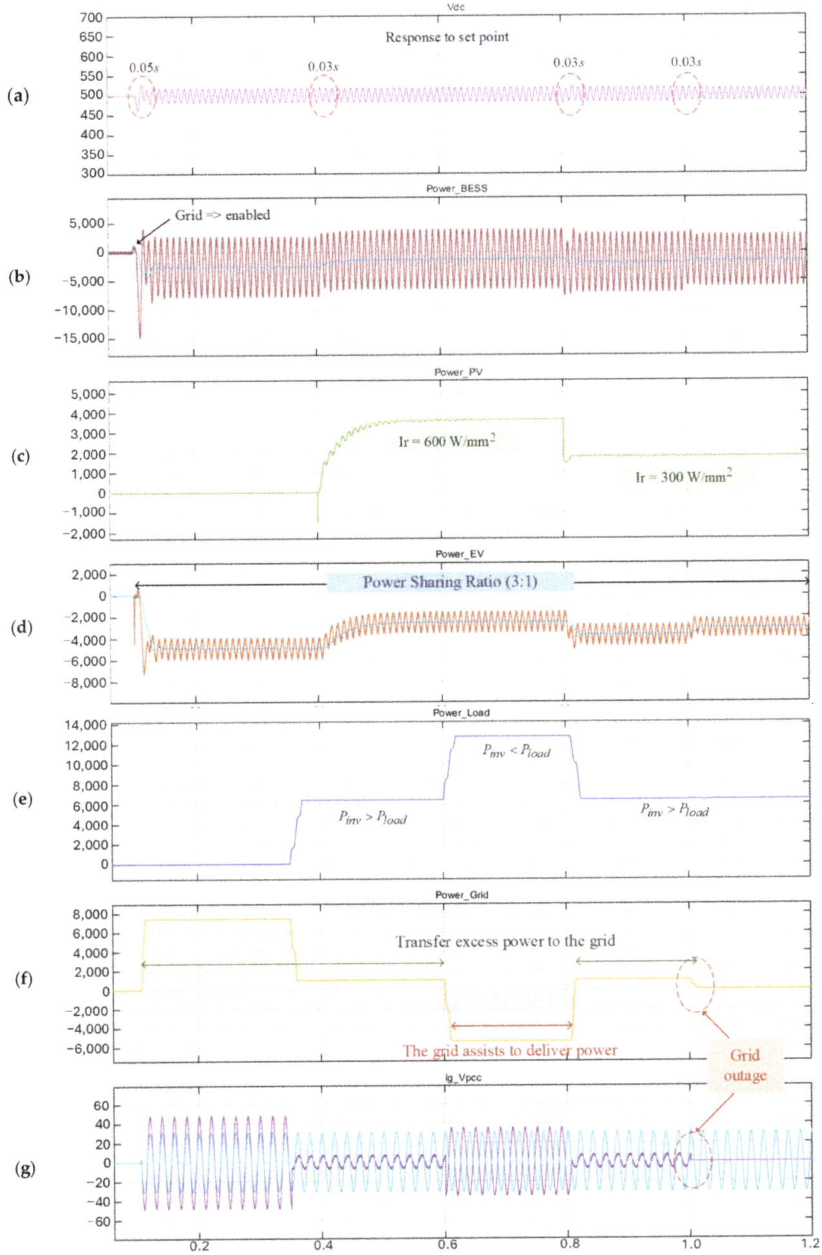

Figure 9. Simulation results in Case 1: top to bottom (**a**) DC link voltage; (**b**) BES power; (**c**) PV power; (**d**) EV power; (**e**) load power; (**f**) grid power; (**g**) grid current and voltage.

The output voltage and current waveforms of the DC-AC inverter are close to pure sinusoidal, with very low waveform distortion. Figure 10 demonstrates the variations in AC side currents, which include grid, load, and inverter currents, during the on- to off-grid transition. It is observed that the system continues to deliver power to load seamlessly, without interruption, when the main grid is disconnected. Meanwhile, the DC-AC inverter and PLL in grid-forming mode successfully establish the system voltage and frequency during off-grid operation.

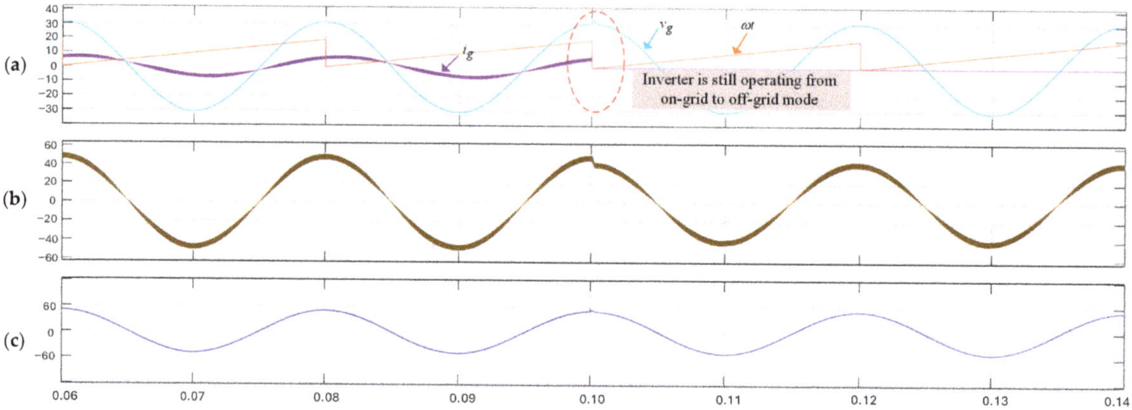

Figure 10. AC side current waveforms in Case 1: top to bottom (**a**) grid current and voltage; (**b**) inverter current; (**c**) load current.

Voltage and current harmonic distortions (HDv and HDi) between the 2nd and 11th orders, in both on-grid and off-grid conditions, are illustrated in Table 4, while the total harmonic distortions of voltage and current (THDv and THDi) are calculated up to the 25th order. It was found that the HD and THD levels, both voltage and current, are relatively low and comply with standards such as EN 50160 [31] (THDv < 8% and HDv of each order < 0.5%).

Table 4. Voltage and current harmonic distortions under on-grid and off-grid operation.

Order	Voltage (% HDv)		Current (% HDi)	
	On Grid	Off Grid	On Grid	Off Grid
2	0.0017	0.0149	0.0208	0.0182
3	0.0006	0.0691	0.0243	0.0660
4	0.0004	0.0032	0.0148	0.0073
5	0.0002	0.0272	0.0190	0.0320
6	0.0001	0.0212	0.0771	0.0246
7	0.0001	0.0176	0.0273	0.0202
8	0.0001	0.0321	0.0062	0.0384
9	0.0000	0.0192	0.1693	0.0196
10	0.0000	0.0379	0.0670	0.0469
11	0.0000	0.0145	0.1684	0.0208
% THD	0.0019	0.1828	0.4707	0.2444

6.2. Case 2: Off-Grid Operation with V2H

The results in Figure 11 demonstrate that the power management in the multi-port converter can successfully deal with fluctuations in PV and local load during the main grid was lost. At times between 0.4 and 0.6 s, the PV power gradually increases to about 4.8 kW while the load demand is roughly 2.16 kW. It was found that at t = 0.56 s, EV and BES can absorb the surplus power from PV generation, which is approximately 2.6 kW, while the EV:BES ratio is 3:1 (EV = 1.73 kW and BES = 0.86 kW). On the other hand, at t = 0.6 s, EV and BES provide supplementary power, the total of which is around 1.7 kW, to assist PV generation in a case where the load demand is increased to 6.4 kW. It is found that the step response of the control system is about 0.03 s.

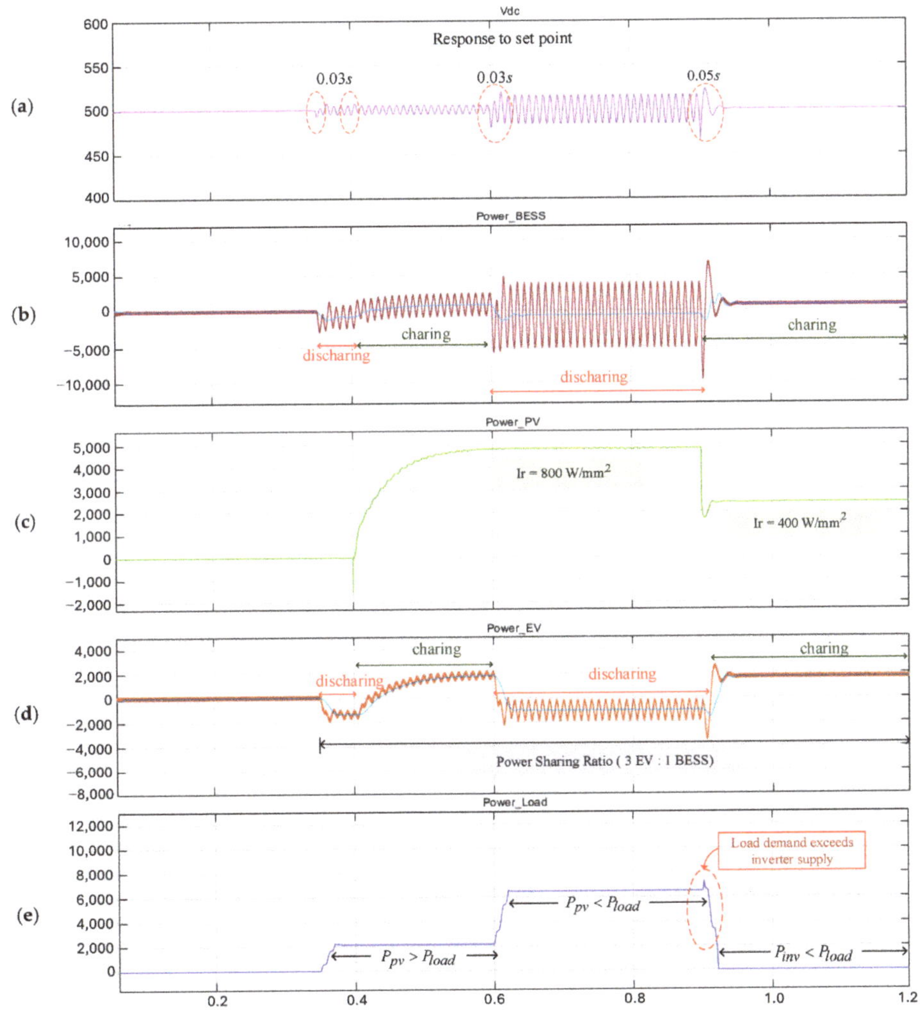

Figure 11. Simulation results in Case 2: top to bottom (**a**) DC link voltage; (**b**) BES power; (**c**) PV power; (**d**) EV power; (**e**) load power.

At t = 0.9 s, PV generation drops rapidly to roughly 2.5 kW, but load demand surges to nearly 12.9 kW, above the DC-AC inverter's rated capacity of 7.5 kW. In consequence, the overcurrent is detected, and the control system quickly disconnects the local load to protect the converter's power circuit. If the PV is still operational, even when the load is detached, the PV's power is then used to charge batteries in EV and BES systems. It was discovered that PV power of 2.4 kW is sent to EV and BES systems at the power-sharing ratio of 3:1 (EV is 1.8 kW and BES is 0.6 kW).

This study only investigated the rapid protection performance when the overload condition occurred. Meanwhile, the automatic reconnection has not yet been explicitly implemented. Once safe conditions have been restored, the load is still manually reconnected, or the DC-AC inverter must be manually activated. The automatic reconnection may become involved when the load demand falls below the inverter-rated capacity while the system voltage and frequency remain within the statutory limits.

The waveforms of AC side currents under the overload condition are presented in Figure 12. It is found that the built-in protection system, which is part of the control system, takes a very short time (less than 0.01 s) to make a decision and issue a command to disconnect the local load when the inverter current is above the predefined threshold.

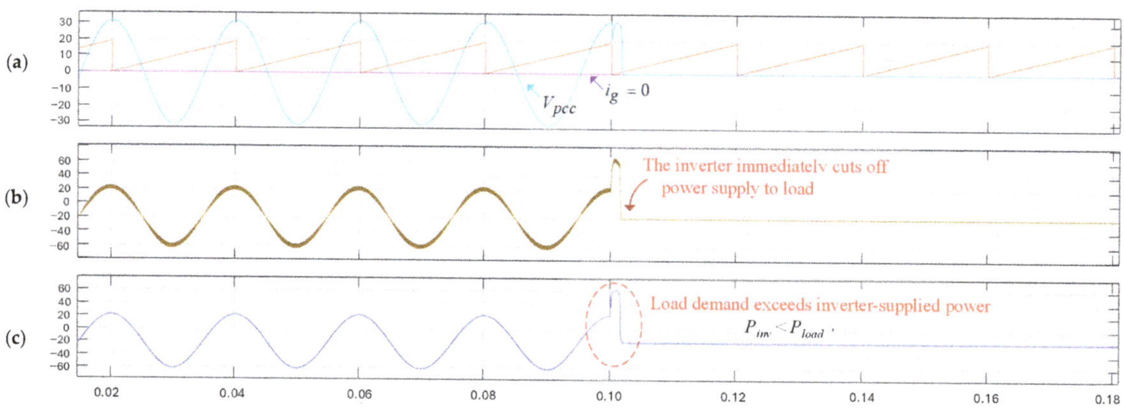

Figure 12. AC side current waveforms in Case 2: top to bottom (**a**) grid current and voltage; (**b**) inverter current; (**c**) load current.

7. Discussion

The proposed multi-port converter demonstrated stable operation in both grid-connected and islanding modes. In Case 1, the system achieved reliable V2G operation, where EV and BES shared power at a 3:1 ratio, based on their SoC levels. During the planned islanding transition, the DC-AC inverter provided a seamless transition to grid-forming mode and maintained a fast transient response to ensure satisfactory voltage and frequency. Voltage and current waveforms remained close to sinusoidal with little distortion, validating the effectiveness of the PLL and inverter control loops.

In Case 2, the multi-port converter enabled V2H functions with SoC-based power sharing between EV and BES. A rapid dynamic response was achieved under various variations in load and PV generation, and the built-in protection scheme disconnected loads within <0.01 s under overcurrent conditions, safeguarding the converter without disrupting power flow continuity. The system also allowed EV and BES to absorb surplus PV power, ensuring efficient power balancing during off-grid operation.

In comparison to conventional dual or three-port systems, the proposed four-port converter offers greater flexibility by simultaneously supporting grid services and residential loads, while ensuring seamless mode transitions and rapid fault protection. These results confirm that the converter with proposed power management strategy can serve as a practical solution for smart grid integration, enhancing reliability, stability, and energy management.

To highlight the advantages of the proposed architecture, the proposed four-port converter is compared with the PV-Integrated Multifunctional Off-Board EV Charger [16], which represents one of the most advanced designs for grid-integrated PV, BES, and EV systems. Both approaches share the objective of enabling bidirectional energy exchange with the grid while supporting household or commercial loads, but their design priorities diverge significantly. A detailed comparison is provided in Table 5, summarizing the key differences in efficiency, dynamic response, battery management, complexity, and safety considerations.

From this comparison, it is evident that the proposed four-port converter achieves a balanced trade-off, combining high efficiency, simple implementation and control, and fast dynamic response with explicit battery lifetime protection, making it a practical candidate for residential V2G/V2H applications. However, without galvanic isolation, extra protection is required to ensure safety.

Table 5. Comparative analysis of the proposed four-port converter and related state-of-the-art solutions.

Feature/Metric	PV-Integrated Multifunctional Off-Board EV Charger [16]	Proposed Four-Port Converter (This Work)
System scale	3-phase 12.6 kVA, laboratory prototype	1-phase 7.5 kVA, high resolution simulation model
Architecture	Transformer-based seven-level cascade H bridge bidirectional DC-AC converter, buck–boost bidirectional DC–DC converters (BES, EV), diode replaces PV boost stage; galvanic isolation included	Non-isolated dedicated four-port: no galvanic isolation; a simple H bridge bidirectional DC-AC converter, interleaved bidirectional DC-DC converter (BES), buck–boost bidirectional DC-DC converter (EV), boost PV converter
Efficiency	93–95% measured in hardware	Higher efficiency due to fewer conversion stages, no isolation transformer, employing interleaved architecture
Dynamic response	Seamless transition between grid-connected and standalone modes; validated experimentally	Fast transition time (0.03–0.05 s), seamless planned islanding transition
Power Quality	Harmonic mitigation, reactive power support, unity power factor (UPF) operation	THD voltage < 8%, compliance with EN50610 [31]
Battery management	Stationary BES used to reduce EV deep cycling; no explicit SoC-based strategy	SoC-based power sharing between BES and EV reduces DoD and current stress, extending battery lifetime
Hardware complexity	High: adaptive notch filter (ANF) and multilevel control	Moderate; simultaneous four-port operation; SoC-based power sharing
Safety	Galvanic isolation provides inherent protection, between grid and proposed EV charger	Non-isolated; efficiency and compactness prioritized, with additional fault detection, grounding protections are required

8. Conclusions

A single-phase four-port conversion system with V2G and V2H functions is proposed for grid-tied PV/BES/EV hybrid energy system. Based on computer simulation, it was discovered that the multi-port converter with novel power management can successfully support grid and home during on- and off-grid operations. The results confirm that the PV/BES/EV hybrid energy system can effectively cope with fluctuations in PV generation and load demand with a rapid response (within 0.05 s). The DC-AC inverter can switch between grid-following and grid-forming modes to continue power supplying without interruptions. When the main grid fails, the bidirectional DC-DC converters in the EV and BES ports serve as vital control devices for power balancing process.

Even though EVs with V2G and V2H functions effectively support the electricity grid and providing backup power, the battery degradation is a concern due to the high DoD. To mitigate the battery degradation, power sharing between EV and BES systems, based on SoC level, is suggested to minimize DoD, resulting in longer battery life. Overall, the proposed system provides a high potential for smart home applications integrating diverse energy sources.

Despite these contributions, several limitations should be acknowledged:

(1) The absence of a detailed battery aging model to quantitatively confirm the extended battery life due to SoC-based power sharing.
(2) The analysis is based on MATLAB/Simulink simulations, and the absence of experimental validation restricts the verification of practical performance under real operating conditions.
(3) The present work does not explicitly address the definition of standardized thresholds for islanding detection and voltage/frequency protection schemes, which may influence the reliability of transitions between grid-connected and islanded operation during unplanned islanding transition.

Future work will therefore focus on the following: (i) incorporating an advanced electrochemical battery aging model to validate the long-term benefits of the SoC-based strategy, (ii) extending this study to a hardware-in-the-loop and laboratory prototype to experimentally verify efficiency and dynamic performance, and (iii) aligning islanding detection and protection schemes with established standards such as IEEE 1547 to ensure seamless transition and safety in practical deployments.

Author Contributions: Conceptualization, C.S. and P.P.; methodology, C.S. and P.P.; software, C.S. and K.P.; validation, C.S., T.K., and P.P.; formal analysis, C.S. and K.P.; investigation, T.K., S.S., and P.P.; resources, C.S., T.K., and S.S.; data curation, C.S. and P.P.; writing—original draft preparation, C.S.; writing—review and editing, P.P.; visualization, C.S. and P.P.; supervision, S.S.; project administration, P.P.; funding acquisition, P.P. All authors have read and agreed to the published version of the manuscript.

Funding: This research was funded by Naresuan University (Thailand), grant number R2567C029.

Data Availability Statement: The data presented in this study are available on request from the corresponding author.

Acknowledgments: This work is part of the "Power flow controller for mobile Electric Vehicle charger integrated with Photovoltaic and High-voltage battery" project (grant number R2567C029) supported by Naresuan University, Thailand. Our gratitude also goes to the Smart Electric Power Conversion Laboratory (Sepcon) and the Centre of Excellence on Energy Technology and Environment (CeTe) for providing the facilities needed to conduct this study.

Conflicts of Interest: The authors declare no conflicts of interest.

Nomenclature

Abbreviations

BES	Battery Energy Storage

DoD	Depth of Discharge
EV	Electric Vehicle
IC	Incremental Conductance (MPPT technique)
MPPT	Maximum Power Point Tracking
P&O	Perturb and Observe (MPPT technique)
PI	Proportional-Integral Controller
PLL	Phase Locked Loop
PWM	Pulse Width Modulation
PV	Photovoltaic
SoC	State of Charge
V2G	Vehicle-to-Grid
V2H	Vehicle-to-Home
Symbols	
C_{batt}	Battery capacitance
C_{dc}	DC bus capacitance
C_{ev}	EV side capacitance
C_{pv}	PV side capacitance
D	Duty ratio of PMW signal
i_{batt}	Reference battery current
i_{ev}	Reference EV current
i_{Lpv}	Inductor current for PV boost converter
i_{Lref}	Reference inductor current for grid-tied inverter
m	Proportional gain of power-sharing ratio
P_{BES}	Power contribution from the BES
$P_{dc,ref}$	DC link power delivered by BES and EV
P_{EV}	Power contribution from the EV
P_{grid}	Excess power flowing back to the grid
P_{inv}	Inverter's power
$P_{inv,rated}$	Inverter's rated capacity
P_{load}	Load demand
P_{PV}	Power from PV
S_1, S_2, S_3, S_4	Switches in 2-phase interleaved bidirectional DC-DC converter
S_5, S_6	Switches in buck–boost bidirectional DC-DC converter for EV
S_7	Switch in DC-DC boost converter for PV
S_8, S_9, S_{10}, S_{11}	Switches in grid-tied full-bridge DC-AC inverter
V_{batt}	BES voltage
V_{dc}	DC link voltage
$V_{dc,ref}$	Reference DC link voltage
V_{ev}	EV voltage
V_{pcc}	Voltage at the point of common coupling
$V_{pv,ref}$	PV voltage reference

References

1. International Energy Agency (IEA). *World Energy Outlook 2024*; IEA: Paris, France, 2024. Available online: https://www.iea.org/reports/world-energy-outlook-2024 (accessed on 14 May 2025).
2. International Energy Agency (IEA). *Global EV Outlook 2025*; IEA: Paris, France, 2025. Available online: https://www.iea.org/reports/global-ev-outlook-2025 (accessed on 28 September 2025).
3. Xu, N.Z.; Chung, C.Y. Reliability Evaluation of Distribution Systems Including Vehicle-to-Home and Vehicle-to-Grid. *IEEE Trans. Power Syst.* **2016**, *31*, 759–768. [CrossRef]
4. Sora, J.; Serban, I.; Petreus, D. Enhancing Microgrid Operation through Electric Vehicle Integration: A Survey. *IEEE Access* **2024**, *12*, 64897–64912. [CrossRef]

5. Lu, W. Home Energy Management Combining a Welfare Calculation and Vehicle-to-House Technique for Grid Stability. *Appl. Comput. Eng.* **2024**, *42*, 208–219. [CrossRef]
6. Jiang, W.; Fahimi, B. Multiport Power Electronic Interface—Concept, Modeling, and Design. *IEEE Trans. Power Electron.* **2011**, *26*, 1890–1900. [CrossRef]
7. Bhattacharjee, A.K.; Kutkut, N.; Batarseh, I. Review of Multiport Converters for Solar and Energy Storage Integration. *IEEE Trans. Power Electron.* **2019**, *34*, 1431–1445. [CrossRef]
8. Panchanathan, S.; Vishnuram, P.; Rajamanickam, N.; Bajaj, M.; Blazek, V.; Prokop, L.; Misak, S. A Comprehensive Review of the Bidirectional Converter Topologies for the Vehicle-to-Grid System. *Energies* **2023**, *16*, 2503. [CrossRef]
9. Tasnim, M.N.; Riana, J.M.K.; Shams, T.; Shahjalal, M.; Subhani, N.; Ahmed, M.R.; Iqbal, A. A Critical Review on Contemporary Power Electronics Interface Topologies to Vehicle-to-Grid Technology: Prospects, Challenges, and Directions. *IET Power Electron.* **2024**, *17*, 157–181. [CrossRef]
10. Acharige, S.S.G.; Haque, M.E.; Arif, M.T.; Hosseinzadeh, N.; Saha, S. A Solar PV Based Smart EV Charging System with V2G Operation for Grid Support. In Proceedings of the 2021 31st Australasian Universities Power Engineering Conference (AUPEC), Perth, Australia, 26–30 September 2021; pp. 1–6. [CrossRef]
11. Nikhare, Y.; Kumar, J.; Samanta, S. A Smart Three-Port Converter for Interconnecting Grid, EV, and Solar-PV for Enhancing System Performance. In Proceedings of the 2023 IEEE Energy Conversion Congress and Exposition (ECCE), Nashville, TN, USA, 29 October–2 November 2023; pp. 39–46. [CrossRef]
12. Lai, C.-M.; Pan, M.-H.; Chen, Z.-M. A Multi-Port Power Converter with High Voltage Gain. In Proceedings of the 2024 6th International Conference on Electrical, Control and Instrumentation Engineering (ICECIE), Pattaya, Thailand, 19–21 January 2024; pp. 1–6. [CrossRef]
13. Savrun, M.M.; Inci, M.; Büyük, M. Design and Analysis of a High Energy Efficient Multi-Port DC–DC Converter Interface for Fuel Cell/Battery Electric Vehicle-to-Home (V2H) System. *J. Energy Storage* **2022**, *45*, 103755. [CrossRef]
14. Firdous, Z.; Mekhilef, S.; Mubin, M.B.; Siddique, M.D.; Elsanabary, A.; Ab Aziz, N.F.; Seyedmahmoudian, M.; Stojcevski, A. Multiport Converters Based EV Charging: Architectures, Challenges, Performance Analysis and Emerging Trends. In Proceedings of the 2023 IEEE International Conference on Energy Technologies for Future Grids (ETFG), Wollongong, NSW, Australia, 3–6 December 2023.
15. Khan, O.; Hredzak, B.; Fletcher, J.E. A Reconfigurable Multiport Converter for Integrated PV/EV/Battery Systems. *IEEE Open J. Power Electron.* **2025**, *6*, 811–829. [CrossRef]
16. Lenka, R.K.; Panda, A.K.; Patel, R.; Guerrero, J.M. PV Integrated Multifunctional Off-Board EV Charger with Improved Grid Power Quality. *IEEE Trans. Ind. Appl.* **2022**, *58*, 5520–5532. [CrossRef]
17. Chaitanya, M.K.; Vinayak, A.S.; Pavan, B.J.G.; Lakshmi, N.N.; Sandeep, K. A Review of Bidirectional DC–DC and DC–AC Converters for V2G and G2V Applications. In Proceedings of the 2023 International Conference on Power, Instrumentation, Energy and Control (PIECON), Chennai, India, 10–12 February 2023.
18. Narayanaswamy, J.; Mandava, S. Non-Isolated Multiport Converter for Renewable Energy Sources: A Comprehensive Review. *Energies* **2023**, *16*, 1834. [CrossRef]
19. Oliveira, C.F.; Afonso, J.L.; Monteiro, V. Fault Analysis of a Non-Isolated Three-Level DC–DC Converter Integrated in a Bipolar DC Power Grid. In Proceedings of the Sustainable Energy for Smart Cities, Cham, Switzerland, 16–18 November 2022.
20. Feretti, P.H.; Tofoli, F.L.; Ribeiro, E.R. Family of Non-Isolated High Step-Up DC–DC Converters Based on the Multi-State Switching Cell. *IEEE J. Emerg. Sel. Top. Power Electron.* **2022**, *10*, 5882–5893. [CrossRef]
21. Anosh, A.; Saji, D.; Jexy, A.; PG, J.; Athikkal, S. Improved Gain Non-Isolated Type DC–DC Converter. In Proceedings of the 2024 1st International Conference on Trends in Engineering Systems and Technologies (ICTEST), Kollam, India, 11–13 April 2024.
22. Collath, N.; Tepe, B.; Englberger, S.; Hepp, J.; Schmitt, J.; Jossen, A.; Hesse, H.C. Aging Aware Operation of Lithium-Ion Battery Energy Storage Systems: A Review. *J. Energy Storage* **2022**, *55*, 105634. [CrossRef]
23. Terkes, M.; Demirci, A.; Gokalp, E. A Comprehensive Review of the Aging Mechanism and Degradation Costs of Fresh and Second-Life Batteries Based on Analytical and Deterministic Methods. *Energy Storage* **2024**, *6*, e661. [CrossRef]
24. Khalid, M.; Panigrahi, B.K. SoC-Based Decentralized Power Management in Multi-BESS-PV for EV Charging Applications. In Proceedings of the 2022 IEEE Global Conference on Computing, Power and Communication Technologies (GlobConPT), New Delhi, India, 23–25 September 2022.
25. Wibisono, A.; Facta, M.; Setiawan, I. An Average Current Control Method in Multiphase Interleaved Bidirectional DC–DC Converter Connected on DC Microgrids. In Proceedings of the 2021 12th International Renewable Engineering Conference (IREC), Amman, Jordan, 14–15 April 2021.
26. Prasatsap, U.; Nernchad, N.; Termritthikun, C.; Srita, S.; Kaewchum, T.; Somkun, S. Comparison of Control Configurations and MPPT Algorithms for Single-Phase Grid-Connected Photovoltaic Inverter. *Adv. Electr. Comput. Eng.* **2023**, *23*, 55–66. [CrossRef]

27. Somkun, S.; Chunkag, V. Unified Unbalanced Synchronous Reference Frame Current Control for Single-Phase Grid-Connected Voltage-Source Converters. *IEEE Trans. Ind. Electron.* **2016**, *63*, 5425–5436. [CrossRef]
28. Srita, S.; Somkun, S.; Kaewchum, T.; Rakwichian, W.; Zacharias, P.; Kamnarn, U.; Thongpron, J.; Amorndechaphon, D.; Phattanasak, M. Modeling, Simulation and Development of Grid-Connected Voltage Source Converter with Selective Harmonic Mitigation: HiL and Experimental Validations. *Energies* **2022**, *15*, 2535. [CrossRef]
29. Preitl, S.; Precup, R.-E. An Extension of Tuning Relations after Symmetrical Optimum Method for PI and PID Controllers. *Automatica* **1999**, *35*, 1731–1736. [CrossRef]
30. Åström, K.J.; Hägglund, T. *PID Controllers*, 2nd ed.; ISA: Research Triangle Park, NC, USA, 1995.
31. *EN 50160*; Voltage Characteristics of Electricity Supplied by Public Distribution Networks. European Committee for Electrotechnical Standardization: Brussels, Belgium, 2022.

World Electric Vehicle Journal

MDPI

Article

Analysis of the Impact of Electromobility on the Distribution Grid

Tomislav Kovačević [1], Ružica Kljajić [1,*], Hrvoje Glavaš [1] and Milan Kljajin [2]

[1] Faculty of Electrical Engineering, Computer Science and Information Technology Osijek, University of Osijek, 31000 Osijek, Croatia; tomislav.kovacevic@student.ferit.hr (T.K.); hrvoje.glavas@ferit.hr (H.G.)
[2] VERN' University, No 82/1 Palmoticeva Street, 10000 Zagreb, Croatia; milan.kljajin@gmail.com
* Correspondence: ruzica.kljajic@ferit.hr

Abstract: This paper analyzes the impact of electromobility on distribution grids and voltage stability. In line with current legislation and the European Commission's plans for the future of electromobility, the aim is to increase the share of electric vehicles to 50% by 2050. However, achieving this target can be challenging due to the characteristics and features of the electric vehicle charging stations and the associated charging methods, which can lead to constraints within the network. The analysis includes the integration of single-phase and three-phase chargers on a radial feeder, as well as the determination of the maximum number of vehicles that can be accommodated on a given feeder without compromising voltage stability. Five scenarios are evaluated using the DigSilent software package to gain a better understanding of the impact of electromobility on the distribution grid.

Keywords: electromobility; voltage stability; charging methods; grid constraints

1. Introduction

The European Commission launched the European Green Deal in 2020, which defines a legally binding target to achieve a 55% reduction in greenhouse gases by 2030 with the aim to achieve a climate-neutral Europe by 2050 [1]. All economic sectors are expected to contribute to this reduction, including transportation, which has been increasing gradually since 1990 and is responsible for 20% of all greenhouse gas emissions in the EU. If the right steps are not taken, it is estimated that the transportation sector's greenhouse gas emissions might rise by over 20% by 2030 and nearly 50% by 2050. In order to meet the shared objectives of lowering greenhouse gas emissions, a significant proportion of new passenger car and cargo vehicle sales must be of electric vehicles (EVs). Additionally, the European Union must expedite the development of electric vehicle charging infrastructure, while the automotive sector employs almost 15 million people in Europe and contributes more than 7% of the EU's GDP, thus being a vital sector [2]. By 2030, fewer than 20% of road traffic will consist of electric vehicles, but 30% of these investments will need to go toward infrastructure to reduce carbon dioxide emissions from traffic, according to the electric charging strategy. This is due to the fact that only 18% of the renewable energy sources that will be deployed by 2030 will be used for EV charging. A 17% share, or 42.8 million electric passenger vehicles (battery electric vehicles, or BEVs, and plug-in hybrid electric vehicles, or PHEVs), 4.4 million electric light delivery vehicles, or 13% of the total number, and 0.3 million electric trucks, or 3.5% of the total number, should be on the road by the end of the decade. In terms of chargers, this translates to 6.8 million chargers for passenger vehicles, 0.7 million for light delivery vehicles, and 0.1 million for trucks and buses. These figures are necessary to achieve a 55% reduction in greenhouse gas emissions by 2030 [2,3].

European directives and national laws contain technical requirements and regulatory regulations for electric car charging stations that are applicable to all EU member states. The section of the Building Act that deals with promoting electromobility and installing charging infrastructure in buildings is in effect in Croatia [4], defining the necessary conditions and terms for EV chargers. The number of charging stations and charger types has significantly increased globally in recent years. Fast and public chargers are the most common forms, offering electric vehicle drivers without a home connection a reliable and safe method of charging. According to [5] the first fully electric EV was registered in Croatia in 2007. The promotion of electromobility, the reduction in vehicle prices and the greater availability of EV and charging stations have led to a significant increase in the number of EVs since 2015. According to same source, in the year 2024, 9987 full EVs and almost 79,000 hybrid vehicles were registered in Croatia. Figure 1 shows a comparison of the number of EVs (all types) in the European Union and in Croatia for the time period between 2013 and 2023 [5,6].

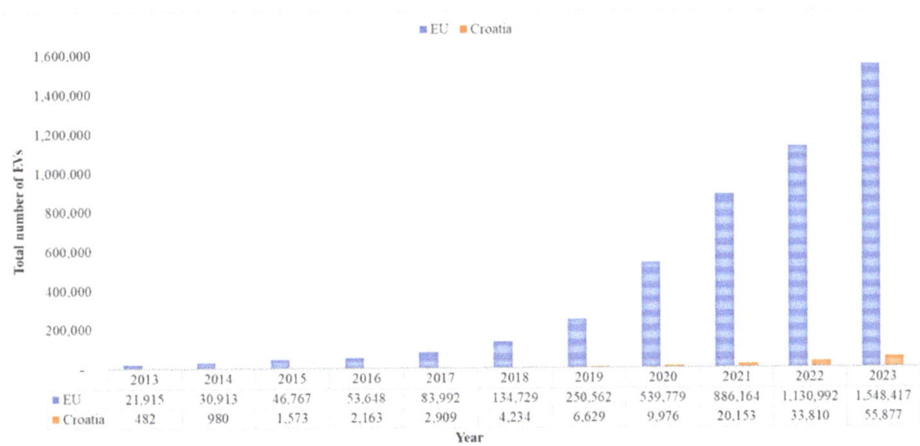

Figure 1. Comparison of total EVs registered in EU and Croatia in 2013–2023.

According to the above data and projections, it is clear that the direction of development will involve a further increase in the number of electric vehicles. However, the current state of the electricity grid and the methods of calculating electricity consumption represent a limitation that will not be easily overcome in a short period of time. This paper analyzes the impact of the increase in the number of EV charging stations on the grid. A passive output without photovoltaic (PV) power plants and with EV integration as provided for by the directives of the European Commission is assumed, corresponding to 50% of electric vehicles. According to the current habits of EV owners, charging is conducted during the night, primarily because EV owners are home during the night and can connect to their chargers, and the secondly because of the price of electricity, which is lower at night. With assumed charging during the night, the rooftop PV power plant that could be integrated in the grid would not provide energy to the grid anyway, and the impact of EVs on the distribution feeder and the voltage conditions could be observed.

In this paper, the impact of single-phase and three-phase EV charger integration on a realistic distribution feeder is analyzed. In scenario 1, which is also a basic scenario, the voltage stability of a passive feeder without integrated EV chargers is observed. Afterwards, in scenarios 2 and 3 it is assumed that the integration level reaches 50%, as aimed by the legislative directives. Finally, in the last two scenarios the maximum number of single-phase and three-phase EV chargers that will not have a negative effect on the voltage profile of the feeder is determined. For each scenario the voltage profiles and the amount of energy taken from the distribution grid are presented, and the results are compared. This paper consists of five chapters, with Introduction as the first chapter, in which the key motivation for this research is presented. The second chapter describes the Materials and Methods used to conduct the research, as well as the key settings for conducting the case study. The third chapter provides the current state of the art and indicates how EV integration affects a distribution grid. In the fourth chapter, a mathematical background regarding voltage calculation in a radial grid is given, while a feeder model and scenarios are presented in the fifth chapter. In the sixth chapter, the results are discussed, and the paper ends with the seventh chapter, containing the conclusions and suggestions for future work.

2. Materials and Methods

The feeder model was based on a real case of radial distribution grid in Croatia, starting from the transformer at the connection with the external grid, including all cables (parameters, lengths, current limits), and modeled according to the actual distribution feeder. The arrangement of consumers, in this case households, corresponded to that of a realistic feeder, and the consumer characteristics were obtained by measurements according to the EN50160 standard over a period of one week. A load profile corresponding to the average working day load was selected. In this case, since the feeder consisted of households, there was no significant deviation in load distribution between working days and weekends, nor were there any significant consumers that would affect the load curve (for example, a small industrial facility).

In addition to the load profile, a harmonic spectrum was assigned to each consumer, which was taken into account during consumer modeling, although harmonic analysis was not conducted in this study. The EV charging load curve was also modeled according to the characteristics of three-phase and single-phase EV chargers obtained from measurements. All data not specified in this paper or in the referenced literature can be obtained upon request from the authors; they were omitted because this is a long-term research project whose results have already been partially presented.

For the analysis, the DIgSILENT PowerFactory 2022 SP3 software package was used, in which the model was created, and quasi-dynamic simulations were performed for the considered 24 h period, observing the voltage profile and the energy exchange with the external grid during that period. EV chargers were distributed so that in two scenarios, the European Commission's requirement for a 50% share of EVs in the transport sector was met, and subsequently, the voltage profile was analyzed for a limited number of vehicles whose simultaneous charging would not compromise voltage quality. All simulations were conducted under the assumption that two electricity tariff rates existed, as is the case in Croatia.

PV integration was not considered in this paper because the focus was on the implementation of the directive that promotes a 50% share of EVs in transport. The analysis was made according to the current electricity charging framework in Croatia (two tariffs) and EV owners' preferences for charging at night due to lower electricity prices and their being at home at night. The price of electricity in Croatia is still affordable, which hardly motivates property owners to integrate PV systems. EV owners are more interested in installing PV systems, but as can be seen from the number of EVs in Croatia, their number is still negligible compared to that in leading EU countries. The analysis of the voltage profiles was made in previous works by the author, where significant PV integration and vehicle charging according to different patterns were assumed, considering daytime, nighttime, and finding the optimal charging plan if the vehicle had the ability to operate the vehicle-to-grid technology (V2G), with or without limiting the battery charge state [7]. However, in comparison to these works, the emphasis is here placed on the network load if the provision of 50% EVs were to be respected with the current state of electricity charging.

3. Integration of EV Charging Stations into the Distribution Grid

Following the requirements of the directives and in an effort to reduce harmful greenhouse gas emissions, the number of EVs is increasing year by year. Given that this is changing the existing power system, it is first necessary to define the legal and regulatory frameworks and enable access for users interested in purchasing an electric vehicle. An overview of possibilities and limitations from the technical, economic and regulatory sides, for the distribution system operator (DSO) and the end consumer, is presented in [8]. In addition, possibilities for utilizing and increasing flexibility with EV integration are proposed, as well as an overview of significant projects in European Union countries. With the increasing number of electric vehicles, different types of vehicle charging are also being developed. There are two basic groups of methods: unidirectional and bidirectional charging [9–11]. Unidirectional charging exclusively directs the energy towards the car and can be carried out as uncontrolled, delayed and controlled. Uncontrolled charging creates the greatest problems in the distribution grid, since it takes place at full power until the battery is fully charged. The negative sides are primarily the consumption of high power at any time of the day, which can lead to thermal loads and stresses on lines and transformers, and increased losses in the grid [12–15].

A second type of unidirectional charging is delayed charging [16], which can have a smaller impact on the grid, since charging takes place at a time when price signals are more favorable (lower electricity prices), thus affecting the total load on the system. However, apart from the adjustable period, there is no possibility of controlling the maximum power; so, in this case, charging is also performed with the highest available power [17]. Due to the lower prices, a second peak load may be created during the day because all owners will decide to charge when the electricity price is lower [18]. A third type, which is also the most favorable among unidirectional charging methods, is controlled charging. In this case, the charging time and the maximum power are controlled, reducing the negative impact on the grid [19–22]. Among unidirectional methods, this appears as the most favorable, with the least impact on the state of the distribution grid.

However, in order to achieve the least dependence on the external grid and at the same time use the full potential of the EV considering its positive impact on the grid, bidirectional charging methods have been developed in which the car battery can serve as an energy storage and provide energy at the moment when it is needed in the system [23,24]. These solutions are known as the vehicle-to-grid (V2G), vehicle-to-home (V2H) or vehicle-to-building (V2B) technologies. In V2B and V2H, energy is used to regulate energy consumption at the household or building level without affecting the external grid and is used for self-powering only in the event of a power outage or reducing dependence on the external grid [25–27]. In these cases, the only communication is between the EV and the building to which it is connected, and these systems do not provide a high level of service flexibility. On the other hand, the V2G concept offers various possibilities such as providing ancillary services, frequency regulation and energy storage, smoothing the daily consumption diagram, reducing the impact of renewable energy sources (RESs) on the grid, reducing dependence on the external grid and reducing the load on lines and transformers by reducing the energy exchange with the external grid. Of all the methods, the V2G technology is also the most common, with the negative aspect of causing accelerated aging of the EV battery. Also, the question arises of how much an individual car owner allows access to and management of his battery and to what extent EV battery discharge is allowed.

The integration of EV chargers can present a significant challenge to the system from the perspective of power quality. If the directive's target—namely, that 50% of vehicles are electric—were met, most owners, given the limited number of public chargers and fixed tariffs (as is the case in Croatia), as well as the low electricity prices, the need of personal time management, and the desire to have their car nearby and available at their convenience, would likely choose to charge their vehicles at home. In such a scenario, three-phase chargers would not cause significant problems for the network; they can be considered simply as an additional load and would not substantially increase the system's asymmetry. On the other hand, single-phase chargers—which constitute the majority of home chargers—can lead to significant asymmetry in the system, since the DSO cannot regulate or determine where and to which phase an owner will connect their charger. In the worst-case scenario, all consumers might be connected to one- or two-phase chargers, thereby increasing unbalance. In Croatia, as in other countries facing similar challenges, periodic system balancing is carried out to equalize the load between phases and mitigate these effects.

Effects of EV Integration in the Distribution Grid

The integration of EVs into the distribution grid has a wide range of impacts on the grid. As with all technologies, EVs have both positive and negative impacts on the grid. In general, their impact is viewed in terms of power quality, the effect on line and transformer loads and that on grid losses. The integration of EV chargers increases the total power in the system, which can lead to an increase in peak loads. EV integration, in addition to its positive impact on the environment in the form of reduced emissions, can also have a positive impact on the operation of the power system. In general, the impact is primarily on the quality of electricity, which is improved, but also on grid management. The first positive effect includes load management and the possibility of demand side management [28]. EVs with their energy storage participate as a variable load and can therefore be used in the way the system is managed to smoothen the load curve. Charging vehicles during the day increases the load on the system during peak periods; however, if charging takes place at night during periods of lower load, the load on the system during the day is flattened, and EVs are used in order to flatten the voltage profile [29]. In addition, an increased load during the day would lead to a decrease in the voltage of the end consumers, while at night the voltages could be significantly higher. By managing EV charging in this way, smaller voltage deviations are achieved during the day, and the system is relieved during the peak period, which means that it is not necessary to ensure significantly higher production power during that period. For prosumer-rich feeders, the opposite happens—voltages increase if a large number of integrated rooftop PV systems are installed, and charging cars reduces the voltages and at the same time ensures less energy exchange with the external grid. Delayed charging thus both affects congestion control and relieves the load on the lines and transformers, which significantly affects their lifespan [30–32]. The increased load caused by the additional consumers in the grid increases the thermal stress of the grid components and the losses. A large number of integrated EVs means higher currents, and the load on the equipment can be brought to thermal limits. The problem can be eliminated by controlled charging, in which the maximum power and charging time of the vehicles are monitored, or by additional investment in the grid to upgrade the existing infrastructure and prepare for a larger number of electric vehicles.

EVs as a variable load can also be used in frequency regulation [33,34]. The frequency is constant if the active power of production and that of consumption are equal at all times. If production is higher than consumption, for example due to a large production from DG that cannot be consumed, the excess energy is stored in the EV battery energy storage system (BESS). In this way, the production/consumption ratios are kept constant, and there will be no frequency deviations. In addition, the EV BESS can be used as an additional energy source in the event of a drop in production. In this sense, EVs can be used as a variable load and enable greater grid flexibility [35]. A further step in increasing flexibility and achieving greater economic benefits is the use of EV BESS for providing ancillary services [36,37]. In addition to active power regulation, due to the way capacitors are designed and implemented as an integral part of the EV BESS, reactive energy regulation can be ensured. Analogous to the active power, the reactive powers of production and consumption must also be equalized in order to keep the voltages within the desired values. In this way, the EV BESS can be used for voltage regulation [38,39] and the management of reactive energy flows. From the above-mentioned considerations, it can be seen that the use of EVs can significantly affect the way the grid is managed, especially in terms of reducing losses and reducing dependence on external energy sources, and directly reduce financial losses for owners with regard to consumption management at times when the price signals are favorable on the market.

However, the use of EVs in improving energy quality (especially if used with the V2G technology) also has negative sides, as stated in the literature [40–42]. This primarily refers to the stochastic behavior of the users depending on price signals and, above all, user habits. Thus, charging most often takes place at night due to the lower prices but also to the fact that users are at home at that time and have the possibility of connecting to their home charger. Connecting to a charger during daytime peak loads will create a higher peak load and therefore a higher thermal load on connecting cables/lines and transformers. On the other hand, if too many users are connected during the night, the same problems can occur because there will be a higher voltage drop and problems in the grid. Due to poor planning and management of the system, as well as to too many integrated sources on the existing infrastructure that cannot handle such a large number of vehicles, the previously mentioned leveling will not occur, but rather another extreme peak loading will occur. The maximum number of vehicles that can be connected without significantly affecting the voltage conditions in the grid was determined [38,43]. The main criteria for determining the maximum number of EVs are based on voltage limitations and allowed voltage deviation, i.e., on terms aiming at maintaining voltage stability. In addition, voltage deviation is strongly correlated with the type of used charger. Three-phase chargers, which are usually of higher power, will load all three phases equally [44]. Single-phase chargers will significantly increase the power of only one phase, and considering that they are consumers of significant power (about 3.8 kW) in relation to household consumption, there will be an increase in the load of the individual phase [45,46]. EV users cannot determine the location or the phase to which they will be connected, as it is a matter of stochasticity, and there may be a significant unbalance that will impair the quality of electricity, with unbalance values above the deviation limits provided by the standard. Closely related to unbalance, harmonic distortion will also be affected due to power electronics circuits.

Each charger has its own harmonic spectrum, and with a smaller number of chargers, the problem will not be so pronounced, as it has been shown that vehicles have little impact on the rest of the grid [47,48].

4. Calculation of the Voltage at the End of the Feeder

According to the European standard EN 50160 [49], which defines the power quality parameters, voltage needs to be kept within the limits of ±10% of the rated voltage. In radial distribution grids, characteristic for rural grids, there are consumers distributed along the feeder. At the beginning of the feeder, there is usually a transformer that defines the voltage of the feeder. In some grids, this transformer can be used to control the voltage. Since the lines are short with a specific X/R ratio, the capacity of the lines can be neglected, and it is assumed that also the angle difference can be neglected. With these neglected terms taken into account, the feeder can be understood as a longitudinal impedance consisting of resistance R_1 and inductance with corresponding reactance X_1 (both expressed in Ω/km), as shown in Figure 2.

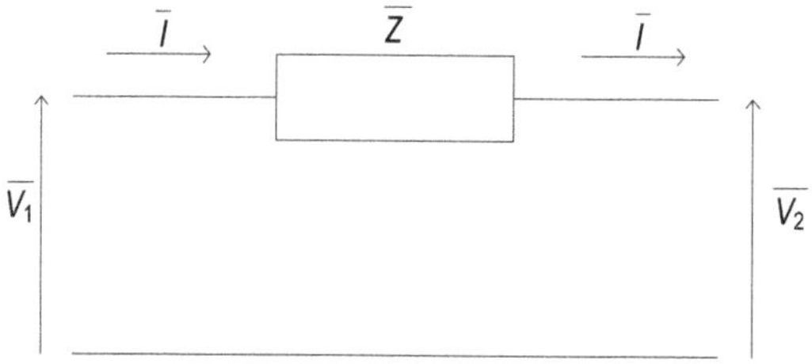

Figure 2. Feeder replacement scheme.

The following calculation of the end voltage is performed based on Ref. [50]. For the feeder shown in Figure 2, the voltage at the beginning of the feeder V_1 (in V or kV) can be calculated using Equation (1)

$$\overline{V_1} = \overline{V_2} + \overline{I}\cdot\overline{Z} \tag{1}$$

where V_2 represents the voltage at the end of the feeder expressed in V or kV, I is the current through the feeder in A or, rarely, for the distribution grid, in kA, and Z is the impedance of the feeder in Ω. The impedance of a feeder with length l (in km) is calculated using Equation (2):

$$\overline{Z} = (R_1 + jX_1)\cdot l = R + jX \tag{2}$$

If the apparent power $\overline{S} = P + jQ$ (in kVA or MVA) is known, for $\overline{V_2} = V_2\angle 0°$, the current can be calculated using Equation (3)

$$\overline{I} = \frac{\overline{S^*}}{\overline{V_2^*}} \tag{3}$$

The voltage drop along the feeder can now be calculated using Equation (4)

$$\overline{\Delta V} = \overline{I}\cdot\overline{Z} = \frac{P_2 - jQ_2}{V_2}\cdot(R + jX) \tag{4}$$

As mentioned before, the angle difference is neglected, and the voltage drop at the end of the feeder consists only of the longitudinal component of the voltage drop, as shown by Equation (5)

$$\Delta V = \frac{P_2\cdot R}{V_2} + \frac{Q_2\cdot X}{V_2} \tag{5}$$

Equation (5) can be written in a general form if the losses are also neglected, in which case, $P_2 = P$, $Q_2 = Q$, and $V_2 = V_n$, where V_n is the rated voltage. However, the feeder does not consist of a single consumer but of a series of n loads integrated along the feeder. In this case, the total apparent power is the sum of the apparent power of each load $\overline{S_i}$ (as shown in Equation (6)):

$$\overline{S} = \overline{S_1} + \overline{S_2} + \cdots \overline{S_i} + \cdots + \overline{S_n} \tag{6}$$

The total power towards the end of the feeder is reduced by the individual consumer powers as follows in Equation (7) for the second load on the feeder

$$\overline{S'} = \overline{S_2} + \cdots \overline{S_i} + \cdots + \overline{S_n} \tag{7}$$

and consequently, the power of the last load is given by the Equation (8)

$$\overline{S'_n} = \overline{S_n} \tag{8}$$

Combining Equations (2) and (5)–(8), the voltage drop at the end of the feeder can be calculated by Equation (9)

$$\overline{\Delta V} = \frac{1}{V_n} \cdot \left(\sum_{i=1}^{n} (R_1 \cdot P'_i \cdot l_i) + \sum_{i=1}^{n} (X_1 \cdot Q'_i \cdot l_i) \right) \tag{9}$$

where P'_i is the cumulative active power of the i-th load, expressed in kW or MW, Q'_i is the cumulative reactive power of the i-th load (in kVAr or MVAr), and l_i is the length of the i-th section of the feeder in the corresponding unit, m or km.

5. Voltage Stability at the End of the Feeder

With the expected projections and set goals regarding the share of EVs in the total number of means of transport, the question arises as to what impact such a level of integration will have on the grid. Vehicle charging must be provided at various locations—office buildings, public parking lots, highways, in front of shopping malls, at rest areas, etc. However, despite the availability of public charging stations, most drivers will charge their cars at home. A large number of chargers integrated into the distribution grid represents a significant increase in load, especially if charging is performed with three-phase chargers of higher power. The primary focus of this study is on the provisions of the directive, specifically, the targeted 50% share of electric vehicles (EVs) in the transport sector, and how achieving this share would affect a realistic distribution network. While it is indeed important and interesting to consider public charging stations in the analysis—especially given spatial constraints such as limited parking availability in residential buildings, where not every EV owner can have a dedicated charging spot, as well as the potential to increase the driving flexibility—this particular work intentionally concentrates on a segment of a radial feeder composed exclusively of single-family homes. Consequently, the need for a significant number of public chargers in this context is minimal. Also, the analyzed feeder represents a typical configuration of a rural grid in Croatia, with old infrastructure.

This work used a model of a real grid (Figure 3)—a distribution low-voltage feeder with a load characteristic of a household. Given that it was a rural feeder, the assumption was that there would be no integrated rooftop PV power plants and that it would not affect the voltage conditions on the feeder. In addition, charging took place during the night; so, the possible integration of the rooftop PV plants would not have an impact on the voltage conditions. Charging at night was considered for two key reasons:

- A two-tariff electricity billing model was used, as currently in effect in Croatia, for which the analysis was conducted, according to which the electricity price is lower during the night and higher during the day.
- EV users are predominantly working individuals who are not at home during the day; therefore, it is not possible to charge the car at home during daytime hours. Even with installed rooftop PV systems, charging could not take place during the peak production period. The high price of rooftop PV power plants and energy storage systems, along with the low price of electricity and a long payback period, will not encourage electric vehicle owners to build a PV power plant.

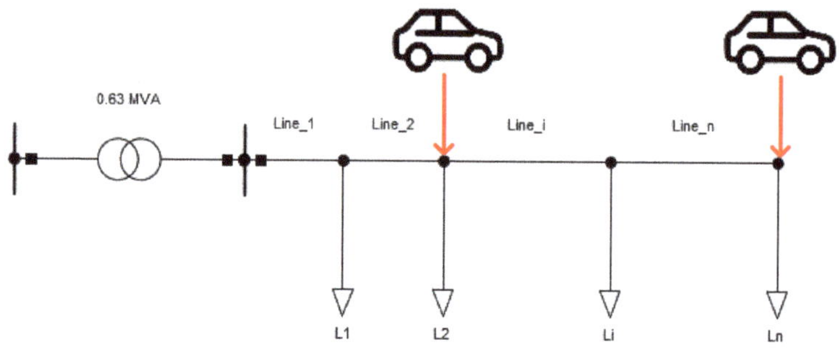

Figure 3. Distribution grid model with integrated EV.

All the simulations were made using DIgSILENT PowerFactory 2022. This research was also a based on previous research conducted by the authors [51–53].

The test grid was modeled with 90 households evenly distributed along the feeder with a three-phase unbalanced load and a peak power of each household of approximately 0.8 kW. The feeder was connected to the distribution grid via a 10/0.4 kV transformer with 0.63 MVA apparent power. The load characteristics were obtained by measurement for a normal working day and are shown in Figure 4.

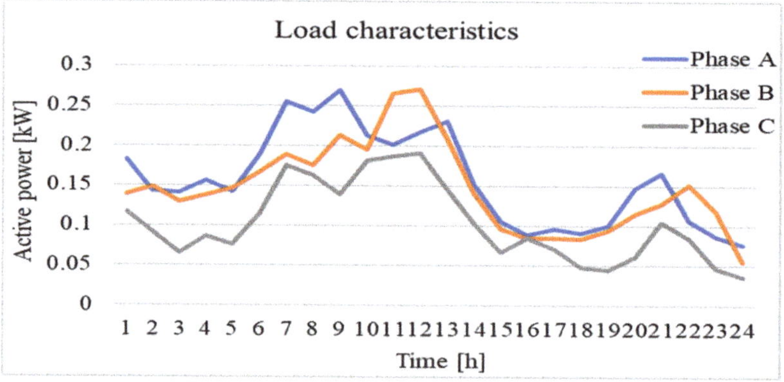

Figure 4. Load characteristics.

Five scenarios were analyzed for different shares of EV chargers and their output power during a 24 h period. In Scenario 1, the voltage profile and energy exchange were calculated for a feeder without EV chargers. This scenario was also used as base scenario for comparison. In Scenario 2, an unbalanced single-phased charger with 3.7 kW was implemented with a share of 50% EVs. The same share of balanced three-phase EV chargers was implemented in Scenario 3. In Scenario 4, the grid was loaded with maximum number of EV chargers. Single-phase 3.7 kW chargers were used in this scenario. In the last scenario, the maximum number of three-phase EV chargers (11 kW) was integrated on a feeder. For all scenarios, the voltage profiles for the end user are shown, as well as the energy exchange. In Scenarios 4 and 5, the number and power of chargers was defined in order to keep the voltages and thermal limitations of the cables within the allowed range. The assumption was that the cars had an empty battery and that charging was conducted over a certain period in all scenarios.

5.1. Scenario 1—Base Scenario

The key feature of the basic scenario was a completely passive load of the feeder, characterized by a large number of consumers with an unbalanced low load characteristic. It was assumed that there were no integrated chargers, and therefore all the energy was used to cover the consumption and losses along the cable. Figure 5 shows the voltage profile along the feeder (for all 90 loads), and it is shown that there was an unbalance, which was almost insignificant, since the voltage between phases was around 0.001 p.u.

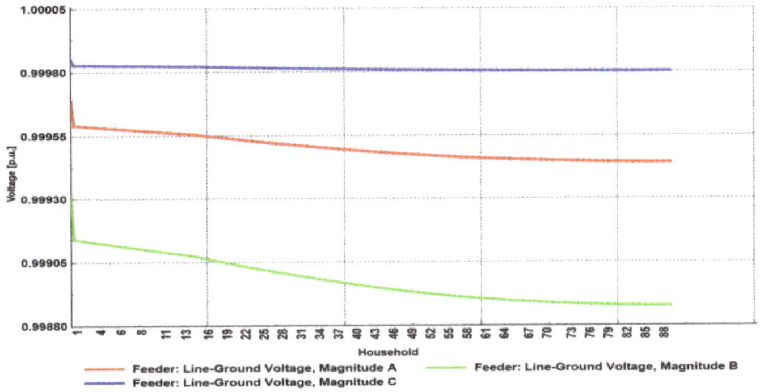

Figure 5. Voltage profile for Scenario 1.

Since all the consumers had a similar load pattern, the load curve and energy exchange during the day were constant (Figure 6). The maximum load was around 12.5 kW, resulting in approximately 300 kWh of energy taken from the grid during the 24 h period.

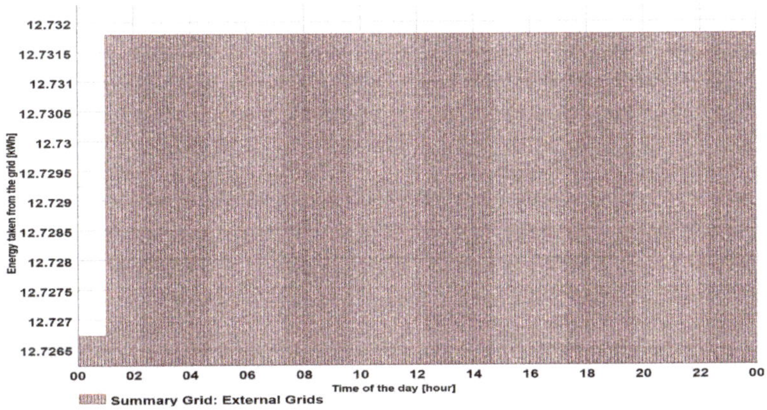

Figure 6. Energy exchange for Scenario 1.

Figure 7 shows the voltage profile for the end consumer located at bus 90. During the whole day, the voltage remained constant and within the specified limits. A voltage unbalance was present but it was within a very narrow range of deviations.

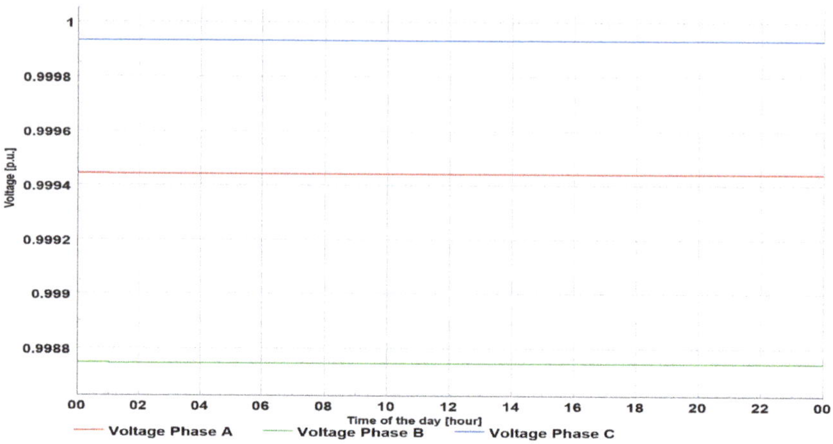

Figure 7. Voltage profile at bus 90, Scenario 1.

5.2. Scenario 2—50% of Single-Phase EV Chargers

In Scenario 2, single-phase EV chargers with an active power of 3.7 kW were integrated, following the expected increase in the number of EVs. We assumed that every second household would have an electric car and charge it via a single-phase charger, with 44 chargers implemented along the feeder. The chargers were evenly distributed on all three phases, and the resulting voltage profile is shown in Figure 8. As can be seen, towards the end of the feeder, there was an increasing unbalance and a voltage drop. When compared to the base scenario, it was observed that the voltage dropped by almost 0.15 p.u., resulting in voltage that was not within the limit defined by power quality standards. Unbalanced load in the phases could be avoided by proper planning and distribution of the EV chargers, but this was hard to accomplish, since EV charger integration was exclusively determined by the behavior and capabilities of the consumers.

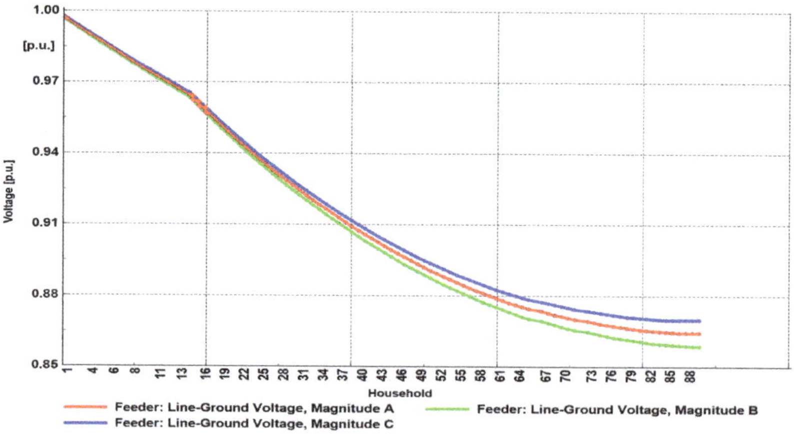

Figure 8. Voltage profile for Scenario 2.

With increased consumption, the total energy taken from the grid also increased, reaching 1003.7 kWh during the 24 h period. The peak power also increased, reaching 138.5 kW during the night period while the EVs were being charged. The total energy taken from the grid is shown in Figure 9. During the day, the total energy was dependent on household consumption and was rather low in comparison with that in the charging periods. A more uniform distribution and lower peak values could be achieved if all consumers would not charge their cars at the same time.

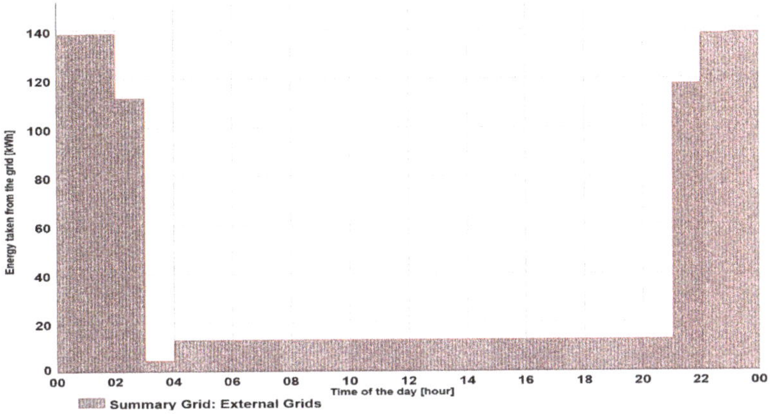

Figure 9. Energy exchange for Scenario 2.

This charging pattern and methods resulted in lower voltages at end consumers, bus 90 (Figure 10). During the day, the voltage was around 1 p.u. for all three phases, since the load was evenly distributed over the phases. The unbalance between phases is negligible. However, when all chargers were connected during the night, the voltage unbalance increased, with a minimum value of 0.855 p. u in phase B, and a maximum value of 0.875 p.u. in phase C.

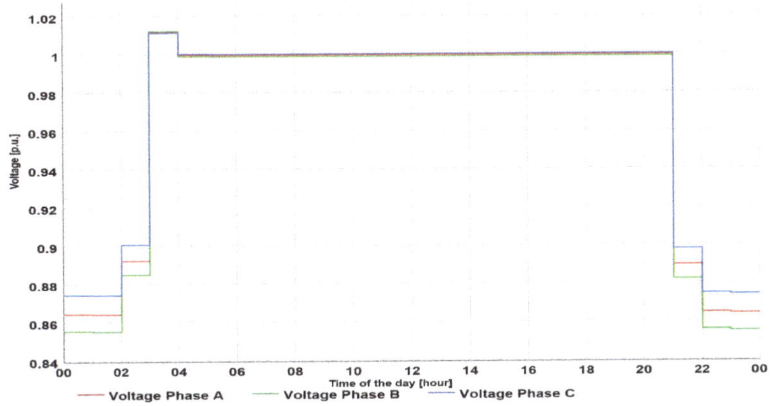

Figure 10. Voltage profile at bus 90, Scenario 2.

The voltage value below 0.9 p.u. was below the limit set by the EN 50160 standard, and such grid operation would not be permitted. With the existing infrastructure and grid loads, it would not be allowed to connect 50% of chargers to this outlet, as expected according to projections on the increase of electromobility.

5.3. Scenario 3—50% of Three-Phase EV Chargers

Similar to Scenario 2, in Scenario 3 it was assumed here that the integration of EV chargers would be according to the projections and expectations about the share of EVs. Accordingly, it was assumed that every second household would have a balanced three-phase EV charger with a power of 11 kW. The number of chargers remained the same as in the previous scenario (44 chargers). The voltage profile is shown in Figure 11, and it can be noticed that the voltage at the end user reached a value of 0.65 p.u. Since the chargers were balanced, the voltage along the feeder as also balanced.

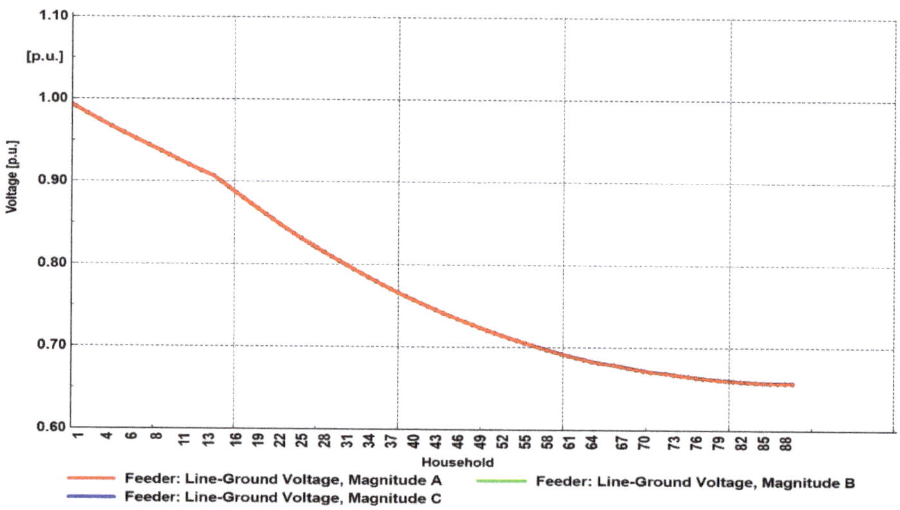

Figure 11. Voltage profile for Scenario 3.

The energy exchange (shown in Figure 12) increased, as the total power in the feeder reached the peak value of 341.6 kW. The total energy taken from the grid reached the value of 2195 kWh for the 24 h period. In comparison with the values for the previous scenarios, this is a huge increase, and the grid would need to be able to deliver this amount of energy. Also, with such increase, the thermal load of the cable would increase its deterioration and aging.

The voltage at the end consumer is shown in Figure 13. As in the previous scenario, the voltage deviated significantly from the lowest value of 0.66 p.u. during the period when charging was at its highest power. From the previously mentioned considerations, it is clear that the integration of three-phase chargers in such a large number should not be allowed for a given feeder.

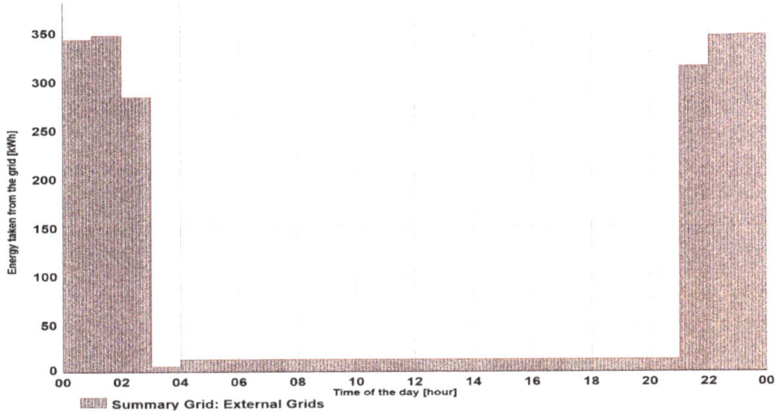

Figure 12. Energy exchange for Scenario 3.

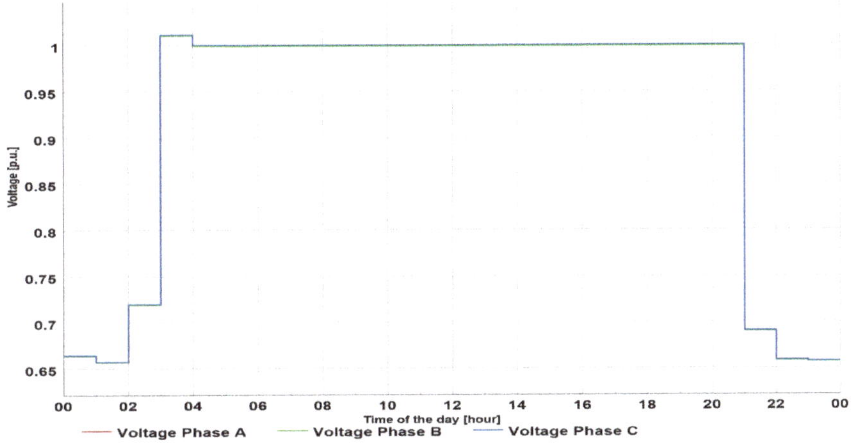

Figure 13. Voltage profile at bus 90, Scenario 3.

5.4. Scenario 4—Maximum Acceptable Number of Single-Phase EV Chargers

As shown in the previous scenarios, a high integration of EV chargers, whether single-phase or three-phase, would result in a significant increase in the load on the system and thus disrupt the voltage conditions. In order to test the maximum limit to which integration is possible, the number of chargers was gradually reduced until the voltage on the last bus reached the limit of 0.9 p.u. The largest number of single-phase chargers at which the voltage remained within the given limits was 22, with the chargers randomly distributed per feeder, which means that the largest possible integration involved 25% of the households with EV charging capability. The chargers were evenly distributed on all three phases, which does not actually reflect a real situation but can serve as a framework for further analysis. Figure 14 shows the voltage profile when the maximum number of single-phase EV chargers were integrated.

The voltage at the end user was lower than in the base scenario, but when compared with Scenario 2, the voltage drops were much lower. In addition, the voltage unbalance was significantly more pronounced. In Scenario 2 the difference between the lowest phase voltage and the highest phase voltage was 0.1 p.u., while in this case the difference was

0.3 p.u.—the lowest voltage was in phase B and amounted to 0.91 p.u., while with a value of 0.94 p.u., the phase C voltage was the highest. Energy taken from the grid is shown in Figure 15. The charging pattern remained unchanged, but the maximum power was reduced, being around 80 kW. In total, the feeder took 686 kWh from the distribution grid.

Comparing the basic scenario and Scenario 4, it is clear that the voltage unbalance and energy taken from the grid increased, which was expected, given the higher power consumption. When Scenario 4 is compared with scenario 2, it is clear that by reducing the number of EV chargers, better voltage conditions would be achieved in the system. Further improvement would be achieved by a better distribution of the chargers by phase and by uniform planning and scheduling of the EV charging pattern during a 24 h period. During the day, the voltage was equal to the rated value (1 p.u.), with low unbalance between phases. However, during the night, due to the increased load, the amount of current through the feeder increased, increasing the feeder's thermal stress and the energy losses.

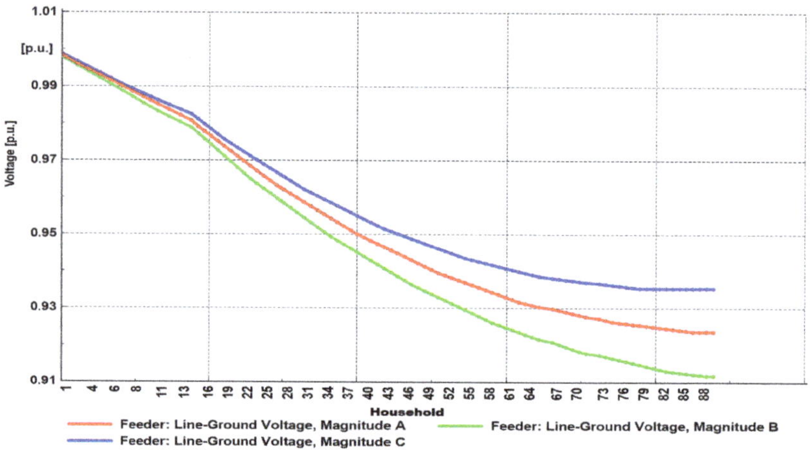

Figure 14. Voltage profile for Scenario 4.

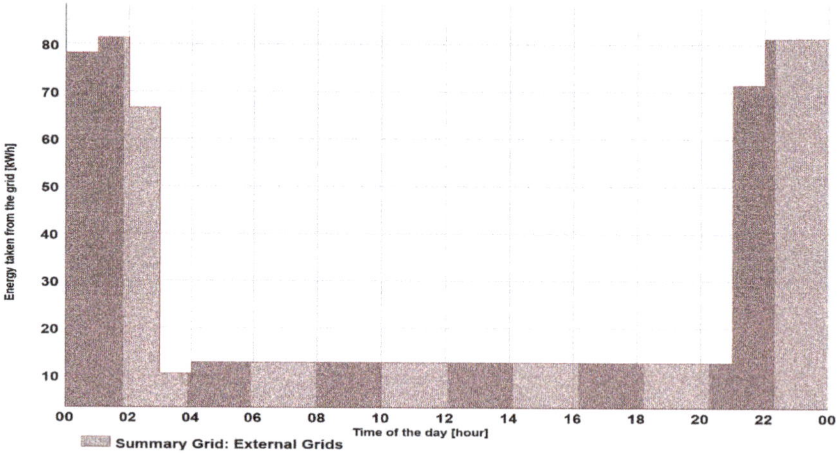

Figure 15. Energy exchange for Scenario 4.

Considering the voltage profile at the end user located on bus 90 (Figure 16), the voltage unbalance was more pronounced than in Scenario 2. In this scenario, the lowest voltage was in phase B (value of 0.91 p.u.), which is in accordance with the values of the feeder's voltage profile.

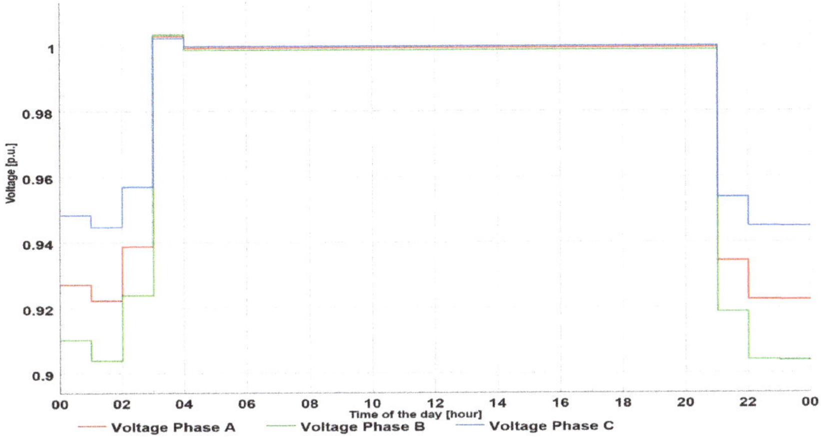

Figure 16. Voltage profile at bus 90, Scenario 4.

5.5. Scenario 5—Maximum Acceptable Number of Single-Phase EV Chargers

In Scenario 5, the same assumptions were used as in Scenario 4, regarding the largest number of three-phase chargers with a power of 11 kW that could be integrated on the feeder without a voltage drop below 0.9 p.u. As shown in Scenario 3, integration at the level of 50% of the households with EVs would lead to an excessive voltage drop. Therefore, as in Scenario 4, the number of chargers was reduced, and these were now symmetrical; ultimately, the total number of chargers that could operate simultaneously was 11, corresponding to 12.5% of the households that could have a three-phase EV charger. The voltage profile for this scenario is shown in Figure 17—the voltages were balanced, as the load was defined by the power of the EV chargers.

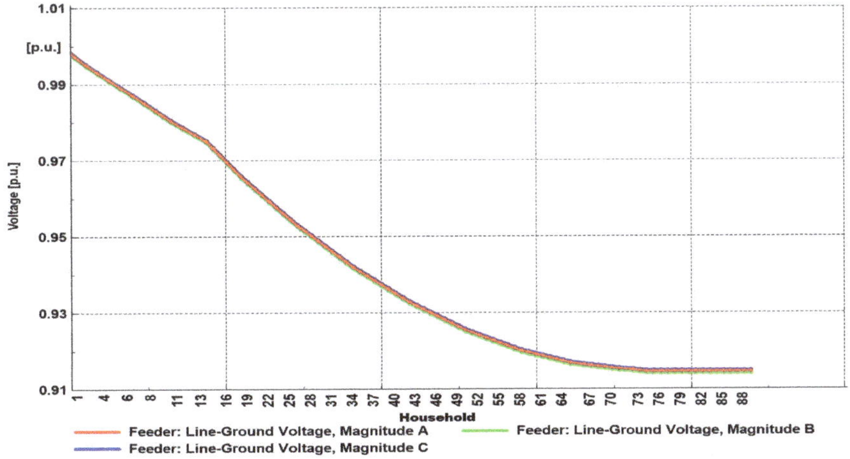

Figure 17. Voltage profile for Scenario 5.

With 11 chargers the total power taken from the distribution grid would increase compared to the base scenario. However, compared to Scenario 3 the power taken from the grid was lower and amounted to 823.94 kWh, with a peak load of 106.1 kW (Figure 18).

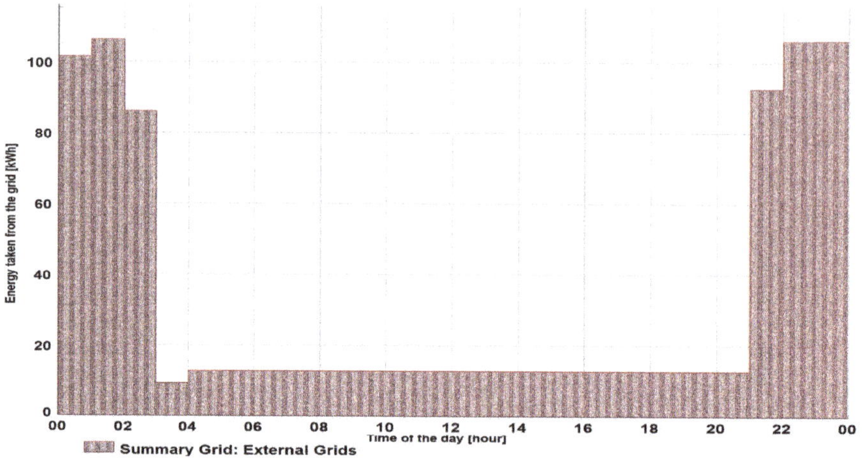

Figure 18. Energy exchange for Scenario 5.

The reduction in peak load from 341.6 kW in Scenario 3 to 106.1 kW in Scenario 5 represents a significant load reduction in the grid. However, this power was still significant and as can be seen in Figure 19, affected the voltage levels at the last consumer. The lowest voltage occurred during the night when the chargers worked at maximum power, but with a limited number of chargers, this value would not fall below 0.91 p.u.

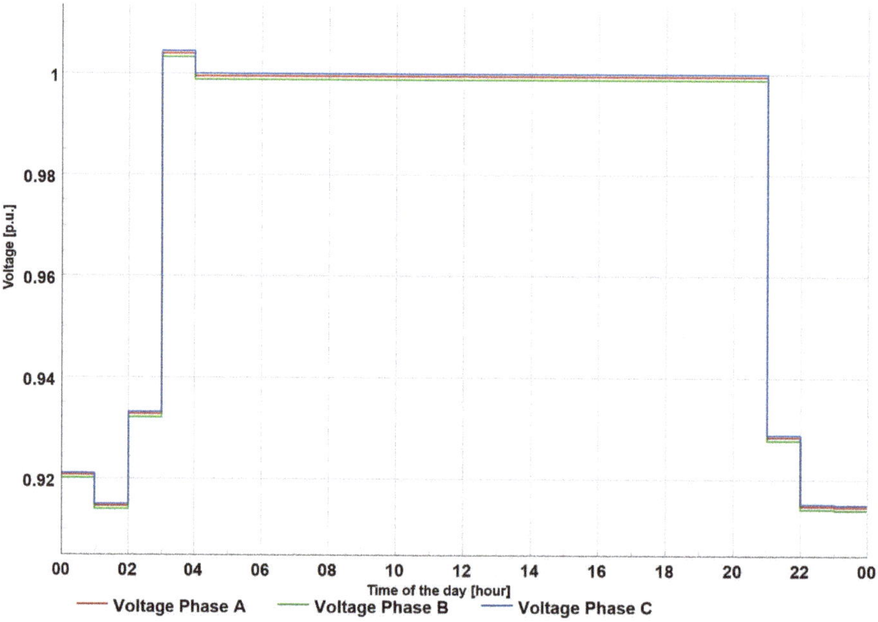

Figure 19. Voltage profile at bus 90, Scenario 5.

5.6. Sensitivity Analysis

In the previous analyses, it was observed that the implementation of EVs could be simply considered as an increase in the load in the system. A sensitivity analysis was made for the case of random daily charging. In this case, there was no integrated rooftop PV power plant, and therefore the voltage profile did not change significantly due to their influence. By connecting an EV with a single-phase charger, stochastically, it could be seen that the unbalance between the phases increased (Figure 20). The profile was obtained for the 12th hour of the day when charging was at its peak.

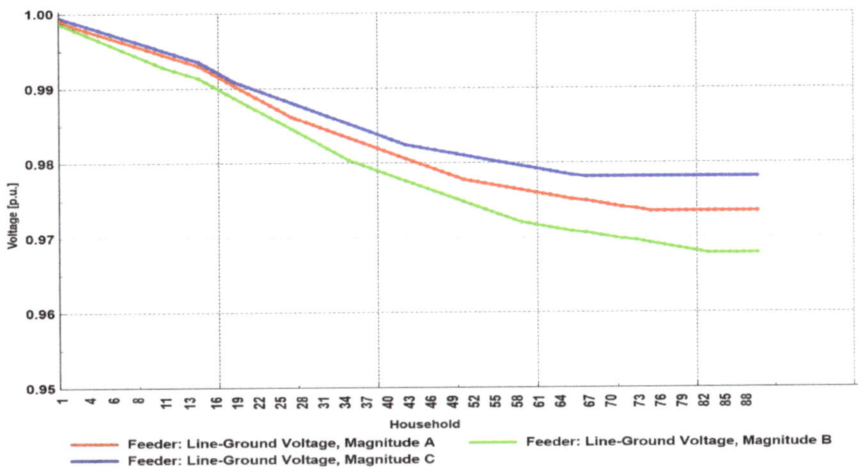

Figure 20. Voltage profile for daytime charging.

In terms of energy exchange, or energy taken from the network, the situation remained the same, since these were the same consumers, and the household load was approximately the same during the day and night, as can be seen in Figure 21.

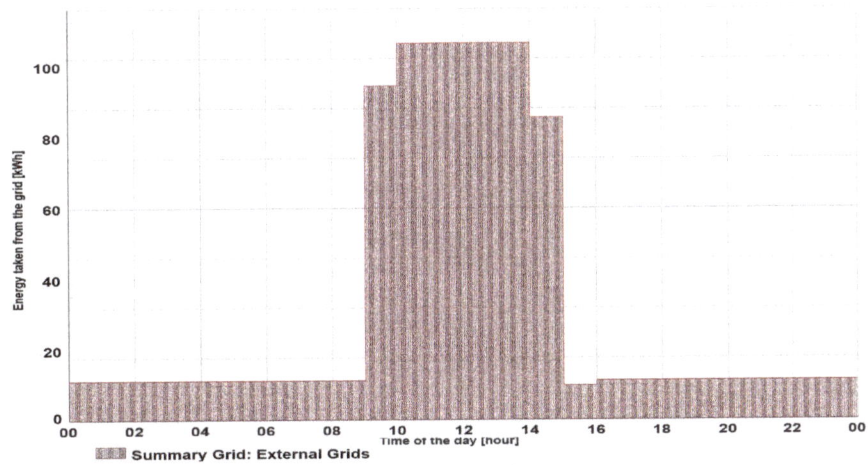

Figure 21. Energy exchange for daytime charging.

The voltage at the last consumer also changed and was the lowest during the car charging period, from the ninth to the fifteenth hour (Figure 22).

Considering the previous cases, it can be seen that by changing the charging time, the load curve in the system was smoothed. Furthermore, any change, for example, delayed EV charging, would reduce the load in the system.

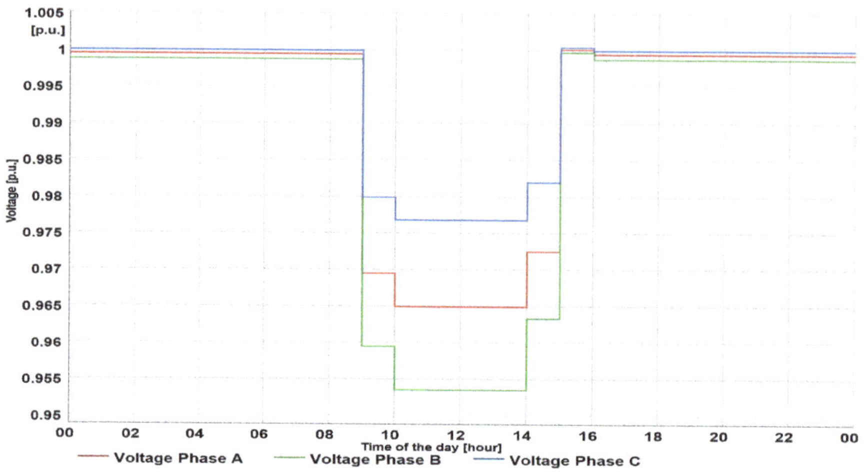

Figure 22. Voltage profile at bus 90 for daytime charging.

6. Result Comparison and Discussion

In the previous five scenarios we compared the voltage conditions along the feeder for different levels of EV charger integration and for two different charger types. Table 1 shows the voltage values at the last consumer, with the lowest phase value given for unbalanced consumers.

Table 1. Voltage at the end of the feeder.

Scenario	Number of Chargers	Peak Power, kW	End Voltage, p.u.
1	0	12.73	0.99
2	44	138.5	0.85
3	44	341.6	0.65
4	22	81.32	0.91
5	11	106.1	0.91

The energy taken from the grid, for each scenario, is shown in Figure 23. The highest energy value was found for Scenario 3 when integrating 44 three-phase EV chargers. When compared to Scenario 1, this was an increase of over 600%. Also, compared to all other scenarios, Scenario 3 with a high integration of three-phase EV chargers should not be allowed under any circumstances, given that the voltage on bus 90 (end user) was significantly impaired.

Slightly better voltage conditions were found with the maximum proposed integration of single-phase EV chargers that had lower maximum power and were distributed evenly on all three phases, which had less impact on the voltage conditions in the feeder.

Scenarios 2 and 3 had in common that although the aim was to equip at least 50% of the households with EVs, the infrastructure was not adapted to meet this goal. Furthermore,

EV charging was still stochastic and dictated by the price signals. A better distribution of the charging pattern during the day, which would result from changing consumer habits, more appropriate electricity prices that would encourage consumers to change their charging patterns, or the integration of rooftop PV plants would allow for a further increase in the number of chargers that can be connected to the feeder. The high peak power used to charge EVs puts a strain on the cables and the transformer and leads to their faster aging—all these are factors that should be taken into account when increasing electromobility.

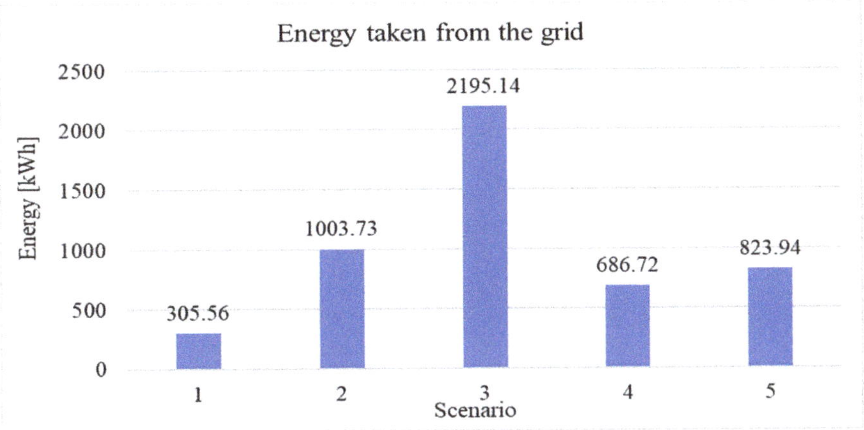

Figure 23. Energy comparison for all scenarios.

In all previous scenarios, the angle differences, harmonics or voltage regulation devices were neglected in all analyses. Given that it was a distribution grid that dealt with specific voltage levels, in this case 0.4 kV, a short distance and a specific R/X ratio, the influence of the angle difference could be justifiably neglected, as the difference was not significant.

Since it was a distribution feeder with a regular two-winding transformer, there was no possibility of voltage regulation. A possible change would be to replace the transformer with a tap changer, which would in that case raise the voltage at the beginning of the feeder and thus improve the voltage conditions in the system, which is not a realistic option in the Croatia nor is its implementation likely in practice. Also, since the loads were defined as ordinary households, the impact of higher harmonics was ignored because according to examples from the literature and according to the authors' available measurement data, EV charging does not contribute significantly to harmonics, which can also be seen in References [47,48]. For further analysis and future research, the following options could be considered:

- Replacing the regular transformer with a regulating transformer;
- Integrating a rooftop PV power plant at the consumer location, which would affect the voltage conditions and power flows in the network, but also possibly using the generated energy for EV charging or storage. In that case, harmonic analysis should also be considered.

According to the analysis, the increased integration of EVs would lead to higher transformer and line loading. From the perspective of the DSO, this poses new challenges, since no consumer can be banned from purchasing an EV or be restricted from charging it in their own household. Based on the analysis of the actual situation, the need for grid maintenance or upgrades, as well as future grid expansion, can be predicted. Currently, the existing overhead lines are being replaced with a self-supporting cable bundle as part of regular maintenance to improve reliability in rural areas and with buried cables in urban areas. In addition to improving the system reliability, the replacement takes into account the increase in household loads and increased electromobility. Since these are complex and expensive capital investments, the process is challenging and slow, and it will take a significant amount of time until all feeders are completely replaced. Another way in which the DSO can contribute to increasing electromobility is by facilitating the construction of rooftop PV power plants—by facilitating the procedure, providing incentives for construction and incentives for the installation of BESS and promoting and encouraging consumer self-sufficiency. A consumer's decision to purchase an electric vehicle will be influenced by numerous factors, such as socio-economic factors, the availability of charging stations, the range of distances they cover, the cost of upgrading charging infrastructure or the cost of a new property purchase in the case of already integrated chargers [54].

Although this paper performed a basic analysis with a uniform distribution of EV chargers, it is clear that there is a maximum number of vehicles that can be charged simultaneously on a given feeder. As shown in other works, for example, [37], in which EVs were integrated into an urban grid, or [41], with their integration into the transmission grid, the feeder topology, available load and voltage levels have a significant impact on the number of integrated vehicles. Also, by implementing complex optimization methods, the maximum number of vehicles that can be integrated and a charging plan can be created. If there is the option of controlling the reactive energy or if the vehicles could provide reactive energy, the number of EVs can significantly increase.

The sensitivity analysis showed that even by changing the charging time, for example by charging during the day or by delayed charging, the voltage profile and the daily load curve would be smoothed. This actually paves the way towards finding the optimal charging time and method that, knowing the actual state and limitations of the network as well as the habits of EV owners, would encourage owners to charge vehicles at certain times when it is convenient from the system's point of view—for example, by offering dynamic pricing.

7. Conclusions

In this paper, a suburban realistic radial feeder with a dominant type of household loads was used as the model for testing different integration rates of electromobility. The consumers were unbalanced, and it was assumed that there was no integrated rooftop PV power plant anywhere on the feeder, which was considered exclusively passive. For a known daily load characteristic, the influence of the integration of EV chargers, single-phase and three-phase, was considered. The level of integration was initially determined by recommendations on the share of electromobility in the amount of 50%. In both cases, for a single-phase charger with a power of 3.7 kW and a three-phase charger with a power of 11 kW, it was shown that such a level of integration was not possible, since the voltages reached the amount of 0.85 p.u. and 0.65 p.u., respectively. Therefore, there was a need to find the largest number of chargers that could be connected to such a feeder without significantly disturbing the voltage limits at the end consumer. By integrating single-phase EV chargers, the peak power was lower, the energy taken from the distribution grid was also lower, and at the same time, a larger number of chargers could be connected. The negative side was that the multiple chargers could be connected to a specific feeder, which could result in increased imbalance. On the other hand, three-phase chargers could reduce the voltage unbalance, but a smaller number of EV chargers could be integrated, with higher peak power. In addition, the stochasticity of consumer behavior needs to be added to the planning of an electromobility analysis. The scenarios were quite roughly defined with their percentages, since the stochasticity and simultaneity factors of charging affect the overall voltage profile and peak load values. The cable and transformer overload was particularly pronounced, since we considered a pure one-way energy transfer, and for further work, the integration of rooftop PV power plants, different patterns and times of EV charging, as well as simultaneity factors of charging should be taken into account.

Author Contributions: Conceptualization, R.K. and T.K.; methodology, R.K. and T.K.; software, R.K.; validation, H.G., R.K. and M.K.; formal analysis, T.K.; investigation, R.K.; resources, H.G. and R.K.; data curation, R.K.; writing—original draft preparation, T.K.; writing—review and editing, H.G.; visualization, M.K.; supervision, M.K.; project administration, M.K.; funding acquisition, R.K. All authors have read and agreed to the published version of the manuscript.

Funding: The APC was funded by the Faculty of Electrical Engineering, Computer Science and Information Technology Osijek.

Data Availability Statement: Test system data are available on request. All other data presented in this study are available on request.

Acknowledgments: This work was supported by the Croatian Science Foundation under the project "Prosumer-rich distribution power network" (project number: UIP-2020-02-5796).

Conflicts of Interest: The authors declare no conflicts of interest.

Abbreviations

The abbreviations used in this study are listed below (in order of appearances)

Symbol	Description
EV	Electric vehicle
EU	European Union
BEV	Battery electric vehicle
PHEV	Plug-in hybrid electric vehicle
PV power plant	Photovoltaic power plant
DSO	Distribution system operator
V2G	Vehicle-to-grid
V2H	Vehicle-to-home
V2B	Vehicle-to-building
RES	Renewable energy sources
BESS	Battery energy storage system

References

1. Fetting, C. The European Green Deal. *ESDN Rep. Dec.* **2020**, 2, 53.
2. Amann, G.; Escobedo Bermúdez, V.R.; Boskov-Kovacs, E.; Gallego Amores, S.; Giannelos, S.; Iliceto, A.; Ilo, A.; Chavarro, J.R.; Samovich, N.; Schmitt, L.; et al. *E-Mobility Deployment and Impact on Grids: Impact of EV and Charging Infrastructure on European T&D Grids–Innovation Needs*; Publications Office of the European Union: Brussels, Belgium, 2022.
3. Schneider Electric. (n.d.). Electric Vehicle Trends. Available online: https://www.electrical-installation.org/enwiki/Electric_Vehicle_trends (accessed on 12 March 2025).
4. The Croatian Parliament, Building Act, Official Gazette 125/19. 2019. Available online: https://narodne-novine.nn.hr/clanci/sluzbeni/2019_12_125_2489.html, (accessed on 12 March 2025).
5. Center for Vehicles of Croatia (CVH). Statistics on the Number of Electric and Hybrid Vehicles (2007–2024). Available online: https://cvh.hr/gradani/tehnicki-pregled/statistika/ (accessed on 13 March 2025).
6. Eurostat, Official Website for European Statistics. Available online: https://ec.europa.eu/eurostat/statistics-explained/index.php?title=Transport_equipment_statistics#Road_transport_equipmentNordic_countries_led_the_new_zero-emission_vehicles_.28ZEV.29_registered_in_2023 (accessed on 17 July 2024).

7. Mišljenović, N.; Žnidarec, M.; Knežević, G.; Topić, D. Electric Vehicles as Flexible Assets for Households and System Operators: Survey Results. In Proceedings of the 2024 International Conference on Smart Systems and Technologies (SST), Osijek, Croatia, 16–18 October 2024; pp. 1–7. [CrossRef]
8. Gonzalez Venegas, F.; Petit, M.; Perez, Y. Active integration of electric vehicles into distribution grids: Barriers and frameworks for flexibility services. *Renew. Sustain. Energy Rev.* **2021**, *145*, 111060. [CrossRef]
9. Habib, S.; Khan, M.M.; Hashmi, K.; Ali, M.; Tang, H. A comparative study of electric vehicles concerning charging infrastructure and power levels. In Proceedings of the 2017 International Conference on Frontiers of Information Technology (FIT), Islamabad, Pakistan, 18–20 December 2017; pp. 327–332. [CrossRef]
10. Habib, S.; Khan, M.M.; Huawei, J.; Hashmi, K.; Faiz, M.T.; Tang, H. A study of implemented international standards and infrastructural system for electric vehicles. In Proceedings of the 2018 IEEE International Conference on Industrial Technology (ICIT), Greenville, SC, USA, 4–8 March 2012; pp. 1783–1788. [CrossRef]
11. Yilmaz, M.; Krein, P.T. Review of charging power levels and infrastructure for plug-in electric and hybrid vehicles. In Proceedings of the 2012 IEEE International Electric Vehicle Conference, Greenville, SC, USA, 4–8 March 2012; pp. 1–8. [CrossRef]
12. Zakaria, A.; Chengyan, D.; Zafar, H.; Djokic, S.Z. After diversity maximum demand and daily load profiles of maximum demand for uncontrolled residential EV charging. In Proceedings of the 2023 IEEE PES Innovative Smart Grid Technologies Europe (ISGT EUROPE), Grenoble, France, 23–26 October 2023; pp. 1–5. [CrossRef]
13. Lojowska, A.; Kurowicka, D.; Papaefthymiou, G.; van der Sluis, L. From transportation patterns to power demand: Stochastic modeling of uncontrolled domestic charging of electric vehicles. In Proceedings of the 2011 IEEE Power and Energy Society General Meeting, Detroit, MI, USA, 24–28 July 2011; pp. 1–7. [CrossRef]
14. Ayala, H.E.V.; Barriga, N.G. Study of the impact of electric vehicles fleets in HV electric power grids based on an uncontrolled charging strategy. In Proceedings of the 2017 IEEE International Autumn Meeting on Power, Electronics and Computing (ROPEC), Ixtapa, Mexico, 8–10 November 2017; pp. 1–6. [CrossRef]
15. Yu, Y.; Shekhar, A.; Mouli, G.C.R.; Bauer, P.; Refa, N.; Bernards, R. Impact of uncontrolled charging with mass deployment of electric vehicles on low voltage distribution networks. In Proceedings of the 2020 IEEE Transportation Electrification Conference & Expo (ITEC), Chicago, IL, USA, 23–26 June 2020; pp. 766–772. [CrossRef]
16. Jäger, G.; Hofer, C.; Füllsack, M. The benefits of randomly delayed charging of electric vehicles. *Sustainability* **2019**, *11*, 3722. [CrossRef]
17. Daziano, R.A. Willingness to delay charging of electric vehicles. *Res. Transp. Econ.* **2022**, *94*, 101177. [CrossRef]
18. Williams, B.; Bishop, D.; Hooper, G.; Chase, J.G. Driving change: Electric vehicle charging behavior and peak loading. *Renew. Sustain. Energy Rev.* **2024**, *189*, 113953. [CrossRef]
19. Pandit, A.K.; Bhalja, B.R.; Seshadrinath, J. EV charging coordination based energy management system for a grid-tied charging station integrated with PV-ESS. In Proceedings of the 2023 IEEE International Conference on Energy Technologies for Future Grids (ETFG), Wollongong, Australia, 3–6 December 2023; pp. 1–6. [CrossRef]
20. Mastoi, M.S.; Zhuang, S.; Munir, H.M.; Haris, M.; Hassan, M.; Alqarni, M.; Alamri, B. A study of charging-dispatch strategies and vehicle-to-grid technologies for electric vehicles in distribution networks. *Energy Rep.* **2023**, *9*, 1777–1806. [CrossRef]
21. Fasugba, M.A.; Krein, P.T. Gaining vehicle-to-grid benefits with unidirectional electric and plug-in hybrid vehicle chargers. In Proceedings of the 2011 IEEE Vehicle Power and Propulsion Conference, Chicago, IL, USA, 6–9 September 2011; pp. 1–6. [CrossRef]
22. Rana, M.M.; Alam, S.M.; Rafi, F.A.; Deb, S.B.; Agili, B.; He, M.; Ali, M.H. Comprehensive review on the charging technologies of electric vehicles (EV) and their impact on power grid. *IEEE Access* **2025**, *13*, 35124–35156. [CrossRef]
23. Liu, C.; Chau, K.T.; Wu, D.; Gao, S. Opportunities and challenges of vehicle-to-home, vehicle-to-vehicle, and vehicle-to-grid technologies. *Proc. IEEE* **2013**, *101*, 2409–2427. [CrossRef]
24. Gamallo, C.; Fraile-Ardanuy, J. Stochastic aggregated battery for vehicle to building applications. In Proceedings of the 2012 IEEE International Energy Conference and Exhibition (ENERGYCON), Florence, Italy, 9–12 September 2012; pp. 749–754. [CrossRef]
25. Ghaderi, A.; Nassiraei, A.A.F. The economics of using electric vehicles for vehicle to building applications considering the effect of battery degradation. In Proceedings of the IECON 2015–41st Annual Conference of the IEEE Industrial Electronics Society, Yokohama, Japan, 9–12 November 2015; pp. 3567–3572. [CrossRef]
26. Marmaras, C.; Corsaro, M.; Xydas, E.; Cipcigan, L.M.; Pastorelli, M.A. Vehicle-to-building control approach for EV charging. In Proceedings of the 2014 49th International Universities Power Engineering Conference (UPEC), Cluj-Napoca, Romania, 2–5 September 2014; pp. 1–6. [CrossRef]
27. Arya, T.; Saxena, D.; Misra, P. Optimizing electricity bill, grid performance, and user satisfaction in a comprehensive home energy management system with electric vehicles and home-to-grid capabilities. In Proceedings of the 2023 International Conference on Electrical, Electronics, Communication and Computers (ELEXCOM), Roorkee, India, 26–27 August 2023; pp. 1–6. [CrossRef]

28. Mohanty, S.; Panda, S.; Parida, S.M.; Rout, P.K.; Sahu, B.K.; Bajaj, M.; Zawbaa, H.M.; Kumar, N.M.; Kamel, S. Demand side management of electric vehicles in smart grids: A survey on strategies, challenges, modeling, and optimization. *Energy Rep.* **2022**, *8*, 12466–12490. [CrossRef]
29. Khemakhem, S.; Rekik, M.; Krichen, L. Double layer home energy supervision strategies based on demand response and plug-in electric vehicle control for flattening power load curves in a smart grid. *Energy* **2019**, *167*, 312–324. [CrossRef]
30. Qian, K.; Zhou, C.; Yuan, Y. Impacts of high penetration level of fully electric vehicles charging loads on the thermal ageing of power transformers. *Int. J. Electr. Power Energy Syst.* **2015**, *65*, 102–112. [CrossRef]
31. Visakh, A.; Selvan, M.P. Seasonal effects of electric vehicle charging on the aging of distribution transformers. In Proceedings of the 2021 13th IEEE PES Asia Pacific Power & Energy Engineering Conference (APPEEC), Thiruvananthapuram, India, 21–23 November 2021; pp. 1–6. [CrossRef]
32. Pavlićević, A.; Mujović, S. Impact of reactive power from public electric vehicle stations on transformer aging and active energy losses. *Energies* **2022**, *15*, 7085. [CrossRef]
33. Jadhav, M.P.; Kalkhambkar, V.N. Frequency regulation by electric vehicle. In Proceedings of the 2018 International Conference on Current Trends towards Converging Technologies (ICCTCT), Coimbatore, India, 1–3 March 2018; pp. 1–6. [CrossRef]
34. Dong, X.; Ma, Y.; Yu, X.; Wei, X.; Ren, Y.; Zhang, X. Secondary frequency regulation control strategy with electric vehicles considering user travel uncertainty. *Energies* **2023**, *16*, 3794. [CrossRef]
35. Shrivastava, S.; Khalid, S.; Nishad, D.K. Impact of EV interfacing on peak-shelving and frequency regulation in a microgrid. *Sci. Rep.* **2024**, *14*, 31514. [CrossRef]
36. Xue, Y.; Chen, Q.; Cai, H.; Xia, M.; Gu, H. Conflict mechanism analysis method for electric vehicles participating in multiple types of power grid ancillary services. In Proceedings of the 2023 International Conference on Power System Technology (PowerCon), Jinan, China, 21–22 September 2023; pp. 1–6. [CrossRef]
37. Zhou, C.; Xiang, Y.; Huang, Y.; Wei, X.; Liu, Y.; Liu, J. Economic analysis of auxiliary service by V2G: City comparison cases. *Energy Rep.* **2020**, *6*, 509–514. [CrossRef]
38. Carrión, M.; Zárate-Miñano, R.; Domínguez, R. Integration of electric vehicles in low-voltage distribution networks considering voltage management. *Energies* **2020**, *13*, 4125. [CrossRef]
39. Chen, J.; Bei, G.; Zhang, Q.; Huang, L.; Liu, K.; Chen, P. Integrated reactive power optimization for distribution systems considering electric vehicle dis-/charging support. *Energy Rep.* **2023**, *9*, 1888–1896. [CrossRef]
40. Yong, J.Y.; Ramachandaramurthy, V.K.; Tan, K.M.; Mithulananthan, N. A review on the state-of-the-art technologies of electric vehicle, its impacts and prospects. *Renew. Sustain. Energy Rev.* **2015**, *49*, 365–385. [CrossRef]
41. Azzopardi, B.; Gabdullin, Y. Impacts of electric vehicles charging in low-voltage distribution networks: A case study in Malta. *Energies* **2024**, *17*, 289. [CrossRef]
42. Rahman, S.; Khan, I.A.; Khan, A.A.; Mallik, A.; Nadeem, M.F. Comprehensive review & impact analysis of integrating projected electric vehicle charging load to the existing low voltage distribution system. *Renew. Sustain. Energy Rev.* **2022**, *153*, 111756. [CrossRef]
43. Zárate-Miñano, R.; Flores Burgos, A.; Carrión, M. Analysis of different modeling approaches for integration studies of plug-in electric vehicles. *Int. J. Electr. Power Energy Syst.* **2020**, *114*, 105398. [CrossRef]
44. Ye, Z.; Li, T.; Low, S. Towards balanced three-phase charging: Phase optimization in adaptive charging networks. *Electr. Power Syst. Res.* **2022**, *212*, 108322. [CrossRef]
45. Farahani, H.F. Improving voltage unbalance of low-voltage distribution networks using plug-in electric vehicles. *J. Clean. Prod.* **2017**, *148*, 336–346. [CrossRef]
46. Mexis, I.; Todeschini, G.; Möller, F.; Meyer, J. Mitigation of voltage unbalance in rural low voltage networks using single-phase BESS inverters. In Proceedings of the CIRED 2021-The 26th International Conference and Exhibition on Electricity Distribution, Online, 20–23 September 2021; pp. 789–794. [CrossRef]
47. Dávila-Sacoto, M.; González, L.G.; Hernández-Callejo, L.; Duque-Perez, Ó.; Zorita-Lamadrid, Á.L.; Alonso-Gómez, V.; Espinoza, J.L. Harmonic Distortion and Hosting Capacity in Electrical Distribution Systems with High Photovoltaic Penetration: The Impact of Electric Vehicles. *Electronics* **2023**, *12*, 2415. [CrossRef]
48. Ahmed, A.; Iqbal, A.; Khan, I.; Al-Wahedi, A.; Mehrjerdi, H.; Rahman, S. Impact of EV charging station penetration on harmonic distortion level in utility distribution network: A case study of Qatar. In Proceedings of the 2021 IEEE Texas Power and Energy Conference (TPEC), College Station, TX, USA, 2–5 February 2021; pp. 1–6. [CrossRef]
49. European Committee for Electrotechnical Standardization. Voltage characteristics of electricity supplied by public electricity networks. In *CENELEC EN 50160*; CENELEC: Brussels, Belgium, 2022.
50. Anagnostopoulos, P.M.; Papathanassiou, S.A. A power flow method for radial distribution feeders with DER penetration. *J. Technol. Innov. Renew. Energy* **2019**, *8*, 1–12. [CrossRef]

51. Kovačević, T. Utjecaj Elektromobilnosti na Distribucijsku Mrežu (The Impact of Electromobility on the Distribution Network). Master's Thesis, Josip Juraj Strossmayer University of Osijek, Faculty of Electrical Engineering, Computer Science and Information Technology, Osijek, Croatia, 2023. Available online: https://repozitorij.etfos.hr/en/islandora/object/etfos:4407/datastream/PDF/view (accessed on 1 May 2025). (In Croatian).
52. Kljajić, R.; Mišljenović, N.; Marić, P. Impact of the Increasing Integration of Photovoltaic Systems and Charging Stations for Electric Vehicles with Vehicle-To-Grid Concept on Voltage Stability of the Distribution Grid. In Proceedings of the IEEE EUROCON 2023–20th International Conference on Smart Technologies, Torino, Italy, 6–8 July 2023; pp. 671–676. [CrossRef]
53. Kljajić, R.; Marić, P.; Mišljenović, N.; Dubravac, M. An Optimized Strategy for the Integration of Photovoltaic Systems and Electric Vehicles into the Real Distribution Grid. *Energies* **2024**, *17*, 5602. [CrossRef]
54. Hoffmann, J.; Szkudlarek, V.; Ludin, D.; Schreier, N.; Mueller Ei Wellbrock, W. Factors Influencing the Purchase of Battery Electric Vehicles (BEVs): An Explorative Study Based on the Analysis of New Registrations and Expert Interviews in Germany. *Teh. Glas.* **2025**, *19*, 359–367. [CrossRef]

World Electric Vehicle Journal

MDPI

Article

Impacts of Electric Vehicle Penetration on the Frequency Stability of Curaçao's Power Network

Daniela Vásquez-Cardona [1], Sergio D. Saldarriaga-Zuluaga [1,*], Santiago Bustamante-Mesa [1,*], Jesús M. López-Lezama [2] and Nicolás Muñoz-Galeano [2]

[1] Departamento de Eléctrica, Facultad de Ingeniería, Institución Universitaria Pascual Bravo, Calle 73 No. 73A-226, Medellin 050036, Colombia; d.vasquez@pascualbravo.edu.co

[2] Research Group on Efficient Energy Management (GIMEL), Department of Electrical Engineering, Universidad de Antioquia (UdeA), Medellin 050010, Colombia; jmaria.lopez@udea.edu.co (J.M.L.-L.); nicolas.munoz@udea.edu.co (N.M.-G.)

* Correspondence: s.saldarriagazu@pascualbravo.edu.co (S.D.S.-Z.); san.bustamante@pascualbravo.edu.co (S.B.-M.)

Abstract: Assessing the impact of electric vehicle (EV) integration on power systems is crucial, particularly regarding frequency stability, which often remains largely unaddressed, especially in developing countries. This paper examines the effects of EV penetration on the frequency stability of Curaçao's power network, an aspect not previously studied for the island. As a key contribution, we present a representative model of Curaçao's power network, adjusting the dynamic models of the speed governors of synchronous machines, using data available to the academic community. Additionally, we analyze the impacts of EVs on the grid's frequency stability under different EV participation scenarios. To achieve this, simulations were conducted considering various EV participation scenarios and different types of chargers to assess their impact on grid stability. The study evaluates key frequency stability metrics, including the rate of change of frequency (RoCoF) as well as the highest and lowest frequency values during the transient period. The results indicated that higher EV penetration can significantly impact frequency stability. The observed increase in the RoCoF and frequency zenith values suggests a weakening of the grid's ability to withstand frequency disturbances, particularly in high-EV-penetration scenarios.

Keywords: electric vehicles; EVs charging methods; load modeling; frequency stability

1. Introduction

Transportation methods have transformed global travel and connectivity. Since the advent of combustion engine vehicles (CEVs), the movement of people and goods over long distances has become faster and more efficient. However, these vehicles have contributed significantly to both air and noise pollution. Emissions from CEVs are associated with numerous health and environmental issues [1]. Road transportation is a significant contributor to carbon dioxide emissions, with cars being a major source due to their heavy reliance on fossil fuels [2,3]. This environmental impact has spurred the rising popularity of electric vehicles (EVs), prompting traditional CEV manufacturers to shift their focus towards EV production and promotion. Additionally, various countries have supported the adoption of EVs through policies and financial incentives, accelerating their integration into the market.

The growing adoption of EVs is expected to have a substantial impact on the power grid. While EVs offer environmental benefits by reducing greenhouse gas emissions and enhancing air quality, they also increase electricity demand [4]. This heightened demand could lead to grid congestion during peak charging periods, potentially causing instability and even blackouts. However, the challenges posed by EV integration can be alleviated through the use of smart charging and vehicle-to-grid (V2G) technology. Smart charging optimizes the timing of EV charging by shifting it to off-peak hours, when electricity demand is lower, while V2G technology enables EVs to supply power back to the grid during times of high demand [5]. These technologies enhance grid stability, bolster resilience, and facilitate the integration of renewable energy sources [6]. As EV adoption grows, it becomes crucial for the power industry to develop innovative strategies to maintain a stable and sustainable grid. Research has already explored areas such as voltage deviations, power system losses, increased peak demand, grid overloads, and harmonics [7–9].

Several studies have been conducted to assess the impact of EVs on modern power systems. In [10], the authors examine how various levels of EV penetration affect power losses, voltage stability, reliability, and economic losses in distribution networks, focusing on the location of charging stations to reduce these impacts. When too many EVs connect to the grid at once, it can cause issues such as congestion and voltage drops, as distribution grids are designed for a specific load. If a large number of EVs connect simultaneously, system security may be compromised and the grid might fail to return to normal levels before significant damage occurs [11]. In [12], the authors perform a sensitivity analysis using Monte Carlo simulations to evaluate the effects of high EV penetration, considering factors like charger types, driving patterns, and grid configuration. Despite providing a comprehensive analysis, the study does not address voltage levels or propose solutions for the identified issues. In [13], a study examines the impact of private EVs on residential electricity prices in Germany, showing that EVs can lower electricity costs for households. However, the study does not propose strategies to address the grid impacts caused by EVs. Rahman et al. [14] review the integration of EV charging loads in low-voltage distribution systems, analyzing voltage profiles as EV penetration increases. They also discuss strategies to optimize grid performance, such as peak load management and increasing renewable energy penetration, but do not suggest solutions for mitigating the impacts of EVs.

There are various methods for modeling the penetration levels of EVs in power distribution networks, which can be divided into three main categories: deterministic approaches, data-driven approaches, and uncertainty-based approaches [15,16]. The first category includes measurement-based load-modeling techniques, such as the approach used in [17], where a ZIP-based model is applied to estimate the load for EV fast-charging stations in AC distribution grids. The second category involves machine learning methods, like k-nearest neighbors [18], linear regression [19], and random forest [20], which analyze large datasets to model the interaction between EVs and power systems. These data-driven methods rely on extensive data to simulate realistic charging behaviors. The third category focuses on approaches that account for the random behavior of EV users, including the use of various probability distributions (such as Gaussian [21], Weibull [22], and lognormal [23]), stochastic processes [24], and Monte Carlo simulations [12]. While all three approaches offer innovative ways to assess the impacts of EVs on distribution networks, data-driven and uncertainty-based methods require significant amounts of data and are computationally demanding. For these reasons, polynomial-based load models, such as ZIP-based methods, offer a promising alternative for analyzing the impact of EVs on distribution systems. Considering the ZIP-based methods, the authors of [25] evaluate the impact of EVs on Curaçao's power grid, analyzing congestion and voltage degradation to determine the grid's hosting capacity.

In addition, several studies have examined the integration of electric vehicles into power systems, focusing on their impact on grid stability and the optimization of charging behavior. In [26], the authors investigate how high load demands from EVs can induce dynamic instabilities in the distribution network, affecting frequency stability. They propose control strategies to mitigate these effects. The effects of large-scale EV charging on power grid stability are analyzed in [27], with a focus on potential frequency fluctuations caused by uncoordinated charging behaviors. In [28], the implications of widespread EV charger deployment on grid stability are examined, including challenges related to frequency and voltage stability. The impact of increased EV penetration on distribution system components, power loss, and grid stability is explored in [29], emphasizing the necessity of effective management to prevent frequency instability. In [30], while the potential benefits of EVs are highlighted, the challenges posed by high charging demands are also addressed, as improper management may lead to grid instability.

In numerous studies examining the impact of EVs on power systems, the specific type of charger is often overlooked. While most research focuses on the overall EV load, the varying categories of chargers, from slow AC chargers to DC fast chargers, significantly influence grid dynamics. EV chargers, primarily composed of power electronic converters, exhibit different behaviors compared to traditional loads in power systems [31]. This study aims to evaluate the frequency stability of Curaçao's power network, building on the work presented in [25], which assessed voltage profiles and component capacity. For this purpose, a ZIP load modeling approach was adopted to represent EV charging stations, following the methodology outlined in [31]. This model is applicable to both static and dynamic studies, and the analysis includes both AC and DC fast charging stations, which have been shown to exert the most significant impact on power systems [32].

As the global transition toward sustainable transportation accelerates, understanding the effects of EV integration on power systems has become increasingly critical. Many studies emphasize steady-state analyses, underscoring the necessity of complementing them with dynamic studies, especially those assessing frequency stability in the context of EV integration. Such research provides essential insights into the challenges and opportunities of incorporating EVs into existing grids. By examining charging patterns, demand fluctuations, stability, and infrastructure needs, researchers can offer valuable guidance to policymakers and utilities. However, in certain regions, including some states and countries, these studies remain unaddressed due to various factors. This paper evaluates the impact of EV penetration on the frequency stability of Curaçao's power network. To the authors' knowledge, no prior studies have specifically examined this aspect on the island. Using the representative network prototype from [25], this research aims to continue and expand previous work, focusing on encouraging the adoption of electric vehicles in Curaçao. The primary features and contributions of this study are as follows:

- The representative prototype of Curaçao's power network presented in [25] is enhanced by adding and adjusting the dynamic models of the speed governors of synchronous machines, with the data made available to the academic community for further studies.
- The impacts of EVs on the frequency stability of the grid are analyzed, considering different scenarios of EV participation, enabling the identification of the potential effects on the island's network.
- One of the main contributions of this study is the implementation of the Inertia Emulation Control Strategy for BESSs through EVs operating in V2G mode using

RoCoF-Based Synthetic Inertia. This technique allows battery energy storage systems (BESSs) present in V2G-capable EVs to contribute to grid stability by providing a synthetic inertia response based on the rate of change of frequency (RoCoF). By dynamically adjusting the power injected or absorbed according to frequency variations, this strategy enhances the system's resilience to disturbances, mitigating the impact of events such as sudden generation loss or massive load disconnections.

The remainder of this document is structured as follows: Section 2 describes the proposed methodology, including the electric power grid representation, EV charging, load modeling, and the effects of EVs on frequency stability. Section 3 describes the results obtained in each of the proposed scenarios. Finally, the conclusions, limitations, and potential directions for future work are presented in Section 4.

2. Methodology to Assess the Effect of EVs in Frequency Stability

2.1. Curaçao's Electric Power Grid Representation

The representative prototype of Curaçao's power network is presented in [25] (see Figure 1). Curaçao, located in the southern Lesser Antilles and part of the Kingdom of the Netherlands, spans around 444 square kilometers and has a population of over 164,000. The island relies heavily on imported energy, primarily oil and natural gas, to meet its electricity and transportation needs. However, recent efforts have focused on adopting renewable energy sources like wind and solar power to reduce dependence on fossil fuels and lower the island's carbon footprint. Although specific data on Curaçao's power grid is not publicly accessible, this study presents a representative model based on information from [33].

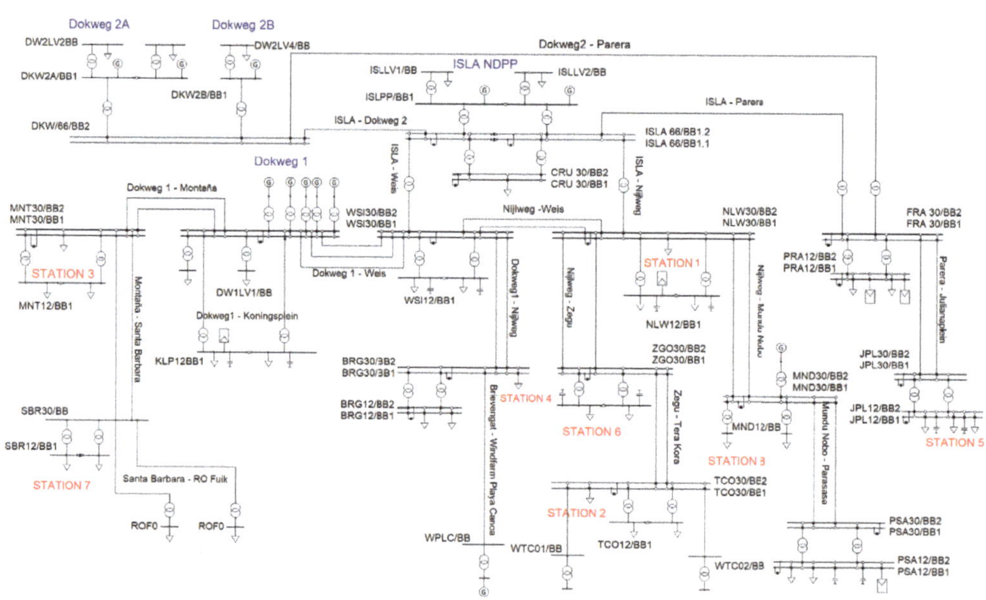

Figure 1. One-line diagram of the power grid under study [25].

The island's energy system primarily consists of fossil fuel power plants, supported by wind and solar installations. Curaçao's grid includes five power plants, two wind farms, 13 substations, 321 km of overhead lines, 1354 km of underground cables, and various transformers. The grid's highest voltage level is 66 kV, which is distributed through eleven 30 kV substations and, subsequently, through twelve 12 kV and two 6 kV substations, where aggregated loads are connected [25].

Curaçao's peak electricity demand is about 130 MW, varying throughout the year. Despite having an installed capacity exceeding peak demand, this capacity is not always fully utilized due to variability in wind and solar power generation. The island's primary power generation includes three main plants: the first plant comprises two sections with a combined capacity of 74.8 MW, the second plant has a capacity of 33.2 MW, and the third plant, consisting of seven smaller generators, has a capacity of 42.8 MW. Together, these fossil fuel power stations contribute a total of 150.8 MW, connected through 66 kV and 33 kV transmission lines.

Wind energy is the second-largest energy source, with three wind farms located on the windward coast. These farms have a combined capacity of 46.5 MW under ideal conditions. Additionally, Curaçao has five solar parks contributing a total of 14 MW, supported by an estimated 760 photovoltaic installations, each with a modeled capacity of 2.8 MW. This renewable energy infrastructure plays a crucial role in supporting the island's electric grid [25].

The speed regulator controller is applied to all synchronous generators in Curacao's networks system. Each synchronous generator is equipped with a speed regulator controller, which is responsible for primary frequency regulation by adjusting the mechanical power input based on frequency deviations. This helps stabilize the system after disturbances such as generation loss or EVs load disconnections. The speed regulator controller depicted in Figure 2 was used for the synchronous thermal generators on the test grid. In this case, Wref represents the reference speed, W is the actual speed, Pref is the reference power, and P is the measured power. Additionally, TA corresponds to the ambient temperature limit, while Pt denotes the turbine output power. The controller parameters are presented in Table 1. Primary frequency regulation, carried out by the speed controller shown in Figure 2, plays a key role in maintaining the dynamic stability of the power system. One of its main advantages is its fast response to sudden imbalances between generation and load, acting within seconds to correct frequency deviations. This helps to dampen initial oscillations and lowers the risk of system failures or blackouts. Since the controller provides a proportional response based solely on local frequency measurements, it operates without the need for centralized coordination, making the system more resilient to local disturbances.

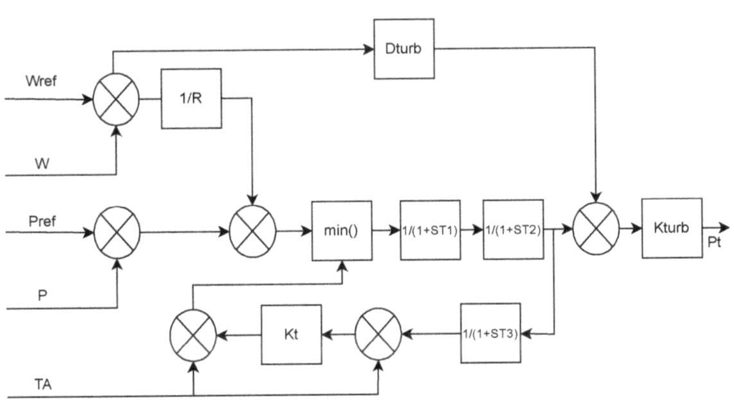

Figure 2. Speed regulator controller.

Table 1. Speed regulator controller parameters.

Parameter	Data
R speed droop [%]	0.047
T1 controller time constant [s]	0.4
T2 actuator time constant [s]	0.1
T3 compressor time constant [s]	3
Kt turbine factor [pu]	2
Kturb turbine [pu]	1
Vmin controller minimum output [pu]	0
Vmin controller maximum output [pu]	1
Dturb frictional losses factor [pu]	1

2.2. Load Modeling and Charging of EVs

EV chargers enable the conversion of alternating current (AC) from the grid into direct current (DC) through power converters [34]. Depending on the charger type, this conversion process can take place either inside or outside the vehicle. There are two main types of EV chargers: onboard and off-board converters.

Onboard chargers are integrated into the vehicle itself, meaning the AC/DC conversion is constrained by the vehicle's size and weight. As a result, the conversion rate is slower, extending the charging time. In contrast, off-board chargers perform the AC/DC conversion externally and inject DC power directly into the EV battery. Since these converters are not restricted by vehicle dimensions, they offer higher efficiency, significantly reducing charging duration [35]. The different charging methods can be classified as indicated below [36,37].

- Level 1 or AC trickle charging: This is the most basic and widely accessible charging method, typically rated up to 3 kW. It operates using a standard household outlet (120 V in North America, 230 V in Europe) without requiring special infrastructure. Due to its low power output, charging times range between 11 and 20 h, depending on battery capacity and state of charge. Level 1 charging is best suited for overnight use, low-mileage drivers, and situations where faster charging is not necessary.
- Level 2 or AC fast charging: This charging method provides a significant improvement in charging speed, with power ratings typically ranging from 7 kW to 22 kW. It requires a dedicated circuit and a specialized charging station, commonly installed in residential garages, workplaces, and public charging locations. Charging times are reduced to approximately 3 to 4 h, making it a more practical solution for daily use. Level 2 charging utilizes the vehicle's onboard charger to convert AC power to DC, balancing efficiency and infrastructure costs while offering a faster alternative to Level 1 charging.
- Level 3 or DC fast charging: This is the most powerful and rapid charging method, delivering between 50 kW and 350 kW. Unlike Level 1 and Level 2, it relies on an off-board charger that performs AC/DC conversion externally before directly supplying DC power to the EV battery. Due to the high voltage and current requirements, Level 3 charging stations are typically located along highways, fleet depots, and urban fast-charging hubs rather than in residential areas. Charging times range from 30 to 60 min, significantly reducing downtime for EV users. However, this method places a substantial demand on the grid, potentially causing voltage deviations, reliability concerns, and increased power losses. Additionally, not all EVs are equipped to handle such high-power charging, and its intensive use may accelerate the aging of distribution network transformers.

The main features of the charging stations are presented in Table 2:

Table 2. Charging level of EVs.

Charging Levels	Level 1	Level 2	Level 3
Location	Domestic	Domestic/ Public	Public
Voltage [V]	120	240	208, 600
Current [A]	11, 16	16, 32, 80	240, 480
Power [kW]	1.4, 1.9	4, 8, 19.2	50, 100
Phase	1 phase AC	1/3 phase AC	3 phase AC or DC

To accurately model the behavior of an EV charging station, the ZIP load model was implemented. This approach provides a mathematical representation of the relationship between power consumption and voltage variations at a load bus. The ZIP model is widely used in both static and dynamic power system studies, as it captures how different types of electrical loads respond to voltage fluctuations. It expresses power as a polynomial function of voltage, incorporating three distinct components: constant impedance (Z), constant current (I), and constant power (P).

In the context of EV charging, the ZIP model is particularly relevant because it reflects the varying electrical characteristics of chargers. Most EV chargers primarily behave as constant power loads (P-component), meaning their power consumption remains stable regardless of voltage changes. However, depending on the type of charger and grid conditions, small contributions from the impedance (Z) and current (I) components may also be present. By utilizing this model, the active power demand of EVs can be accurately simulated, allowing for a more precise assessment of their impact on grid stability and performance.

Equation (1) indicates the polynomial load model, where P_0 is the active power for a supply voltage of 1 p.u. Furthermore, p_1, p_2, and p_3 correspond to model parameters. If any of them draws close to 1, it indicates that the load operates as a constant impedance, constant current, or constant power, respectively. Finally, the independent parameter V indicates the supply voltage in p.u.:

$$P = P_0\left[p_1\hat{V}^2 + p_2\hat{V} + p_3\right] \quad (1)$$

Table 3 presents the parameters of Equation (2) that characterize each EV charging station, which were estimated using the least-squares method as described in [31].

Table 3. EV charging modeling parameters.

Charging Type	p_1	p_2	p_3
DC super-fast charging	−0.1326	0.1816	0.951
DC fast charging	0.0620	−0.2199	1.156
AC fast charging	0.0034	−0.1199	1.086

The load model used in this work is based on the model proposed in [31] to represent the dynamic behavior of EV chargers. To improve the accuracy of the load modeling, the ZIP model parameters were carefully estimated using various detailed EV models and validated with simulated data. The least-squares estimation method, as referenced in [31], was employed to calibrate the model, ensuring that it accurately reflected the dynamic behavior of EV charging stations under different grid conditions. Additionally, the implementation of the ZIP model accounted for the voltage-dependent nature of EV loads, allowing for a more comprehensive representation of their impact on power system stability. Since most EV chargers exhibit power electronics-based behavior, their response to voltage fluctuations is primarily governed by the P-component of the ZIP model. However, variations in grid voltage and charger design can introduce small Z and I components, which were incorporated into the analysis to improve realism. This approach ensured that the load dynamics in the network were realistically represented, enabling a more accurate assessment of frequency stability under different operating conditions.

2.3. EVs in Frequency Stability

The frequency stability of the power system is crucial for ensuring the reliable operation of the electrical grid [38]. In this context, the impact of V2G integration on grid frequency becomes a significant concern. Firstly, from a demand perspective, the charging load of a large number of EVs can lead to a substantial increase in the total grid load, especially during peak charging periods [38]. Secondly, with the growing share of renewable energy in the power system, the high randomness and decentralization of energy generation and consumption further complicate grid frequency management. Beyond the overall frequency decline caused by increased load, the volatility of EV charging demand is also a concern that cannot be overlooked. As a new type of load, EV charging demand is highly variable, which can cause a sharp increase in grid load over a short period. If this short-term load fluctuation is not effectively managed and regulated by the grid, it may lead to rapid frequency fluctuations. These short-term frequency fluctuations can pose a challenge to the stable operation of the power grid, particularly when the grid's regulation capacity is limited [38].

According to [28], one of the key findings in this area is the variation in the behavior of the RoCoF in the Great Britain system. The disconnection of the DC circuit, where EVs are charged, can lead to high RoCoF when the DC circuit is heavily loaded. The company that conducted the study (Sygensys) reported this phenomenon and identified that the increasing EV load could exacerbate the RoCoF, with varying impacts throughout the system. The variation in the RoCoF is correlated with the physical location of the generator. For EV charging, this means that the observable frequency during significant generation losses depends not only on the size of the loss but also on where the load is fed from the transmission system. The disconnection of V2G-enabled EVs during this time due to the RoCoF could worsen the disturbance and potentially lead to a blackout. The results from Great Britain serve as a reminder that transmission planning personnel must understand oscillatory modes and identify whether the RoCoF increases due to large amounts of inverter-based resources (IBRs) and converter-based resources could trigger generation trips. As an increasing RoCoF raises global risks, planners need to pay special attention to the impact of EV charging when exporting to the grid in V2G modes or when EV chargers draw power from the grid in charging modes. The conclusions of [28] are based on comparisons of the RoCoF at various levels of EV penetration.

2.4. Inertia and Frequency Dynamics in Power Systems

Inertia plays a fundamental role in the dynamic response of power systems following a disturbance. The inertia constant, typically denoted H, quantifies the kinetic energy stored in the rotating masses of generators and turbines. This stored energy helps to momentarily balance the power deficit caused by sudden generation losses or load changes, thereby resisting abrupt frequency deviations. The following section introduces the mathematical definition of the inertia constant and its relationship to system frequency, power imbalance, and the rate of change of frequency (RoCoF), forming the basis for inertia estimation and system stability analysis.

The inertia constant H of a rotating system, or H_i in the case of an individual generator, characterizes the kinetic energy stored in its rotating mass, represented by E_{c0}. It is defined as the amount of time (in seconds) required to replenish this stored energy when the system operates at its nominal mechanical speed ω_{sm} and nominal apparent power S_{base}. The overall inertia of the system is given by Equation (2) [39]:

$$H = \frac{E_{c0}}{S_{base}} = \frac{J\omega_{sm}^2}{2S_{base}} \tag{2}$$

where J is the total moment of inertia in kg·m^2, ω_{sm} is the angular velocity of the system in radians per second (rad/s), and S_{base} denotes the selected base power in megavolt-amperes (MVAs).

The dynamic relationship between the frequency f_i and the power imbalance ΔP_i at the terminals of generator i is described by Equation (3):

$$\frac{2Hi}{f_n} \cdot \frac{df_i}{dt} = p_{m,i} - p_{e,i} = \Delta p_i \qquad i = 1, 2, \ldots, N \tag{3}$$

where $p_{m,i}$ is the mechanical power supplied by the turbine (in per unit), $p_{e,i}$ is the electrical power output (in per unit), Δp_i denotes the power imbalance between generation and load (in per unit), H_i is the inertia constant (in seconds), f_i is the frequency observed by element i (in Hz), f_n is the nominal system frequency (in Hz), and df/dt represents the RoCoF (in Hz/s).

If the time at which the disturbance occurs is known, $t = t_0$, along with the corresponding power imbalance $\Delta p_i(t_0)$ and the RoCoF df_i/dt then the inertia contributed by generator i can be estimated using Equation (4):

$$\hat{H}_i = \frac{\Delta p_i(t_0) f_n}{2\frac{df_i}{dt}|_{t=t_0}} \tag{4}$$

Equation (4) is valid only within a short time window following the disturbance, prior to the influence of other dynamic factors such as control actions or system damping effects [40]. Since the RoCoF represents the time derivative of frequency, it can be approximated by evaluating the slope between two consecutive frequency measurements, as shown in Equation (5):

$$\frac{df_i}{dt}|_{t=t_0} \cong \frac{f_i(t_0^+) - f_i(t_0^-)}{t_0^+ - t_0^-} \tag{5}$$

where $f_i(t_0^+)$ and $f_i(t_0^-)$ correspond to the frequency data taken at t_0^+ and t_0^-, respectively.

Figure 3 illustrates the frequency response of a power system following a generation disturbance. The initial sharp decline in frequency is a direct result of an imbalance between generation and demand. When generation suddenly falls short of the load, the system experiences a power deficit. As a consequence, the kinetic energy stored in the rotating masses of synchronous machines is used to supply the shortfall, causing the system to slow down and the frequency to drop.

The lowest point reached after the disturbance is known as the nadir (56.84 Hz), marking the moment of greatest energy deficit. Subsequently, the frequency recovers as primary control mechanisms (such as governors) act to restore balance. This leads to a temporary overshoot known as the zenith (58.88 Hz), occurring after 50 s in this case. Eventually, the system reaches a new steady-state frequency of approximately 58.62 Hz, which reflects the new operating point at a lower speed due to the sustained imbalance.

These three key points—nadir, zenith, and steady-state frequency—are commonly used metrics to characterize the frequency behavior of the system after a disturbance.

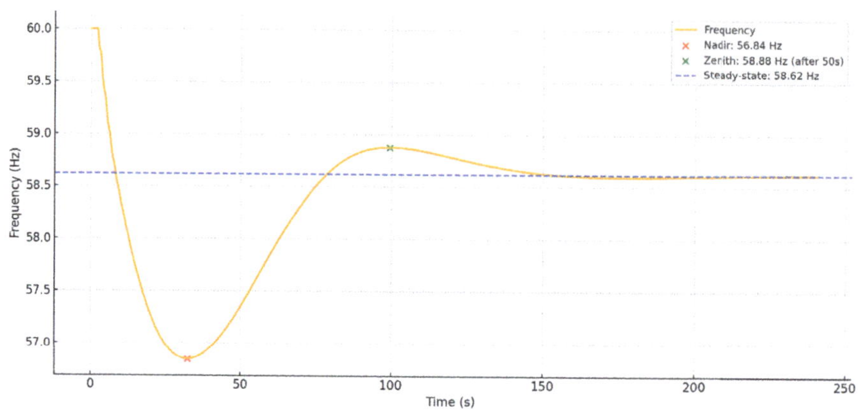

Figure 3. Frequency response after generation loss showing nadir, zenith, and steady state.

2.5. Inertia Emulation Control Strategy for BESS in EVs with V2G Using RoCoF-Based Synthetic Inertia

One of the main contributions of this study is the implementation of the Inertia Emulation Control Strategy for battery energy storage systems (BESSs) through electric vehicles (EVs) operating in V2G mode, using RoCoF-based synthetic inertia. This technique enables BESSs integrated into V2G-capable EVs to support grid stability by providing a synthetic inertia response based on frequency variation. By dynamically adjusting the injected or absorbed power according to frequency changes, this strategy enhances the system's resilience to disturbances, mitigating the impact of events such as sudden generation loss or massive load disconnections.

Conventional battery energy storage systems (BESSs) do not inherently provide frequency response. Therefore, additional control strategies are required to establish a link between grid frequency and the power output of the BESS, allowing it to contribute to frequency regulation during disturbances in the AC system. In this context, [41,42] presents two approaches for delivering synthetic inertia through BESSs to address frequency deviations or active power imbalances: one-shot RoCoF control and continuous RoCoF control.

Continuous RoCoF control in BESSs utilizes real-time measurements of the instantaneous RoCoF on the grid side, offering high adaptability to frequency disturbances. However, this approach requires more sophisticated control mechanisms to effectively filter out harmonics, as instantaneous RoCoF measurements can amplify noise.

In the context of BESSs, once the grid frequency is measured its derivative with respect to time can be obtained to estimate the RoCoF. Due to the inherently noisy nature of this signal, it must be properly filtered before being scaled by a gain known as the synthetic inertia constant. This process enables the BESS to emulate inertia response. The behavior is represented by Equation (6):

$$\Delta p = 2H_{syn} \cdot \frac{df_{sys}}{dt} \quad (6)$$

where Δp represents the inertial active power of the machine, H_{syn} is the inertial of the machine, and f_{sys} is the frequency of the system.

Figure 4 illustrates the control strategy used for inertia emulation in BESSs. The RoCoF is first filtered to remove high-frequency noise. The resulting signal is then multiplied by $2H_{\text{syn}}$, where H_{syn} represents the synthetic inertia constant of the machine. This product yields the inertial active power contribution, Δp, which reflects the amount of power the BESS injects or absorbs in response to frequency deviations. This value is then added to the reference power setpoint, P^*, to compute the total active power output, P. In this context, f_{sys} represents the system frequency. This control loop enables the BESS to support grid stability by dynamically adjusting its power output according to frequency variations [43].

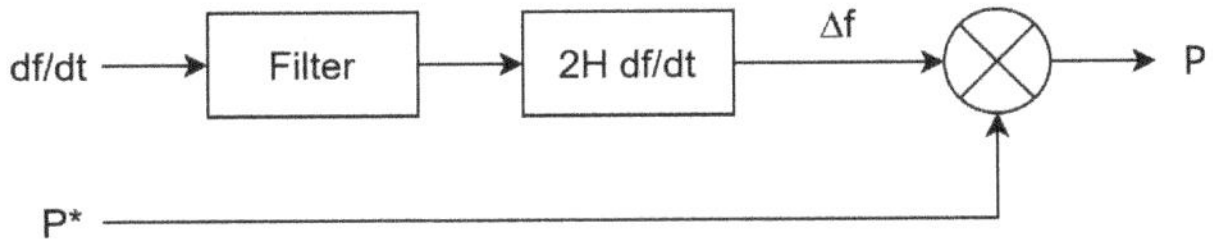

Figure 4. Inertial response-based power control scheme.

3. Tests and Results

The research analyzed three different demand conditions: peak, average, and minimal, which corresponded to electricity consumption levels of 149.81 MW, 104.3 MW, and 74.5 MW, respectively. In each scenario, a designated proportion of the total vehicle fleet was assumed to be electric, with all EVs simultaneously drawing power from the grid. The penetration levels analyzed ranged from 0.5% to 4.5%, increasing incrementally by 0.5%. Table 4 provides a detailed breakdown of these values. In this context, CL refers to the charging level. For instance, the first row illustrates that a single Charging Level 3 station with a 0.5% EV penetration results in a 9 MW increase in load. When EV penetration reaches 1%, the load rises to 17 MW, with further increments following the same pattern. The corresponding increase in the number of EVs at each penetration level is shown in the final row of Table 4.

Table 4. Different levels of EV penetration and their corresponding load increment in MW.

Stations	CL	0.5%	1%	1.5%	2%	2.5%	3%	3.5%	4%	4.5%
1	3	9	17	22.5	30	47	57.5	78.8	90	112
2	3	2.5	5	11.25	15	16.25	20	21.9	25	30.5
3	3	4.25	8.75	15.25	20.35	22.85	27.5	32.8	37.5	39
4	3	4.25	8.75	12.75	17	23.75	28.5	30.6	35	35
5	2	0.19	0.39	0.42	0.55	0.61	0.74	0.87	0.99	1.25
6	2	0.18	0.37	0.41	0.55	0.61	0.74	0.85	0.97	1.06
7	2	0.18	0.37	0.41	0.55	0.61	0.74	0.65	0.74	0.77
8	2	0.18	0.37	0.41	0.55	0.61	0.5624	0.49	0.56	0.56
Vehicles	-	412	824	1235	1647	2059	2471	2883	3294	3706

Eight EV charging stations were planned for installation at specific locations on the island, highlighted in red in Figure 1. The first station corresponded to a DC superfast charging station, while stations two, three, and four were standard DC fast charging stations. The remaining four stations were AC fast charging stations. Stations one through four were linked to 30 kV substations, whereas the last four were connected to 12 kV substations. A total of nine EV penetration scenarios were evaluated for each demand condition. The objective of the study was to assess the impact of EV integration on frequency stability.

To analyze the most critical scenario, all the EVs were assumed to be connected to the system simultaneously.

For the frequency stability analysis, the guidelines outlined in [28] were followed, examining two scenarios across all demand conditions and EV penetration levels. The first scenario analyzed the system's frequency response to the abrupt disconnection of EVs. The second scenario evaluated the consequences of generation loss during peak EV load periods.

In the first case, the analysis focused on assessing the quality of the frequency response by examining the RoCoF and the highest frequency value recorded during the transient period (the zenith). The results for this scenario are summarized in Table 5.

For the second case, the evaluation centered on the RoCoF and the lowest frequency value observed during the transient period (the nadir). The corresponding results are presented in Table 6.

Table 5. Zenith and RoCoF in response to disconnection of EVs.

Penetration	Scenario	Zenith [Hz]	RoCoF [Hz/s]
0.5	Min	50.50693	0.03602914
	Med	50.16299	0.155229015
	Max	50.15784	0.151769669
1	Min	50.98678	0.098481038
	Med	50.50058	0.118902698
	Max	50.3098	0.297885475
1.5	Min	51.21816	0.180201263
	Med	50.85184	0.252023892
	Max	50.54709	0.3315703
2	Min	51.57346	0.426008253
	Med	51.13135	0.436815178
	Max	50.64446	0.550821925
2.5	Min	51.59436	0.528325375
	Med	51.14047	0.543081728
	Max	50.84851	0.815877353
3	Min	51.59731	0.616722722
	Med	51.18087	0.747387495
	Max	51.02416	0.984772071
3.5	Min	51.62149	0.790971889
	Med	51.32756	1.013406901
	Max	51.24755	1.21121712
4	Min	51.67122	0.879590863
	Med	51.48829	1.365407428
	Max	51.41541	1.374188469
4.5	Min	51.81297	1.323338664
	Med	51.68066	1.366393577
	Max	51.62029	1.58852428

According to the results presented in Table 5, an increase in EV penetration within Curaçao's electrical grid may compromise frequency stability during simultaneous EV disconnection events. This is evident when analyzing the behavior of the zenith and the RoCoF, as these parameters showed a rising trend with higher EV penetration levels. Such an increase poses a risk to the grid, indicating a reduced ability to withstand frequency disturbances.

When evaluating the zenith and the RoCoF across different demand conditions, it becomes clear that these indices experienced a consistent rise in all scenarios, as illustrated in Figures 5 and 6. Additionally, Figures 5 and 6 show that the zenith values tended to peak during low-demand conditions, whereas the RoCoF reached its highest levels in maximum-demand scenarios.

Table 6. Nadir and RoCoF in response to disconnection of generation.

Penetration	Penetration	Nadir [Hz]	RoCoF [Hz/s]
0.5	Min	49.92492	0.051707884
	Med	49.92632	0.052818324
	Max	49.93891	0.059892333
1	Min	49.93457	0.053583502
	Med	49.93753	0.05479839
	Max	49.94096	0.058455619
1.5	Min	49.93661	0.060952099
	Med	49.93949	0.056027933
	Max	49.94332	0.056118979
2	Min	49.934523	0.06099
	Med	49.94147	0.059724673
	Max	49.94559	0.052825397
2.5	Min	49.94155	0.060257918
	Med	49.94436	0.055640167
	Max	49.94855	0.051969854
3	Min	49.94399	0.057153236
	Med	49.94694	0.053596122
	Max	49.95122	0.049775663
3.5	Min	49.94737	0.055400175
	Med	49.95041	0.051123869
	Max	49.95466	0.048752845
4	Min	49.95005	0.054293655
	Med	49.95311	0.052685571
	Max	49.95735	0.04792151
4.5	Min	49.95343	0.051175993
	Med	49.95608	0.045278491
	Max	49.96075	0.044602425

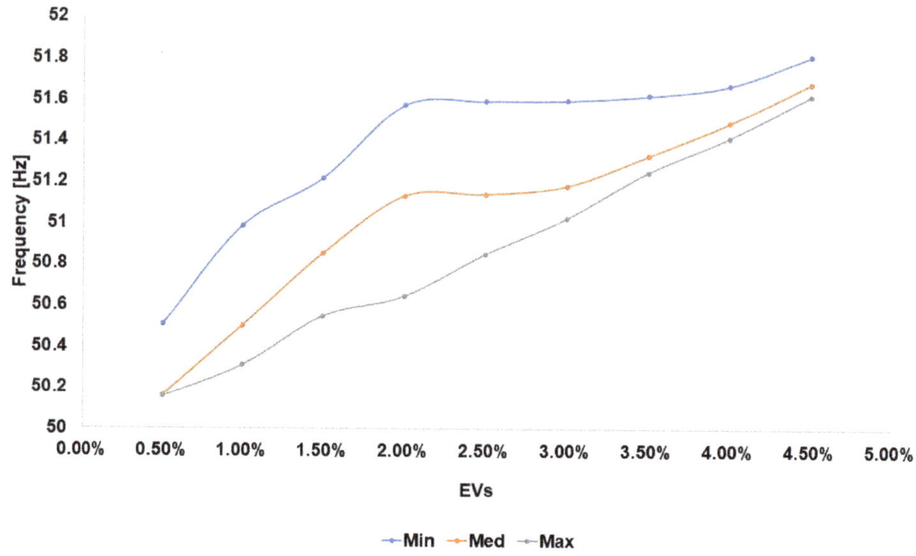

Figure 5. Zenith in response to disconnection of EVs.

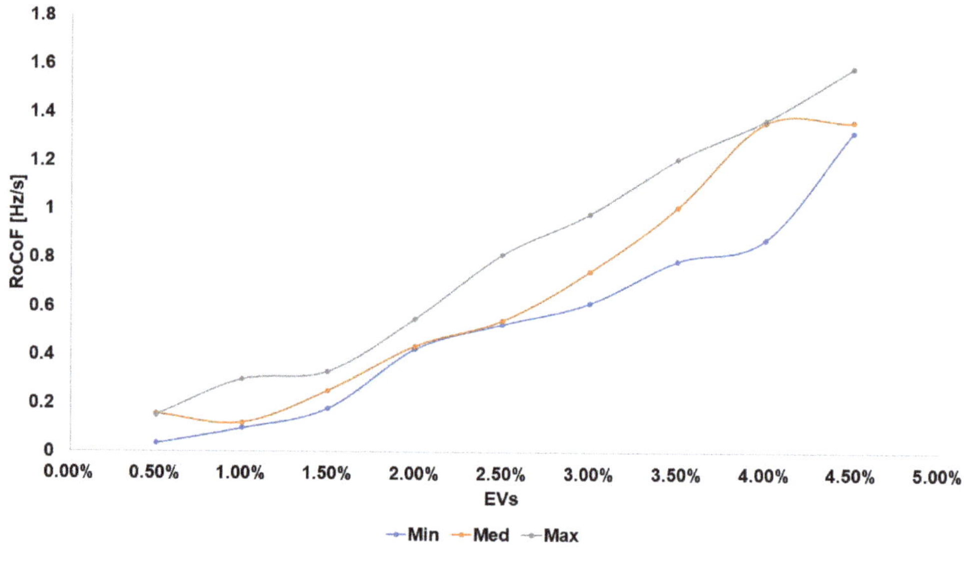

Figure 6. RoCoF in response to disconnection of EVs.

According to the results presented in Table 6, in the case of generation loss, the risk is not as critical at the moment of the event because an increase in generation leads to a rise in system inertia. However, the risk becomes more pronounced when system inertia is low due to the prevalence of renewable energy plants.

By analyzing the behavior of the nadir and the RoCoF, it is evident that the former tends to increase, while the latter decreases as EV penetration rises. This trend suggests an improvement in the grid's ability to withstand frequency disturbances, assuming a dispatch primarily composed of thermal generation plants. Figures 7 and 8 also illustrate that nadir values tend to be higher in maximum-demand scenarios (max), whereas RoCoF values decrease under the same conditions due to greater system inertia.

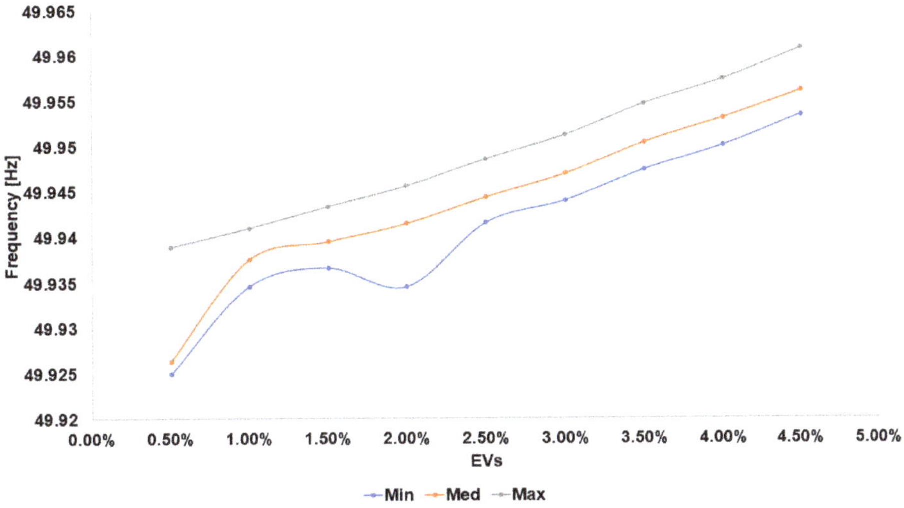

Figure 7. Nadir in response to disconnection of generation.

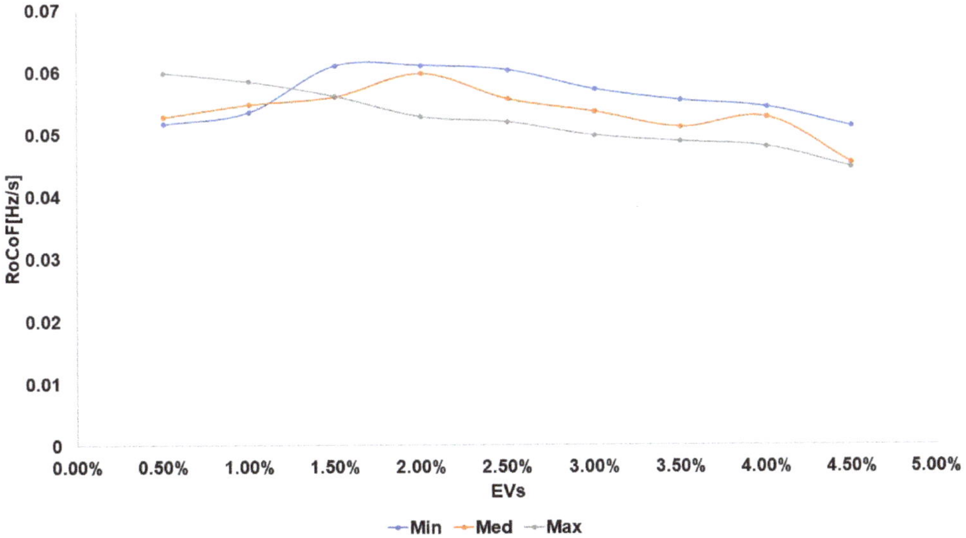

Figure 8. RoCoF in response to disconnection of generation.

Figures 9 and 10 illustrate the frequency response during disturbance events. Specifically, Figure 9 depicts a frequency event triggered by the sudden disconnection of EVs, whereas Figure 10 represents a frequency event following an abrupt loss of generation.

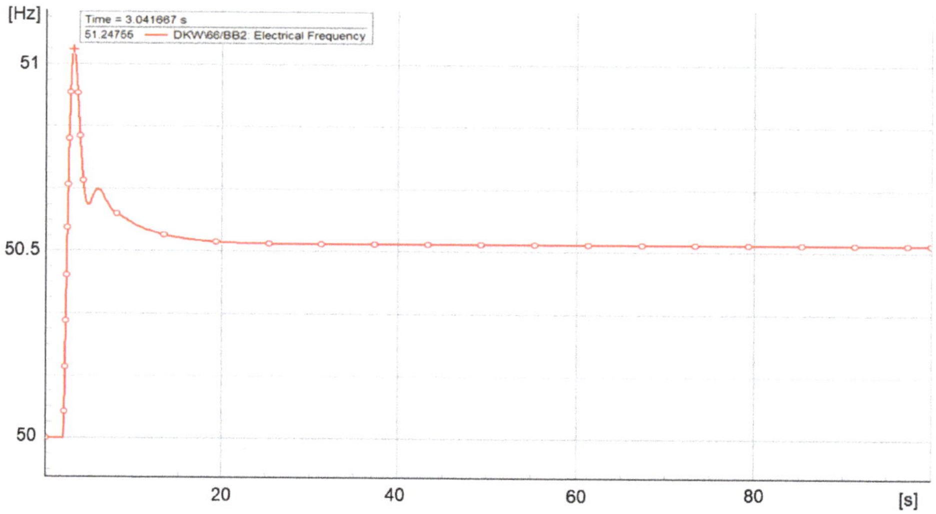

Figure 9. Frequency event after a sudden disconnection of EVs.

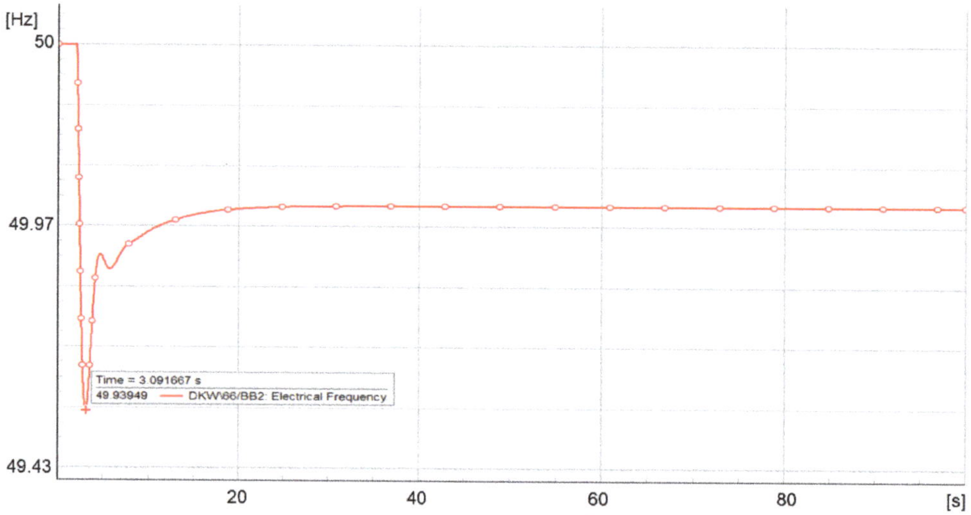

Figure 10. Frequency event after a sudden disconnection of generation.

Figure 11 illustrates the frequency response of a power system under different conditions. Frequency 1 represents the system without any fault and remains stable at 50 Hz, reflecting the ideal operating behavior. In contrast, Frequency 2 shows the system's response to a fault without any inertial support, resulting in a sharp and prolonged frequency dip that highlights the system's vulnerability and lack of resilience. Frequency 3, however, includes an inertial response provided by electric vehicles or battery energy storage systems (BESSs) leading to a noticeably smaller frequency drop and faster recovery.

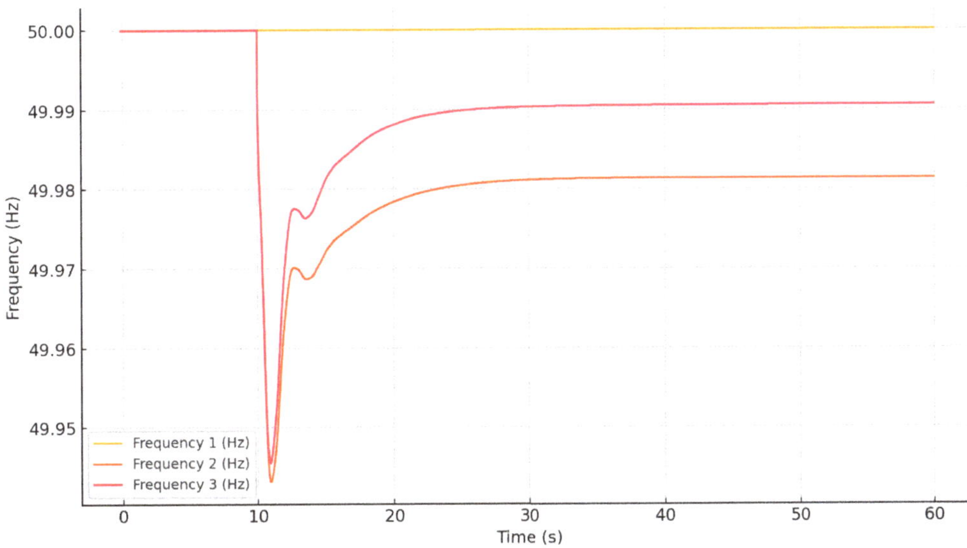

Figure 11. Frequency response comparison under fault conditions with and without inertial support.

4. Conclusions

This study assessed the impact of electric vehicle (EV) penetration on the frequency stability of Curaçao's power network. By conducting a series of simulations, various levels of EV participation were examined in scenarios of minimum, average, and peak demand. The key findings highlight both the potential risks and benefits associated with EV integration into the power grid.

The results indicate that increasing EV penetration can significantly affect frequency stability, particularly during simultaneous EV disconnection events. The observed rise in the RoCoF and frequency zenith values suggests a weakening of the grid's ability to withstand frequency disturbances, especially under high EV penetration scenarios. These findings emphasize the importance of implementing robust control mechanisms and demand management strategies to counteract potential instability.

Conversely, the study also found that in scenarios involving generation loss, a higher EV presence—when effectively regulated—can enhance system inertia. The increase in frequency nadir values and the reduction in the RoCoF in these cases suggest that a well-regulated EV integration strategy, potentially incorporating vehicle-to-grid (V2G) technology, could enhance frequency stability by acting as a controllable energy resource.

To facilitate large-scale EV adoption while preserving power system stability, it is essential to implement coordinated charging strategies, enhance grid flexibility, and establish advanced regulatory frameworks. Future research should explore the dynamic behavior of EV charging infrastructure under real-time conditions and assess the role of smart grids in optimizing frequency response to contingencies. We would also highlight, for future research, areas such as incorporating more detailed EV models, adaptive frequency control mechanisms, effect of ultra-fast charger and stochastic analyses to enhance the accuracy of frequency impact assessments.

Author Contributions: Conceptualization, D.V.-C., S.D.S.-Z. and S.B.-M.; methodology, D.V.-C., S.D.S.-Z. and S.B.-M.; software, D.V.-C., S.D.S.-Z. and S.B.-M.; validation, D.V.-C., S.D.S.-Z., S.B.-M., J.M.L.-L. and N.M.-G.; formal analysis, D.V.-C., S.D.S.-Z. and S.B.-M.; investigation, D.V.-C., S.D.S.-Z., S.B.-M., J.M.L.-L. and N.M.-G.; resources, D.V.-C., S.D.S.-Z., S.B.-M., J.M.L.-L. and N.M.-G.; data curation, D.V.-C. and S.D.S.-Z.; writing—original draft preparation, D.V.-C.; writing—review and editing, D.V.-C., S.D.S.-Z., S.B.-M., J.M.L.-L. and N.M.-G.; visualization, D.V.-C., S.D.S.-Z., S.B.-M., J.M.L.-L. and N.M.-G.; supervision, S.B.-M., S.D.S.-Z., J.M.L.-L. and N.M.-G.; project administration, S.D.S.-Z., S.B.-M., J.M.L.-L. and N.M.-G.; funding acquisition, S.B.-M., S.D.S.-Z., J.M.L.-L. and N.M.-G. All authors have read and agreed to the published version of the manuscript.

Funding: This work was funded by the Institución Universitaria Pascual Bravo, Proyecto de Investigación "Regulación de frecuencia eléctrica empleando resortes eléctricos en sistemas con alta penetración de energías renovables", código PCT00036.

Data Availability Statement: The original contributions presented in this study are included in the article. Further inquiries can be directed to the corresponding author.

Acknowledgments: The authors would like to acknowledge the Institución Universitaria Pascual Bravo, Proyecto de Investigación "Regulación de frecuencia eléctrica empleando resortes eléctricos en sistemas con alta penetración de energías renovables", código PCT00036.

Conflicts of Interest: The authors declare no conflicts of interest.

References

1. Sanguesa, J.A.; Torres-Sanz, V.; Garrido, P.; Martinez, F.J.; Marquez-Barja, J.M. A Review on Electric Vehicles: Technologies and Challenges. *Smart Cities* **2021**, *4*, 372–404. [CrossRef]
2. Directorate-General for Mobility and Transport (European Commission). *EU Transport in Figures: Statistical Pocketbook 2022*; Publications Office of the European Union: Luxembourg, 2022. [CrossRef]
3. Gruetzmacher, S.B.; Bento Vaz, C.; Ferreira, A. Sustainability performance assessment of the transport sector in European countries. *Rev. Fac. Ing. Univ. Antioq.* **2021**, *1*, 42–52. [CrossRef]
4. Qiu, D.; Wang, Y.; Hua, W.; Strbac, G. Reinforcement learning for electric vehicle applications in power systems: A critical review. *Renew. Sustain. Energy Rev.* **2023**, *173*, 113052. [CrossRef]
5. Alsharif, A.; Tan, C.W.; Ayop, R.; Al Smin, A.; Ali Ahmed, A.; Kuwil, F.H.; Khaleel, M.M. Impact of electric Vehicle on residential power distribution considering energy management strategy and stochastic Monte Carlo algorithm. *Energies* **2023**, *16*, 1358. [CrossRef]
6. Saberi-Beglar, K.; Zare, K.; Seyedi, H.; Marzband, M.; Nojavan, S. Risk-embedded scheduling of a CCHP integrated with electric vehicle parking lot in a residential energy hub considering flexible thermal and electrical loads. *Appl. Energy* **2023**, *329*, 120265. [CrossRef]
7. Gomez, J.; Morcos, M. Impact of EV battery chargers on the power quality of distribution systems. *IEEE Trans. Power Deliv.* **2003**, *18*, 975–981. [CrossRef]
8. Tripathi, S.; Singh, V.P.; Kishor, N.; Pandey, A. Load frequency control of power system considering electric Vehicles' aggregator with communication delay. *Int. J. Electr. Power Energy Syst.* **2023**, *145*, 108697. [CrossRef]
9. Yang, Z.; Yang, F.; Min, H.; Tian, H.; Hu, W.; Liu, J. Review on optimal planning of new power systems with distributed generations and electric vehicles. *Energy Rep.* **2023**, *9*, 501–509. [CrossRef]
10. Deb, S.; Tammi, K.; Kalita, K.; Mahanta, P. Impact of Electric Vehicle Charging Station Load on Distribution Network. *Energies* **2018**, *11*, 178. [CrossRef]
11. Lopes, J.A.P.; Soares, F.J.; Almeida, P.M.R. Integration of electric vehicles in the electric power system. *Proc. IEEE* **2010**, *99*, 168–183. [CrossRef]
12. Stiasny, J.; Zufferey, T.; Pareschi, G.; Toffanin, D.; Hug, G.; Boulouchos, K. Sensitivity analysis of electric vehicle impact on low-voltage distribution grids. *Electr. Power Syst. Res.* **2021**, *191*, 106696. [CrossRef]
13. Kühnbach, M.; Stute, J.; Gnann, T.; Wietschel, M.; Marwitz, S.; Klobasa, M. Impact of electric vehicles: Will German households pay less for electricity? *Energy Strategy Rev.* **2020**, *32*, 100568. [CrossRef]
14. Rahman, S.; Khan, I.A.; Khan, A.A.; Mallik, A.; Nadeem, M.F. Comprehensive review & impact analysis of integrating projected electric vehicle charging load to the existing low voltage distribution system. *Renew. Sustain. Energy Rev.* **2022**, *153*, 111756. [CrossRef]

15. Zuluaga-Ríos, C.D.; Florián-Ceballos, D.F.; Ángel Rojo-Yepes, M.; Saldarriaga-Zuluaga, S.D. Review of Charging Load Modeling Strategies for Electric Vehicles: A Comparison of Grid-to-Vehicle Probabilistic Approaches. *Tecnura* **2021**, *25*, 108–125. [CrossRef]
16. Rojo-Yepes, M.Á.; Zuluaga-Ríos, C.D.; Saldarriaga-Zuluaga, S.D.; López-Lezama, J.M.; Muñoz-Galeano, N. A Novel Neuro-Probabilistic Framework for Energy Demand Forecasting in Electric Vehicle Integration. *World Electr. Veh. J.* **2024**, *15*, 493. [CrossRef]
17. Gil-Aguirre, J.; Perez-Londoño, S.; Mora-Flórez, J. A measurement-based load modelling methodology for electric vehicle fast-charging stations. *Electr. Power Syst. Res.* **2019**, *176*, 105934. [CrossRef]
18. Li, X.; Zhang, Q.; Peng, Z.; Wang, A.; Wang, W. A data-driven two-level clustering model for driving pattern analysis of electric vehicles and a case study. *J. Clean. Prod.* **2019**, *206*, 827–837. [CrossRef]
19. Frendo, O.; Graf, J.; Gaertner, N.; Stuckenschmidt, H. Data-driven smart charging for heterogeneous electric vehicle fleets. *Energy AI* **2020**, *1*, 100007. [CrossRef]
20. Gerossier, A.; Girard, R.; Kariniotakis, G. Modeling and Forecasting Electric Vehicle Consumption Profiles. *Energies* **2019**, *12*, 1341. [CrossRef]
21. Sun, K.; Sarker, M.R.; Ortega-Vazquez, M.A. Statistical characterization of electric vehicle charging in different locations of the grid. In Proceedings of the 2015 IEEE Power Energy Society General Meeting, Denver, CO, USA, 26–30 July 2015; pp. 1–5. [CrossRef]
22. Li, G.; Zhang, X. Modeling of Plug-in Hybrid Electric Vehicle Charging Demand in Probabilistic Power Flow Calculations. *IEEE Trans. Smart Grid* **2012**, *3*, 492–499. [CrossRef]
23. Khoo, Y.B.; Wang, C.H.; Paevere, P.; Higgins, A. Statistical modeling of Electric Vehicle electricity consumption in the Victorian EV Trial, Australia. *Transp. Res. Part Transp. Environ.* **2014**, *32*, 263–277. [CrossRef]
24. Jiang, H.; Ren, H.; Sun, C.; Watts, D. The temporal-spatial stochastic model of plug-in hybrid electric vehicles. In Proceedings of the 2017 IEEE PES Innovative Smart Grid Technologies Conference Europe (ISGT-Europe), Turin, Italy, 26–29 September 2017; pp. 1–6. [CrossRef]
25. Robles-Lozano, G.; Saldarriaga-Zuluaga, S.D.; Zuluaga-Ríos, C.D.; López-Lezama, J.M.; Muñoz-Galeano, N. Assessing the Impacts of Electric Vehicle Penetration in Curaçao's Power Network. *World Electr. Veh. J.* **2023**, *14*, 231. [CrossRef]
26. Podder, A.K.; Sadamoto, T.; Chakrabortty, A. Co-Optimization of EV Charging Control and Incentivization for Enhanced Power System Stability. *arXiv* **2024**, arXiv:2405.00947.
27. Li, Y.; Jenn, A. Impact of electric vehicle charging demand on power distribution grid congestion. *Proc. Natl. Acad. Sci. USA* **2024**, *121*, e2317599121. [CrossRef]
28. NERC. *Potential Bulk Power System Impact of Electric Vehicle Chargers*, 1st ed.; North American Electric Reliability Corporation: Atlanta, GA, USA, 2024.
29. Ibrahim, R.A.; Gaber, I.M.; Zakzouk, N.E. Analysis of multidimensional impacts of electric vehicles penetration in distribution networks. *Sci. Rep.* **2024**, *14*, 27854. [CrossRef]
30. Štogl, O.; Miltner, M.; Zanocco, C.; Traverso, M.; Starý, O. Electric vehicles as facilitators of grid stability and flexibility: A multidisciplinary overview. *WIRE Energy Environ.* **2024**, *13*, e536. [CrossRef]
31. Tian, H.; Tzelepis, D.; Papadopoulos, P.N. Electric Vehicle Charger Static and Dynamic Modelling for Power System Studies. *Energies* **2021**, *14*, 1801. [CrossRef]
32. Dharmakeerthi, C.; Mithulananthan, N.; Saha, T. Impact of electric vehicle fast charging on power system voltage stability. *Int. J. Electr. Power Energy Syst.* **2014**, *57*, 241–249. [CrossRef]
33. Bulbaai, R.R. Toward 100% Sustainable Energy Production and a Structural Decrease in Energy Demand: Curaçao, as a Case Study of Small Island Developing States. Ph.D. Thesis, University of Twente, Enschede, The Netherlands, 2019.
34. Collin, R.; Miao, Y.; Yokochi, A.; Enjeti, P.; Jouanne, A. Advanced Electric Vehicle Fast-Charging Technologies. *Energies* **2019**, *12*, 1839. [CrossRef]
35. Brenna, M.; Foiadelli, F.; Leone, C.; Longo, M. Electric vehicles charging technology review and optimal size estimation. *J. Electr. Eng. Technol.* **2020**, *15*, 2539–2552. [CrossRef]
36. Dericioglu, C.; Yirik, E.; Unal, E.; Cuma, M.; Onur, B.; Tumay, M. A review of charging technologies for commercial electric vehicles. *Int. J. Adv. Automot. Technol.* **2018**, *2*, 61–70.
37. Saldarriaga-Zuluaga, S.D.; López-Lezama, J.M.; Zuluaga Ríos, C.D.; Villa Jaramillo, A. Effects of the Incorporation of Electric Vehicles on Protection Coordination in Microgrids. *World Electr. Veh. J.* **2022**, *13*, 163. [CrossRef]
38. Yang, Y.; Wang, W.; Qin, J.; Wang, M.; Ma, Q.; Zhong, Y. Review of vehicle to grid integration to support power grid security. *Energy Rep.* **2024**, *12*, 2786–2800. [CrossRef]
39. Kundur, P. *Power System Stability And Control*; McGraw-Hill, Inc.: Chicago, IL, USA, 1994; p. 1167.

40. Gonzalez-Longatt, F.; Chikuni, E.; Stemmet, W.; Folly, K. Effects of the synthetic inertia from wind power on the total system inertia after a frequency disturbance. In Proceedings of the IEEE Power and Energy Society Conference and Exposition in Africa: Intelligent Grid Integration of Renewable Energy Resources (PowerAfrica), Johannesburg, South Africa, 9–13 July 2012; pp. 1–7. [CrossRef]
41. Attya, A.B.; Vickers, A. Operation and Control of a Hybrid Power Plant with the Capability of Grid Services Provision. *Energies* **2021**, *14*, 3928. [CrossRef]
42. Kosmecki, M.; Rink, R.; Wakszyńska, A.; Ciavarella, R.; Di Somma, M.; Papadimitriou, C.N.; Efthymiou, V.; Graditi, G. A Methodology for Provision of Frequency Stability in Operation Planning of Low Inertia Power Systems. *Energies* **2021**, *14*, 737. [CrossRef]
43. Gonzalez-Longatt, F.; Chikuni, E.; Rashayi, E. Effects of the Synthetic Inertia from wind power on the total system inertia after a frequency disturbance. In Proceedings of the 2013 IEEE International Conference on Industrial Technology (ICIT), Cape Town, South Africa, 25–28 February 2013; pp. 826–832. [CrossRef]

World Electric Vehicle Journal

MDPI

Article

User Cost Minimization and Load Balancing for Multiple Electric Vehicle Charging Stations Based on Deep Reinforcement Learning

Yongxiang Xia [1], Zhongyi Cheng [1], Jiaqi Zhang [1] and Xi Chen [2,3,*]

[1] School of Communication Engineering, Hangzhou Dianzi University, Hangzhou 310018, China; xiayx@hdu.edu.cn (Y.X.); horizonczy@hdu.edu.cn (Z.C.); zjqzzz@hdu.edu.cn (J.Z.)
[2] China Electric Power Research Institute, Beijing 100192, China
[3] GEIRI North America, San Jose, CA 95134, USA
* Correspondence: xc@ieee.org

Abstract: In the context of global energy conservation and emission reduction, electric vehicles (EVs) are essential for low-carbon transport. However, their rapid growth challenges power grids with load imbalances across networks and increases user charging costs. To address the issues of load balancing across large-scale distribution networks and the charging costs for users, this paper proposes an optimization strategy for EV charging behavior based on deep reinforcement learning (DRL). The strategy aims to minimize user charging costs while achieving load balancing across distribution networks. Specifically, the strategy divides the charging process into two stages: charging station selection and in-station charging scheduling. In the first stage, a Load Balancing Matching Strategy (LBMS) is employed to assist users in selecting a charging station. In the second stage, we use the DRL algorithm. In the DRL algorithm, we design a novel reward function that enables charging stations to meet user charging demands while minimizing user charging costs and reducing the load gap among distribution networks. Case study results demonstrate the effectiveness of the proposed strategy in a multi-distribution network environment. Moreover, even when faced with varying levels of EV user participation, the strategy continues to demonstrate strong performance.

Keywords: electric vehicle; distribution networks; load balancing; deep reinforcement learning; Markov decision process

1. Introduction

In the global effort to address climate change and reduce greenhouse gas emissions, electric vehicles (EVs) are playing an increasingly significant role, offering immense potential to enhance sustainability and environmental friendliness in urban areas [1]. With advancements in technology and proactive policy support, EVs are gradually becoming the mainstream market choice. According to statistics, the market share of EVs reached approximately 18% in 2023, compared to 14% in 2022 and only 2% in 2018 [2]. However, the widespread adoption of EVs has had a significant impact on the distribution of loads within power grids, thereby affecting the operation of distribution networks [3] and the charging costs for users. On the one hand, the charging behavior of EV users is highly uncertain; uncoordinated charging during peak periods may exacerbate the load on the distribution network, threatening the safe operation of the grid [4,5] and increasing user charging costs. On the other hand, as an important adjustable resource within the distribution network [6,7], EVs offer flexibility in terms of charging location and the controllability of the charging process. If EV charging behavior can be effectively guided and their flexibility fully utilized, it can not only mitigate the negative impact on the operation of distribution networks but also promote load balancing among distribution networks [8–10] and reduce user charging costs [11,12].

In recent years, significant progress has been made in academic research on EV charging behavior, which can be broadly divided into two categories. The first category focuses on optimizing the selection of charging stations for EVs to achieve effective coordination of charging behavior [11–16]. In this area of research, scholars guide users to specific charging stations based on different optimization objectives. For instance, in [11], a trip planning scheme for EVs aimed at minimizing user travel time and energy consumption was proposed, with consideration given to communication security between users and the dispatch center. In [12], a road speed matrix acquisition and recovery algorithm was developed based on crowd-sensing technology to optimize travel time and economic costs for EV users. Sortomme et al. [13] and Sun et al. [14] proposed EV dispatch strategies and graph game models, focusing on power grid losses and charging delays in the traffic network, respectively. Although these studies are abundant, most of them focus on minimizing grid losses or user charging costs, with less attention paid to the impact among distribution network loads. Chen et al. [15] introduced the issue of load balancing across multiple microgrids for the first time and proposed a charging station allocation strategy considering spatial load balancing. Based on [15], Xia et al. [16] further optimized the strategy by considering the load conditions of multiple microgrids during user service. However, these studies primarily focus on the selection of charging stations, while relatively neglecting the flexibility of EV users' charging behavior within the stations.

Another category of research focuses on how to implement orderly charging control within charging stations for large-scale EVs. In recent years, this area of study has attracted widespread attention from scholars [17–23], with research methods covering heuristic algorithms [17], integer programming [18,19], the Alternating Direction Method of Multipliers (ADMM) [20], and model predictive control (MPC) methods [21–24]. For example, Liu et al. [17] used a genetic algorithm to solve the optimal charging scheduling problem under the constraint of a limited number of chargers. Sun et al. [18] achieved orderly charging within charging stations by solving two integer programming problems under the constraint of distribution network line capacity. Virgilio Pérez et al. [24] adjusted the charging power within charging stations based on grid information, electricity prices, and user preferences. By employing the MPC method, they effectively reduced user costs and mitigated load fluctuations in the distribution network. These methods typically optimize decision variables based on the global state of the environment and are model-based, requiring accurate modeling [17–19] or prediction of EV behavior [21–24]. However, in practice, accurately modeling the randomness of EV user arrivals or predicting their arrival outcomes is highly challenging. This makes these traditional model-based methods less effective for real-time scheduling of EVs.

In recent years, advancements in artificial intelligence (AI) technologies have significantly contributed to the optimization of EV charging management. Deep learning has been widely applied in predicting EV charging demand to optimize grid scheduling and resource allocation [25]. Furthermore, the efficient operation of smart grids relies on the coordinated optimization of communication, sensing, and computation. Studies have shown that the HTed SC method can significantly reduce communication, sampling, and computational loads in distribution networks, thereby improving system computational efficiency [26]. Meanwhile, the combination of deep learning and reinforcement learning has further enhanced reinforcement learning's decision-making capability, enabling it to autonomously learn optimal strategies in complex environments. Deep Reinforcement Learning (DRL) is a model-free, online algorithm that can autonomously learn optimal strategies based on past and current environmental information, without requiring assumptions about the stochastic models of uncertain future events. Since DRL does not rely on explicitly modeling uncertainty, it can efficiently make decisions in dynamic environments. Based on this feature, numerous researchers have employed DRL to study orderly charging within charging stations [27–30]. Additionally, in the broader field of power grid optimization, multi-agent deep reinforcement learning (MADRL) has been utilized for adaptive reconfiguration of distribution networks, improving cost efficiency and security [31]. The successful application of this approach further demonstrates the potential of DRL in achieving efficient and real-time scheduling in complex power grid environments. For instance, Wan et al. [27] approached the problem from the user's perspective, considering the uncertainties in travel and electricity prices, and used DRL to formulate the optimal charging strategy for users. In [28], the problem was considered from the perspective of charging stations, combining orderly charging with station pricing, and reinforcement learning was used to increase the overall revenue of the station. In [29], a fixed-dimensional matrix was used to anchor the state space of reinforcement learning, effectively addressing the dimensionality issue in DRL. Following [29], Chen et al. [30] further fixed the size of the action space, making it independent of the number of EVs. Despite the significant progress made in these studies, most research scenarios have been limited to a single charging station, failing to fully consider the interactions among the distribution networks where the charging stations are located.

From the existing literature, it can be observed that in the research on EV charging station selection, the load balancing algorithm proposed in [15] creatively utilizes EVs to address the issue of load imbalance across multiple microgrids, providing a solid foundation for research on spatial load balancing. However, it does not account for the charging flexibility of EV users once they arrive at the charging stations, which limits the overall system performance. In contrast, studies on the orderly charging of EVs successfully apply DRL to solve the orderly charging problem in a single EV charging station. Nevertheless, these studies fail to consider the impact of charging behavior on other distribution networks. To complement the previous work and fully exploit the flexibility of EV charging, while achieving both the minimization of charging costs and the load balancing across multiple distribution grids, this paper proposes a novel optimization strategy for the EV charging process. This strategy comprehensively considers the flexibility of EVs in both the selection of charging stations and the charging process within the stations, dividing the optimization of the charging process into two stages: charging station selection and in-station charging optimization.

The main contributions of this paper are listed as follows:

1. Unlike previous studies that focused primarily on either charging station selection or on orderly charging for EVs within a single charging station, this paper considers both processes simultaneously. By fully leveraging the flexibility of EVs, this approach enhances their contribution to load balancing across distribution networks and reduces charging costs for EV users.
2. This paper designs a novel DRL reward function that considers both EV users and power grids. This reward function can guide charging stations to implement an EV power distribution strategy that minimizes user charging costs while optimizing load balancing across distribution networks.

3. The case study analyses demonstrate that this strategy remains effective even under varying levels of user participation.

The remainder of this paper is organized as follows: Section 2 introduces the charging station allocation model and the charging resource scheduling model, and formulates the orderly charging problem within charging stations as a Markov Decision Process (MDP) problem. Section 3 presents the EV charging station selection algorithm based on the Load Balancing Matching Strategy (LBMS) and the charging resource allocation algorithm based on the Double Deep Q Network (DDQN). Section 4 compares the system performance under different algorithms. We conclude this paper in Section 5.

2. System Model and MDP

2.1. Charging Station Allocation

In this paper, we consider a city with M distribution networks, which are uniformly distributed across different regions. Each distribution network adopts a hybrid model integrating distributed energy resources (DERs) and energy storage systems (ESSs), offering greater independence and stability compared to traditional power grids [32]. These distribution networks can flexibly integrate renewable energy sources and optimize energy management through intelligent scheduling systems, thereby improving energy utilization efficiency. Moreover, the energy storage system can be combined with vehicle-to-grid (V2G) technology, enabling EVs not only to act as power loads for charging but also to feed energy back into the grid when needed. This integration further enhances the flexibility and balancing capability of the power system. Additionally, by incorporating AI technologies, these distribution networks can dynamically adjust energy distribution and charging scheduling using reinforcement learning, intelligent optimization, and data-driven prediction, further improving the system's adaptability and overall operational efficiency. Each distribution network is equipped with a large aggregator that connects all charging facilities within the area, thereby unifying the management of all charging facilities in the distribution network into a single huge charging station. The model employed in this paper is a time-series model, in which time is divided into several equidistant time windows of fixed duration ΔT, and at the beginning of each time window, a number of EV users make charging requests.

Due to the variation in the areas served by different distribution networks (such as industrial and residential zones), there are significant differences in peak electricity demands across these regions. As a result, the base load of the distribution networks fluctuates considerably over time, with notable load disparities among different networks. Figure 1 illustrates these load variations, reflecting the differences in the load levels across distribution networks at different times. Moreover, the distribution of EVs adjusts dynamically based on the current load conditions of the distribution networks, with vehicle locations changing in response to load variations. At every moment, the load among the distribution networks has a minimum and maximum value, denoted as $P_{valley}(t)$ and $P_{peak}(t)$, respectively. Specifically, $P_{valley}(t)$ represents the valley power load of distribution networks at time t, and $P_{peak}(t)$ represents the peak power load of distribution networks at time t.

When the battery levels of EV users are low, they will search for a charging station to park and charge until their charging needs are met or the predetermined parking time is reached. The distance between the EV user and the charging station plays a crucial role in the selection process. Without proper guidance, users tend to choose the nearest charging station, and this uncoordinated charging behavior may result in users gravitating toward distribution networks with higher loads, potentially affecting load balancing among grids.

To ensure the load balancing of the power grids, it is essential to provide reasonable guidance in the selection of charging stations for EV users. In this context, the vehicle scheduling center should guide EV users to recommended charging stations, thereby optimizing charging behavior. The core objective of the scheduling center is to ensure the safe operation of the power grids. Therefore, when recommending charging stations, the scheduling center must consider not only the geographical locations of stations to ensure that users can reach stations with their remaining battery capacity, but also the current electrical loads of various distribution networks to balance the system's demand and supply. This guidance and scheduling strategy within the system is critical for the management of EV charging and contributes to maintaining the stability and safety of the power grids.

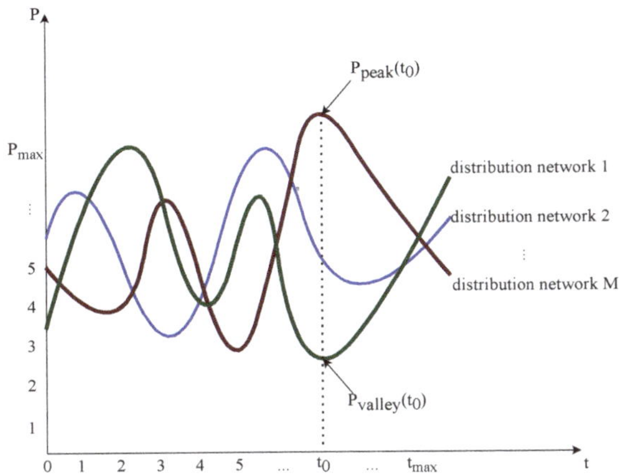

Figure 1. Load variation across different distribution networks over time.

Specifically, when EV i requests for charging at time t, the scheduling center receives information regarding the EV's current position (x_i, y_i), as well as its initial state of charge (SOC) S_i^{ini}. The scheduling center calculates the distances from the location of EV i to various charging stations. In this paper, a matrix $\boldsymbol{R}$ is utilized to store the distances from EV users to each charging station, and matrix $\boldsymbol{R}$ is defined as

$$\boldsymbol{R} = \begin{bmatrix} R_{1,1} & R_{1,2} & \cdots & R_{1,M} \\ R_{2,1} & R_{2,2} & \cdots & R_{2,M} \\ \vdots & \vdots & \ddots & \vdots \\ R_{N,1} & R_{N,2} & \cdots & R_{N,M} \end{bmatrix} \tag{1}$$

where $R_{i,j}$ represents the distance from EV i to charging station j, and N denotes the number of EVs requesting charging.

Subsequently, the scheduling center calculates the distance that EV i can travel based on its initial SOC S_i^{ini}, denoted as R_i. This distance R_i must satisfy inequality constraint

$$R_i \leq R^{max} S_i^{ini} \tag{2}$$

where R^{max} represents the maximum driving range of the EV when fully charged. By calculating $R^{max} S_i^{ini}$, the maximum distance that EV i can travel under its initial battery charge state can be determined. Combining the matrix $\boldsymbol{R}$ with this formula, the scheduling center identifies a set of candidate charging stations for EV i, denoted as $\mathbf{Cand}_i$. This candidate set $\mathbf{Cand}_i$ is selected using the formula

$$\mathbf{Cand}_i = \left\{ V_j^G | R_{i,j} \leq R^{max} S_i^{ini} \right\} \tag{3}$$

where V_j^G represents charging station j that satisfies the above formula. The scheduling center will recommend an appropriate charging station j for the EV from this candidate set.

In this paper, the primary objective of recommending charging stations for EVs is to reduce load disparities among distribution networks. To evaluate the load distribution across the distribution networks at time t, we introduce the Spatial Load Valley-to-Peak Ratio (SLVTPR), denoted as $\eta(t)$. Its calculation is based on the total load of the entire distribution network, which includes both the base load (e.g., residential or industrial demand) and the additional load induced by EV charging within charging stations. $\eta(t)$ is defined by the equation

$$\eta(t) = \frac{P_{valley}(t)}{P_{peak}(t)} \times 100\%. \tag{4}$$

The SLVTPR reflects the degree of disparity among peak and valley loads across distribution networks. A higher SLVTPR indicates more balanced loads among the distribution networks, signifying smaller extreme differences and more stable grid operation. Therefore, to achieve load balancing among distribution networks, the LBMS strategy is used in the allocation of charging stations. Specifically, the scheduling center needs to allocate EVs to the charging station with the minimal power load within their reachable range at time t, which can be expressed as

$$j = \arg\min_j P_{j,t}, \, j \in \mathbf{Cand}_i \tag{5}$$

where $P_{j,t}$ represents the sum of the loads of the distribution network where charging station j is located at time t.

After determining the charging station, the scheduling center will calculate the time at which the user arrives at charging station j. In this paper, it is assumed that the EV travels at a constant speed on the road, denoted as v. Therefore, the travel time $T_{i,j}^{tra}$ for EV i can be calculated using the equation

$$T_{i,j}^{tra} = \frac{R_{i,j}}{v}. \tag{6}$$

Based on the travel time $T_{i,j}^{tra}$ and the time t_i^0 when EV i initiates the charging request, the charging start time T_i^{str} can be determined as

$$T_i^{str} = t_i^0 + T_{i,j}^{tra}. \tag{7}$$

Using the aforementioned methods, the scheduling center assigns EVs to appropriate charging stations.

2.2. Charging Resource Scheduling

When EV i arrives at charging station j, the operator needs to arrange charging resources for the EV. In this section, we analyze the charging arrangements for EVs at charging station j.

In this paper, to facilitate more convenient control of EV charging behavior, the charging service provided by the station follows an on–off strategy. This strategy implies that EVs connected to the charging piles will only operate in either full-power charging or zero-power charging modes.

Assuming that the number of charging piles at the charging station is sufficient, EVs can immediately connect to a charging pile and begin charging upon arrival at the station. Let I_t represent the set of EVs arriving at the station at time t, J_t the set of EVs that arrived before time t and have not yet completed their charging requirements, and $K_t = I_t \cup J_t$ the set of EVs requiring charging at time t. For any EV i, once it connects to a charging pile, the charging station can immediately access the user's charging demand d_i, the estimated parking time p_i, and the start time of charging T_i^{str}.

In practical operations, the charging station operator makes decisions about charging EVs within the station only at the beginning of each ΔT. Therefore, in this section, time is discretized into T^{tot} time slots, with each time window having a length of ΔT.

The smart charging agent (hereafter referred to as the agent) accesses user information at the start of each ΔT, and since the agent makes decisions at discrete time points, it needs to calculate the charging time slots required by the user in the continuous charging condition (hereafter referred to as the required charging time slots) $c_{i,t}$ at time t and the estimated parking time slots $e_{i,t}$ at time t. In order to ensure that the user can complete the charging demand within the available parking time slots, this paper rounds up the required charging time slots and rounds down the estimated parking time slots during discretization. Assuming that the time interval between the two decision times is ΔT, the required charging time slots $c_{i,t}$ and the estimated parking time slots $e_{i,t}$ are given as

$$c_{i,t} = \left\lceil \frac{d_i}{\Delta T p} \right\rceil, \ \forall i \in I_t \tag{8}$$

$$e_{i,t} = \left\lfloor \frac{p_i}{\Delta T} \right\rfloor, \ \forall i \in I_t \tag{9}$$

where p represents the charging power of the charging station.

It can be observed that over time, both $c_{i,t}$ and $e_{i,t}$ will gradually decrease until $c_{i,t}$ or $e_{i,t}$ decreases to 0.

Therefore, the information of the EV is transformed into a three-dimensional vector $(c_{i,t}, e_{i,t}, T_i^{str})$. The agent needs to find the appropriate $c_{i,t}$ time slots for user charging before the parking time slots $e_{i,t}$ expires. The charging urgency $\Delta T_{i,t}^{emr}$ at time t for each user can be calculated as

$$\Delta T_{i,t}^{emr} = \frac{c_{i,t}}{e_{i,t}}, \forall i \in K_t. \tag{10}$$

From the above equation, it can be seen that $\Delta T_{i,t}^{emr}$ is a value between 0 and 1. At every moment, the agent calculates the urgency of charging for each EV. As $\Delta T_{i,t}^{emr}$ increases, the user's charging demand becomes more urgent. The agent can adjust the charging order based on the urgency of the users, ensuring that the charging station meets the users' charging demands while achieving the desired optimization objectives.

2.3. Markov Decision Process

By analyzing the aforementioned scenario within the charging station, it can be concluded that at the beginning of each time slot, the agent determines the charging schedule for the current time slot based on observations of the past and current charging time slots and estimated parking time slots of EVs. This decision, in turn, affects the remaining charging demand in future time periods. Therefore, the environment of orderly charging for EVs within a charging station exhibits Markov properties, and the solution to the optimal decision-making process is referred to as the solution to the MDP. Typically, the MDP model is represented by a four-tuple $\{S, A, P, R\}$, where the elements of the tuple denote the state of the environment, the actions taken by the agent, the transition probabilities of the environment, and the rewards obtained by the agent, respectively. Through continuous interaction with the environment, the agent gradually learns the optimal policy.

2.3.1. State

To address the curse of dimensionality in DRL, this paper employs a two-dimensional matrix to discretize EV feature information, as illustrated in Figure 2. In Section 2.2 , an EV's feature can be represented as $(c_{i,t}, e_{i,t}, T_i^{str})$. By ignoring T_i^{str}, the user feature information within a charging station can be simplified to $(c_{i,t}, e_{i,t})$. The agent only needs to access the user feature information within the charging station, count the number of users with the same (c, e), and store these data in the corresponding position of the matrix to complete the discretization process. Finally, by concatenating the matrix elements with the current time t, the state s_t can be formally expressed as follows:

$$s_t = \left[N_t^{1,1}, N_t^{1,2}, \ldots, N_t^{1,E^{max}}, N_t^{2,1}, N_t^{2,2}, \ldots, N_t^{c,e}, \ldots, N_t^{C^{max},E^{max}}, t\right] \quad (11)$$

where $N_t^{c,e}$ represents the total number of EVs with remaining charging time slots c and expected parking time slots e at time t. C^{max} denotes the maximum charging time for EVs, and E^{max} represents the maximum parking time for EVs.

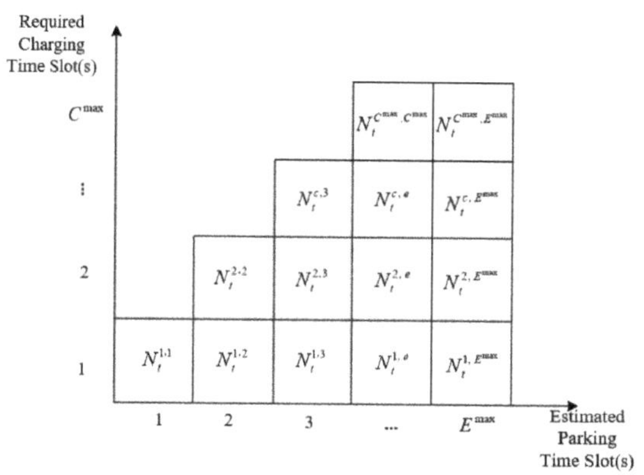

Figure 2. Information of EVs within charging station.

Moreover, since $c \leq e$ (because once c exceeds e, even if the EV continues charging during the remaining parking period, it will still be unable to meet the charging demand), and both the charging demand and parking time for EVs are bounded, the dimensionality of the state space S_{dim} can be computed as

$$S_{dim} = C^{max}E^{max} - \frac{C^{max}(C^{max}-1)}{2} + 1 \quad (12)$$

It can be seen that through this method, the state dimensionality can be fixed. This state representation preserves necessary information while reducing the complexity of the state space, thereby providing a more efficient computational framework for the agent's decision-making process.

The aforementioned analysis of the state is specific to a particular charging station. In this paper, considering there are M large charging stations, the same analytical method can be applied to determine the state of charging station j at time t.

2.3.2. Action

In this paper, the action $a_{j,t}$ is defined as the ratio of the number of charging EVs $N_{j,t}^{c}$ to the total number of EVs $N_{j,t}^{tot}$ at station j at time t. Therefore, the action selected by the agent must satisfy the constraint

$$0 \leq a_{j,t} = \frac{N_{j,t}^{c}}{N_{j,t}^{tot}} \leq 1 \tag{13}$$

It can be seen that the action $a_{j,t}$ only indicates the number of charging EVs but does not specify which EVs should be served at time t. Therefore, it is necessary to determine a charging order to select which users will be provided with service. In this paper, an optimal Least Laxity First (LLF) charging strategy based on the urgency of user demand is used to determine which EVs should be prioritized for charging operations.

In (10), $\Delta T_{i,t}^{emr}$ was defined to represent the urgency of users' charging demands. For users with larger $\Delta T_{i,t}^{emr}$, the gap between the required charging time and the estimated parking time is smaller, indicating a higher urgency. These users must maintain a fully charged state as much as possible before leaving the charging station; otherwise, there is a risk of not meeting their charging needs. Therefore, these users should be prioritized for charging. For users with the same $\Delta T_{i,t}^{emr}$, their charging urgency is identical, so users with smaller $e_{i,t}$ (remaining parking time) should be prioritized. This is because, for users with the same $\Delta T_{i,t}^{emr}$, the risk of not fulfilling their charging demand is the same. However, for users with shorter remaining parking times, the available charging time slots are fewer, and the risk of not meeting their charging needs is higher, so it is crucial to prioritize meeting the needs of these users.

Therefore, by combining the action with the LLF charging strategy, the agent can determine which users should be provided with charging services at time t.

2.3.3. State Transition

Given s_t and a_t, and the system state will be transited to s_{t+1} with the probability $\boldsymbol{P}$ as follows:

$$\boldsymbol{P}(s_t, s_{t+1}) = \mathbf{Pr}(s_{t+1} \mid s_t, a_t) \tag{14}$$

Based on s_t and a_t, EVs previously stationed at the charging station can transition into one of three states: (1) delayed charging, (2) charging, or (3) charging completion. Meanwhile, at this decision-making moment, additional EVs may arrive at the charging station and wait for charging. The state transition probabilities are influenced not only by the actions of EVs already at the station but also by the charging demands and parking durations of newly arriving EVs. Since these factors are dynamically changing and inherently unpredictable, the state transition probabilities cannot be explicitly formulated. Therefore, in the subsequent sections, a DRL-based approach will be introduced to learn and approximate these transition probabilities through interaction with the environment.

2.3.4. Reward

The optimization objectives of this paper are twofold. The first optimization objective is to minimize the charging cost for users. Therefore, the optimization objective for charging station j can be expressed as

$$r_{j,t}^{cost} = a_{j,t} N_{j,t}^{tot} p_{j,t} \tag{15}$$

where $p_{j,t}$ represents the price of charging station j at time t.

The second optimization objective is to minimize the load disparity among multiple distribution networks, which can be expressed as

$$r_{j,t}^{disp} = a_{j,t}N_{j,t}^{tot}\left(\frac{P_{j,t}^{ini}}{P_{t}^{tot}} - \frac{1}{M}\right) = a_{j,t}N_{j,t}^{tot}l_{j,t} \quad (16)$$

where $P_{j,t}^{ini}$ represents the baseline load of the distribution network j at time t when no EV users are charging, P_t^{tot} represents the total baseline load of all distribution networks at time t when no EV users are charging, and $l_{j,t}$ represents the difference between the proportion of the baseline load of distribution network j at time t in the total baseline load and the proportion of the ideal load for the distribution network.

Due to the significant differences in the quantitative scales of charging station prices and the load disparity among grids, it is necessary to normalize both factors in order to jointly optimize user charging costs and the load disparity among distribution networks. The normalization formula is given as

$$x^{nor} = \frac{x^{max} - x_t}{x^{max} - x^{min}} \quad (17)$$

where x^{nor} represents the normalized value, x_t represents the value of the element at time t, x^{max} represents the maximum value of the same type of element, and x^{min} represents the minimum value of the same type of element.

According to (17), by normalizing both the charging price and the spatial load disparity among distribution networks to the same level, the charging price $p_{j,t}$ in (15) is normalized to $p_{j,t}^{nor}$, and the difference between the load proportion of the distribution network and the ideal load proportion $l_{j,t}$ in (16) is normalized to $l_{j,t}^{nor}$. Thus, the total immediate reward $r_{j,t}$ of charging station j can be expressed as

$$r_{j,t} = \alpha a_{j,t}N_{j,t}^{tot}p_{j,t}^{nor} + \beta a_{j,t}N_{j,t}^{tot}l_{j,t}^{nor} \quad (18)$$

where α and β represent the weight factors in the reward function for minimizing user charging costs and balancing the load among distribution networks, respectively, ranging from 0 to 1.

The goal of agent is to maximize the total reward value for all decision points, and each decision point's reward value includes both the immediate reward at the current moment and the future reward. Therefore, at each time step, the state-action function $Q_j^{\pi}(s,a)$ can be computed as

$$Q_j^{\pi}(s,a) = \mathbb{E}\left[\sum_{k=0}^{\infty} \gamma^k r_{j,t+k} \mid s_t = s, a_t = a\right] \quad (19)$$

where $\mathbb{E}$ represents the expectation function, γ denotes the discount factor, and r_{t+k} represents the reward the agent will receive at future time steps. $Q_j^{\pi}(s,a)$ represents the cumulative reward the agent can obtain by taking action a in state s, and then following an optimal policy thereafter. The discount factor γ is used to reduce the weight of future rewards relative to immediate rewards, indicating how much the agent values future rewards. A higher γ focuses on long-term gains, while a lower γ emphasizes short-term rewards. To prevent the agent from only considering immediate rewards, γ is typically set to a value close to 1. The value function $Q_j^{\pi}(s,a)$ measures the quality of taking action a in state s, influencing the agent's selection of actions. The DRL algorithm is ultimately about selecting the best action from all available actions to maximize the total reward value.

3. The DDQN DRL Algorithm for Charging Station

The Deep Q Network (DQN) is a technique that combines deep learning with reinforcement learning. It builds upon the traditional Q-learning algorithm by incorporating deep neural networks to approximate the Q-function in Q-learning, enabling DQN to handle high-dimensional state space problems. Additionally, DQN introduces techniques such as experience replay and fixed Q-targets, which have enabled DQN to achieve significant advancements in the field of DRL. However, similar to traditional Q-learning, DQN updates the Q-values by selecting the maximum Q-value of the next predicted state. Specifically, the formula for updating the target Q-values in DQN is

$$q_{target} = r_t + \gamma \max_{a'} Q(s', a') \tag{20}$$

where s' represents the next state, and a' represents all possible actions in state s'. It can be observed that this method of updating the Q-value assumes that the maximum Q-value corresponds to the best future return. However, in practical applications, due to various objective factors such as estimation errors, sampling errors, or noise, this maximization operation often leads to an overestimation of the Q-value. Given the inherent approximation errors of neural networks and the uncertainty of the environment, this bias can be further amplified. The issue of overestimation can cause the learned policy in the DQN algorithm to deviate from the optimal policy, significantly reducing the algorithm's performance. Furthermore, overestimation can also affect the convergence of the algorithm, causing it to converge on a suboptimal policy, thereby negatively impacting overall effectiveness.

To address the overestimation problem in DQN, DDQN was proposed. DDQN mitigates overestimation by decoupling the max operation in the target Q-value calculation into two parts: action selection based on the training network and value estimation based on the target network. This effectively reduces overestimation, leading to more accurate and reliable target Q-value estimates, thereby improving the overall quality of the policy and the stability of the algorithm. The process of the DDQN algorithm within the charging station is presented in the following section and the specific training algorithm of DDQN is shown in Figure 3.

First, two neural networks with the same structure but independent parameters are initialized, one as the training network $Q(\theta)$ and the other as the target network $Q(\bar{\theta})$. The parameters of the training network are initialized as θ, and the same parameters are used in the target network, i.e., $\theta = \bar{\theta}$. Then, the total number of training episodes E^{tot} and the target network parameter update frequency B are set.

Next, the initial state s_0 of the EV charging station is generated, and the training network selects an action using the ϵ-greedy strategy, i.e.,

$$a_t = \begin{cases} \arg\max Q(s, a \mid \theta), & \lambda \geq \epsilon \\ random a_t \in A, & \lambda < \epsilon \end{cases} \tag{21}$$

where ϵ represents the exploration rate, a value that gradually decreases from 1 to 0; λ is a random number between 0 and 1; and A is the set of actions available for the agent to choose from.

After selecting action a_t, the load $P_{j,t}$ of charging station j is updated as

$$P_{j,t} = P_{j,t}^{ini} + a_t p \tag{22}$$

Finally, the agent chooses which users to charge at time t based on the urgency of user demand using the LLF charging strategy. The agent then calculates the immediate reward r_t for the selected action. Next, the agent obtains the next state s_{t+1} at the following time step. The tuple (s_t, a_t, r_t, s_{t+1}) is stored in the experience replay buffer. Finally, the agent randomly samples a batch (s_t, a_t, r_t, s_{t+1}) from the experience replay buffer and updates the parameters of the training network according to the equations

$$q_{target} = r_t + \gamma \max_{a'} Q(s', \arg\max_{a'} Q(s', a'; \theta); \bar{\theta}) \tag{23}$$

$$L(\theta) = \mathbb{E}\left[(q_{target} - Q(s, a \mid \theta))^2 \right] \tag{24}$$

$$\theta = \theta - \mu \nabla_\theta L(\theta) \tag{25}$$

where μ is the learning rate, and ∇ represents the gradient operator.

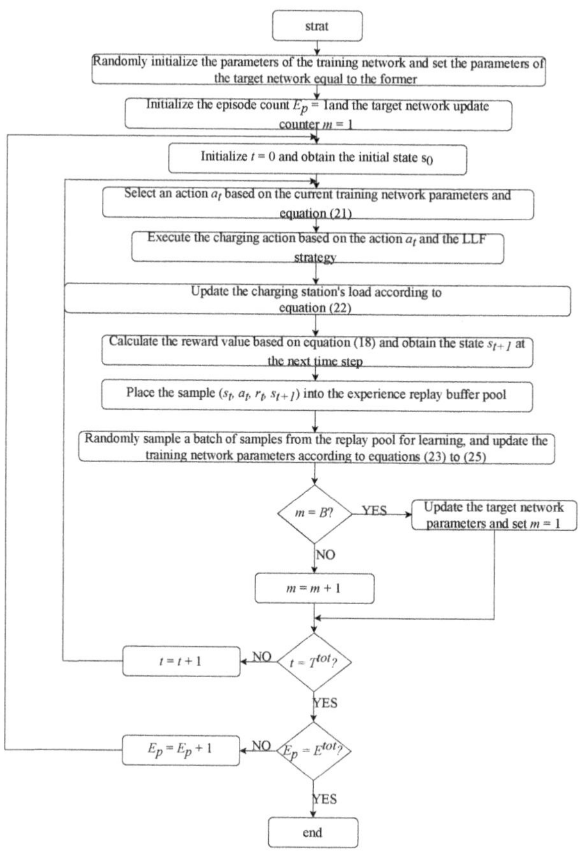

Figure 3. The specific training algorithm of DDQN.

By comparing (20) and (23), we can observe the key differences between the DDQN algorithm and the DQN algorithm. In DDQN, the update of the target Q-value relies on the cooperation of two networks: the training network is used for action selection, while the evaluation of the selected action's value is performed by the target network. This decoupling strategy effectively mitigates the overestimation issue commonly encountered in DQN algorithms. The agent evaluates the difference between the predicted Q-value and the target Q-value using (24) and computes the gradient of the loss function with respect to the network parameters using (25), which is then used to update the parameters of the training network. Once the time reaches T^{tot}, the current training episode ends, and the next episode begins. This process is repeated until the predefined number of training episodes E^{tot} is completed. During each episode, after every B training steps, the parameters of the training network are copied to the target network.

4. Simulation and Analysis

4.1. Parameter Setting

In this study, the basic load of the power distribution network and the hourly electricity price are based on real-world data. The base load data for the distribution network are derived from actual demand data in California and Nevada, USA [33]. After scaling according to a certain ratio, the data are randomly assigned to the distribution network. Then, according to the basic load of the distribution network, the time-of-use electricity price is formulated [17]. This paper adopts the DC charging method, and the distribution networks are evenly distributed in the city. The number of EV users requesting charging is based on the setting in [15]. At the beginning of each time slot, a certain number of users request charging. However, the distribution of these requests is positively correlated with the load levels of the respective distribution network areas. Specifically, areas with higher loads tend to have a greater number of users requesting charging. The DDQN training process was conducted on a PC with an i9-10900k CPU at 3.70 GHz. The program was implemented in Python 3.7 and simulated using TensorFlow 1.13. The MDP formulation in this paper is similar to that in [30], and therefore, some DDQN parameters were set with reference to its configuration. However, given the larger scale and higher complexity of our problem, certain hyperparameters we re adjusted and validated through multiple simulations. The final parameter settings are presented in Table 1.

In each simulation, we compare four different charging strategies, which are detailed as follows:

- greedy: This charging strategy only considers LBMS. In this strategy, EVs begin charging immediately upon arrival.
- DRL_Space: The optimization objective of the DDQN algorithm is to minimize the load imbalance among distribution networks.
- DRL_Cost: The optimization objective of the DDQN algorithm is to minimize the users' charging costs.
- DRL_Both: The optimization objective of the DDQN algorithm is to simultaneously minimize the users' charging costs and the load imbalance among distribution networks.

Table 1. Parameter settings.

Parameter	Value
Maximum charging time slots for EVs	7
Maximum parking time slots for EVs	12
Time window	00:00–23:55

Table 1. *Cont.*

Parameter	Value
Scope of city	30 × 30 km
State dimension	64
Action space	[0,0.1,0.2,...,1]
Number of hidden layers in network	2
Number of neurons per hidden layer	128
Activation function	ReLU
γ	0.95
μ	0.0005
Batch size	64
B	5
E^{tot}	80,000
ΔT	15
T^{tot}	96
M	4
p	20 kW
v	30 km/h
R^{max}	300 km

4.2. Convergence of the DRL Algorithm

Using the parameter settings from Table 1, we configure the neural network and conduct a comparative analysis of the convergence performance of the DDQN algorithm under different optimization objectives. Figures 4–6 illustrate the convergence behavior of the algorithm under the DRL_Space, DRL_Cost, and DRL_Both strategies when the number of EVs requesting charging is set to 1250.

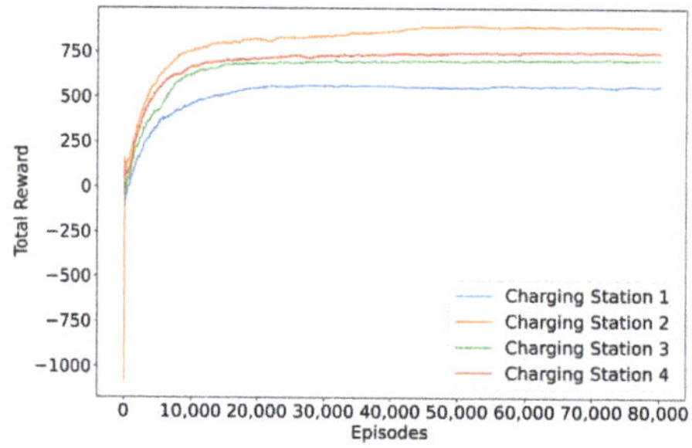

Figure 4. Convergence diagram of DDQN algorithm under DRL_Space ($\alpha = 0$, $\beta = 1$) charging strategy.

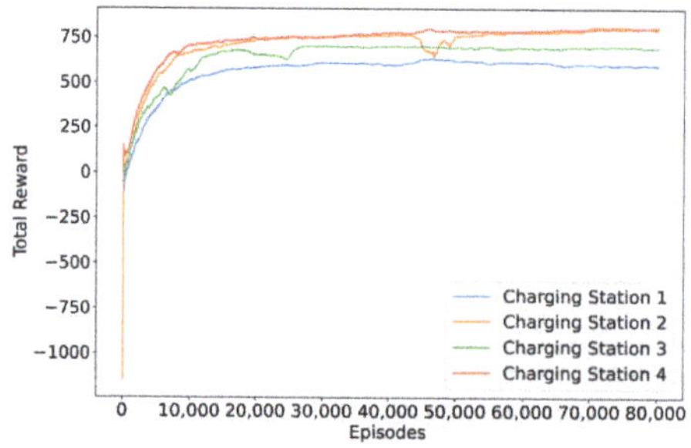

Figure 5. Convergence diagram of DDQN algorithm under DRL_Cost ($\alpha = 1$, $\beta = 0$) charging strategy.

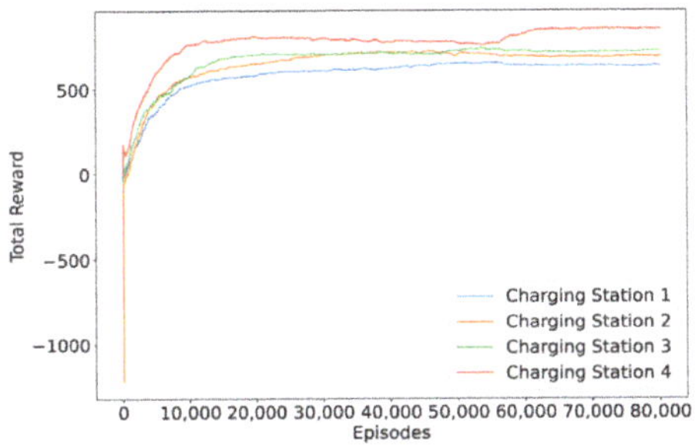

Figure 6. Convergence diagram of DDQN algorithm under DRL_Both ($\alpha = 0.5$, $\beta = 0.5$) charging strategy.

These figures reveal distinct learning patterns under different strategies. At the early stages of training, the high exploration rate leads the agent to attempt a variety of actions, resulting in significant fluctuations in the total reward, which gradually increases as training progresses. As training continues, the exploration rate gradually decreases, shifting the agent's behavior from exploration to exploitation, causing the reward values to stabilize and eventually converge. The final convergence of the reward values demonstrates that the proposed MDP formulation exhibits stable performance under the DDQN algorithm.

However, as observed from the three figures, the final converged reward values vary across different strategies. Additionally, convergence alone does not indicate that the agent has learned the optimal policy. Therefore, in the following sections, we will further validate and evaluate the learned policies to assess their effectiveness in achieving the desired optimization goals.

4.3. Performance of EV Charging Station Allocation

In this study, we first employ the LBMS to guide EV users in selecting charging stations. Therefore, analyzing the EV charging distribution at different time slots across various charging stations is crucial, as it directly reflects whether the proposed station allocation strategy effectively promotes load balancing among distribution networks. Figures 7–9 illustrate the distribution of EVs in each charging station at different time slots under the DRL_Space, DRL_Cost, and DRL_Both strategies, respectively, when the number of EVs is set to 1250.

As shown in the figures, the proposed charging station allocation strategy effectively guides users to charging stations with lower loads. Regardless of the specific strategy employed, a consistent trend is observed—when a charging station has a higher load, the number of EVs charging at that station is relatively low, whereas when the station has a lower load, more EVs tend to charge there. This pattern emerges because the LBMS strategy prioritizes recommending charging stations with lower current loads. Through this scheduling mechanism, EV users are directed to stations in less-loaded distribution networks, effectively reducing the load disparity among distribution networks and further enhancing overall load balancing. This trend demonstrates that the LBMS strategy successfully distributes EV charging demand, preventing excessive charging concentration in high-load areas.

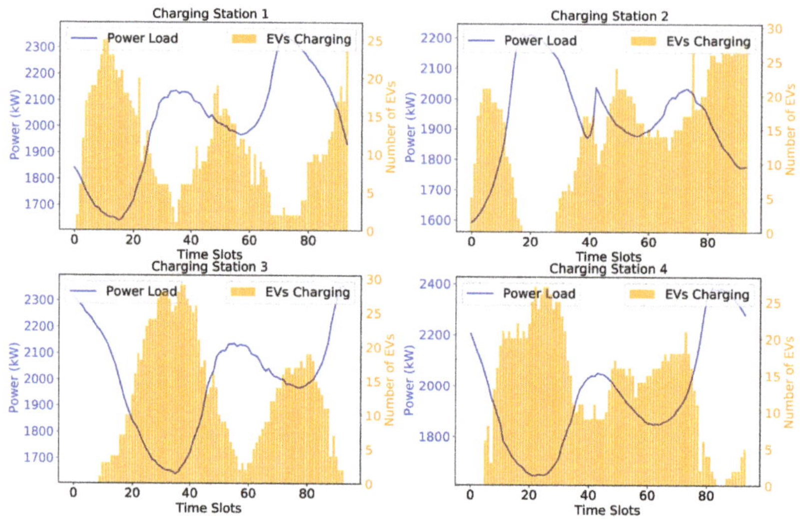

Figure 7. Distribution of EVs across charging stations at each time slot under the DRL_Space ($\alpha = 0$, $\beta = 1$) charging strategy.

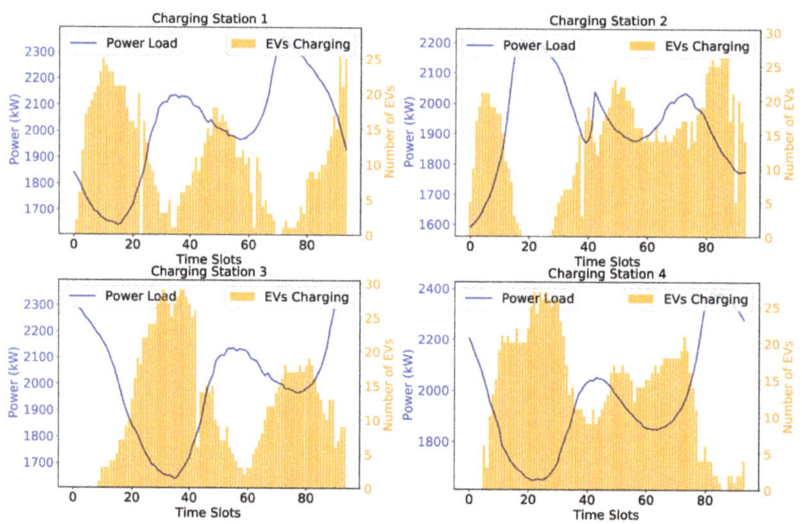

Figure 8. Distribution of EVs across charging stations at each time slot under the DRL_Cost ($\alpha = 1$, $\beta = 0$) charging strategy.

However, it is worth noting that although all three figures exhibit a trend toward load balancing, the number of EVs in the same charging station varies across different time slots. This variation primarily stems from the impact of different in-station coordinated charging strategies on EV scheduling. Different strategies not only influence the load fluctuation patterns of charging stations but also have distinct effects on user charging costs. Therefore, in the following analysis, we will further explore the specific impact of different in-station coordinated charging strategies on charging station load and user costs, as well as the influence of varying weight factors on overall system performance.

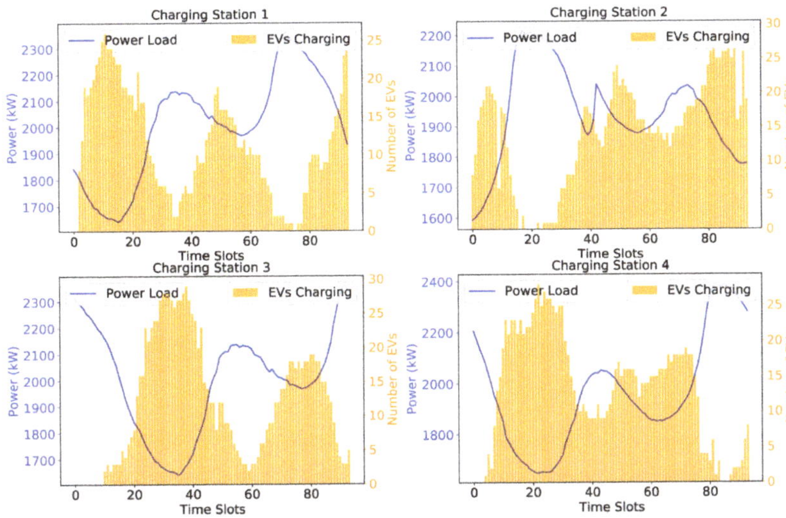

Figure 9. Distribution of EVs across charging stations at each time slot under the DRL_Both ($\alpha = 0.5$, $\beta = 0.5$) charging strategy.

4.4. Performance of Spatial Load Balancing

To compare the performance of different algorithms in spatial load optimization, we comprehensively evaluate the impact of EV charging on spatial load using the average SLVTPR and the average spatial load standard deviation. Specifically, the calculation of the average SLVTPR is

$$\bar{\eta} = \frac{1}{T^{tot}} \sum \eta(t) \tag{26}$$

$\bar{\eta}$ only reflects the extreme differences among distribution networks, making it difficult to capture the overall load disparity among all distribution networks. Therefore, it is necessary to introduce the average spatial load standard deviation to better show the load differences across all distribution networks. Specifically, the calculation of the average spatial load standard deviation is

Figures 10 and 11 depict the performance of $\bar{\eta}$ and $\bar{D}$ in the distribution network under different numbers of EVs requesting charging. From these figures, a comparison between the greedy strategy and the DDQN-based strategies under different optimization functions reveals that, although the greedy strategy can achieve spatial load balancing to a certain extent, it is not as effective as the DRL_Space strategy and DRL_Both strategy. This is because the greedy strategy essentially represents an LBMS approach that only considers spatial load balancing. This strategy selects charging stations based on the distribution network load at the time a charging request is submitted. However, due to the delay effect in user travel, by the time the user actually arrives at the charging station, the load at that station may have changed and may no longer be the lowest among all stations. Moreover, the greedy strategy prioritizes immediate charging, which, although facilitating spatial load balancing to a certain extent, does not account for the dynamic evolution of future loads. Consequently, its scheduling performance remains suboptimal and fails to achieve a globally optimal allocation.

$$\bar{D} = \frac{1}{T^{tot}} \sum D(t) \tag{27}$$

where $D(t)$ is the standard deviation of load among distribution networks at time t.

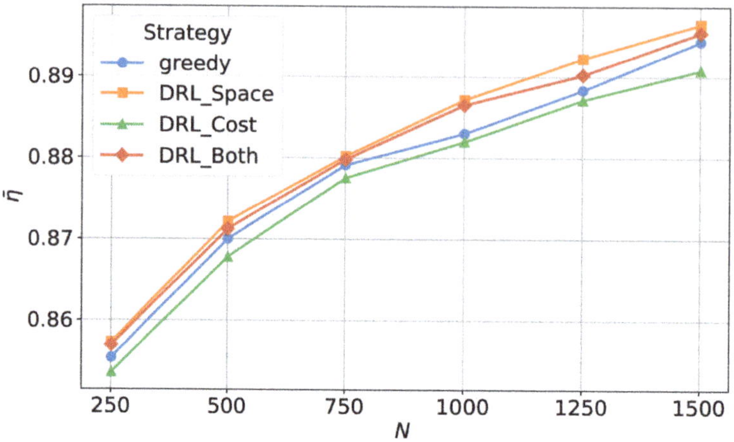

Figure 10. Valley-to-peak ratio of space load under different numbers of EVs.

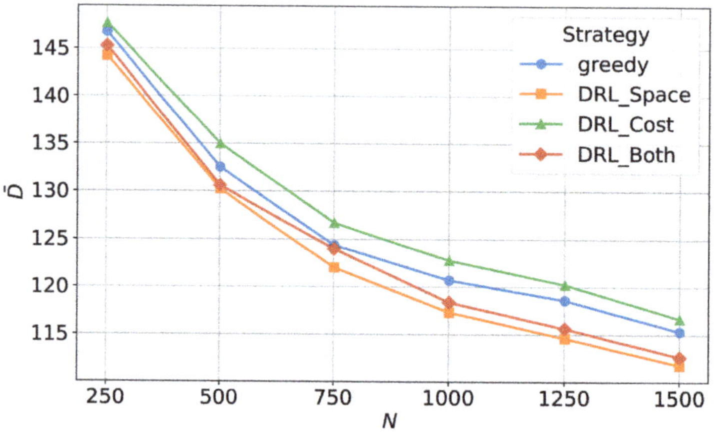

Figure 11. Standard deviation of space load under different numbers of EVs.

The DRL_Space strategy and DRL_Both strategy demonstrate better performance in terms of spatial load balancing. Although the DDQN algorithm uses the same charging station allocation strategy as the greedy strategy when assigning charging stations to users, it optimizes the charging power selection based on the base load conditions of the distribution network at the time the user arrives. Under this charging power allocation strategy, the actions taken by the agent lead to a more balanced load across the distribution networks. On the other hand, the DRL_Cost strategy does not consider spatial load balancing in its charging strategy, which results in less effective load balancing compared to the other three strategies.

4.5. Performance of Charging Cost

As the primary participants in the charging process, EV users are primarily concerned with charging costs. Uncoordinated charging can lead to some users charging during periods of high electricity prices, resulting in increased costs. Additionally, certain users may be unable to complete their charging needs within their parking time, leading to "default" situations, where the charging stations must compensate these users, significantly impacting the costs of the charging station as well. Therefore, it is necessary to compare the user charging costs across all charging stations under the four different charging strategies. In this paper, we use the following formula to calculate the total charging cost for EV users:

$$Cost = \sum_{t=1}^{T^{tot}} \sum_{j=1}^{M} N_{j,t}^{c} p_{j,t} \tag{28}$$

From Figure 12, it is evident that, when comparing the greedy strategy with those based on the DDQN algorithm, the greedy strategy results in the highest user charging costs. This is because the greedy strategy fulfills the user's charging needs as quickly as possible, neglecting the flexibility users may have during their parking period, which leads to increased charging costs.

In contrast, although the DDQN-based strategy relies on the LBMS strategy during the charging station allocation phase, it fully exploits user charging flexibility in the in-station charging phase. This allows it to dynamically adapt to varying load conditions and electricity price fluctuations, intelligently adjusting charging power to optimize overall charging scheduling. As a result, the user charging costs under the three DDQN algorithms are lower than those under the greedy strategy. However, different reward functions have varying impacts on user charging costs. For the DRL_Space strategy, the agent does not consider information about electricity prices and, therefore, cannot adjust charging power based on the prices at the charging stations. Consequently, the performance in terms of user charging costs is not as favorable as in the other two DDQN algorithms that incorporate cost considerations in their reward functions.

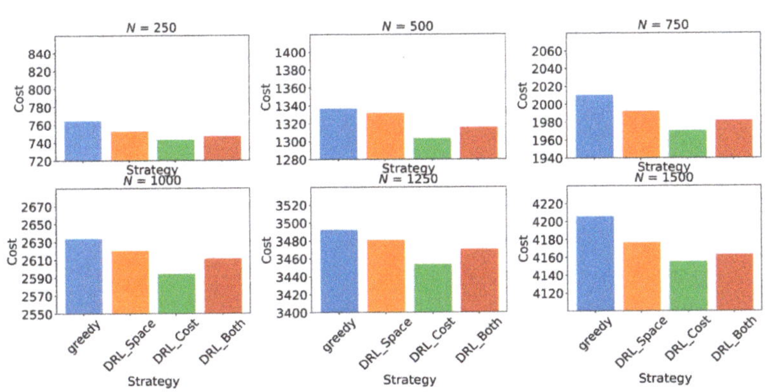

Figure 12. Comparison of users' charging costs under different numbers of EVs.

For the DRL_Both strategy, the reward function design includes consideration of the users' charging costs, which results in lower charging costs compared to the DRL_Space strategy. However, this strategy must balance between minimizing charging costs and the spatial load standard deviation, leading to higher costs relative to the DRL_Cost strategy.

In the case of the DRL_Cost strategy, the input information to the agent consists solely of the charging prices at each time slot. After multiple iterations, the agent selects a charging strategy that minimizes the user's charging costs. As a result, this reward function yields the lowest user costs.

4.6. Comprehensive Performance of Cost and Spatial Load Balancing

The above results focus on comparing the system performance from the perspective of spatial load balancing or user charging costs, making it difficult to reflect the overall performance of the DDQN algorithm under different optimization functions. To comprehensively evaluate the overall system performance under the three reward functions, we first normalize the charging costs and spatial load standard deviations obtained from the DDQN algorithm with different reward functions using (17). This yields the normalized charging station costs $Cost^{norm}$ and the normalized spatial load standard deviations D^{norm} as

$$Cost^{norm} = \frac{Cost^{max} - Cost_t}{Cost^{max} - Cost^{min}} \tag{29}$$

$$D^{norm} = \frac{D^{max} - D_t}{D^{max} - D^{min}} \tag{30}$$

Then, by adding $Cost^{norm}$ and D^{norm}, we obtain the normalized total result $Total^{norm}$, which comprehensively considers both factors. The corresponding expression is

$$Total^{norm} = Cost^{norm} + D^{norm} \tag{31}$$

For example, when the number of EVs is 1000, the charging costs under the DRL_Space, DRL_Cost, and DRL_Both strategies are 2620.19, 2594.48, and 2611.23, respectively, and the corresponding spatial load standard deviations are 117.32, 122.78, and 118.36, respectively. Using (29) and (30), the $Cost^{norm}$ values under the three reward functions are 0, 1, and 0.35, and the D^{norm} values are 1, 0, and 0.81, respectively. Summing them up using (31), the $Total^{norm}$ values are 1, 1, and 1.16. Similarly, the total normalized results can be obtained for other EV numbers, comprehensively considering both factors.

As shown in Figure 13, both the DRL_Space strategy and DRL_Cost strategy have a $Total^{norm}$ value of 1. Under the DRL_Cost strategy, this approach selects a charging power that minimizes the charging price for EVs, disregarding the load distribution across the power grids. In contrast, the DRL_Space strategy prioritizes charging power that balances the load among the power grids, overlooking the users' charging costs. Both strategies focus on only one aspect, which limits the overall performance of the algorithm.

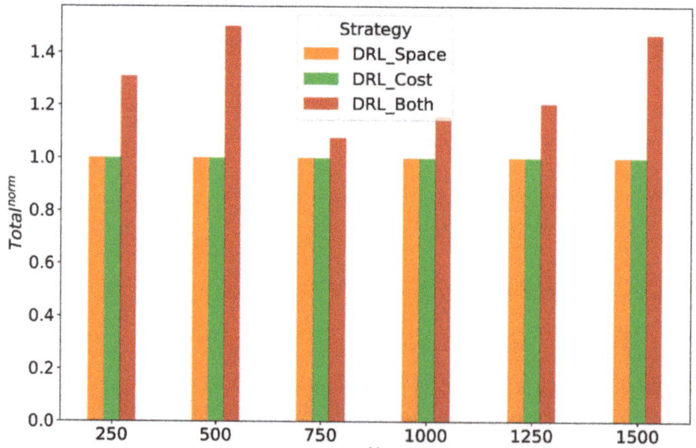

Figure 13. Comparison of normalized results under different numbers of EVs.

However, under the DRL_Both strategy, $Total^{norm}$ is the highest. This is because, this strategy considers both user charging costs and load balancing when selecting the charging power at each time slot. After balancing these two factors, the strategy selects actions that can both reduce user charging costs and minimize the load disparity among distribution networks. Therefore, this strategy achieves the highest reward value, and the system's overall performance is the most balanced.

Table 2 presents the impact of different weights on system performance when there are 1250 EVs. As observed from the table, with the increase in α (user charging cost weight), user charging costs gradually decrease, while the spatial load standard deviation increases. This happens because as α increases, the agent prioritizes minimizing user charging costs during optimization, with less emphasis on balancing the load across the distribution networks. Therefore, when α = 1, the agent fully focuses on optimizing user charging costs, leading to the highest charging cost and the poorest load balancing.

Table 2. Impact of different weights on system performance with 1250 EVs.

Different Weight	$\bar{\eta}$	$\bar{D}$	Cost
$\alpha = 0,\ \beta = 1$	89.23%	114.62	3480.74
$\alpha = 0.3,\ \beta = 0.7$	89.15%	114.93	3474.34
$\alpha = 0.5,\ \beta = 0.5$	89.03%	115.62	3470.37
$\alpha = 0.7,\ \beta = 0.3$	89.01%	116.63	3466.26
$\alpha = 1,\ \beta = 0$	88.72%	120.24	3453.79

4.7. Performance of Weight Factor Sensitivity

Conversely, as β (load balancing weight) increases, the agent focuses more on balancing the load across the distribution networks during optimization. With the increase in β, user charging costs gradually rise, while the spatial load standard deviation decreases. This is because, when the agent places more emphasis on balancing the load among the distribution networks, it actively schedules some users to charge during time periods with more uneven spatial load, thereby achieving load balancing among the distribution networks. However, this adjustment may lead to higher charging costs for some users. Consequently, when β is high (corresponding to higher charging costs in the table), the agent's scheduling strategy can significantly reduce the spatial load standard deviation.

The above analysis demonstrates the varying impacts of different weights on system performance, but it is not straightforward to select a single optimal strategy. Therefore, a comprehensive evaluation of the DDQN algorithm's performance under different weights is necessary. In this section, we continue to apply the normalization method from Section 4.6 to the results obtained under different weights, and the outcomes are shown in Figure 14.

Figure 14 illustrates the impact of different weights on overall system performance under varying numbers of vehicles. From the figure, it is evident that when $\alpha = 1$ or $\alpha = 0$, the system's overall performance is the worst. In both of these cases, although the optimization objectives are individually achieved at their optimal values, the failure to account for the other factor simultaneously leads to a decline in overall system performance. However, when $\alpha = 0.3$ or $\alpha = 0.7$, the system's overall performance is significantly better than when considering only a single factor. This is because, during the optimization process, the introduction of an additional factor forces the agent to balance both optimization goals. However, due to the weighting scheme, the agent inevitably favors one aspect over the other. Thus, despite some improvements, the optimal balance has not been fully achieved.

When $\alpha = 0.5$, the system achieves the best overall performance across different vehicle numbers. This is because, in this configuration, the system does not prioritize either user charging costs or load balancing across the distribution networks but seeks a balance between the two. Although, in this case, the user charging costs and spatial load standard deviation are not optimal, they represent the most balanced solution, indicating that the proposed strategy is a well-balanced optimization approach.

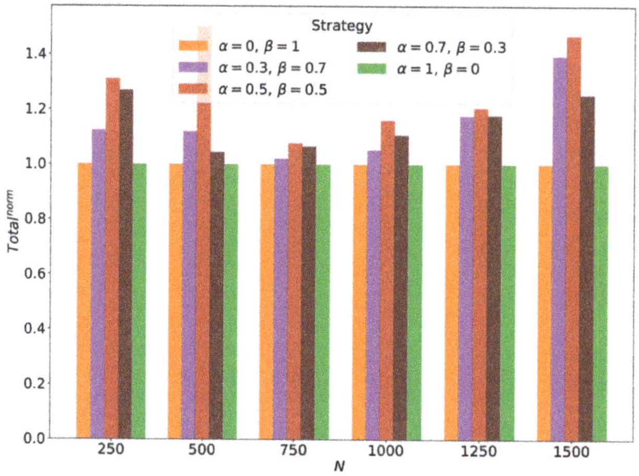

Figure 14. Comparison of normalized results under varying EV counts and weights.

In summary, the trade-off between α and β weights directly influences the system's optimization objectives. When α is higher, the system tends to prioritize minimizing user charging costs, which may result in imbalanced load distribution; whereas when β is higher, the system focuses more on optimizing load balancing across the distribution networks, which may lead to sacrificing some user charging cost efficiency. When α and β are equal, the system exhibits the strongest overall performance. Therefore, a reasonable setting of the α and β weight ratio can achieve a better balance between load balancing and user cost optimization.

4.8. The Impact of Different User Participation Levels on System Performance

In practical applications, the level of user participation plays a crucial role in charging station scheduling and the overall performance of the system. Not all users strictly follow the scheduling center's recommendations when selecting a charging station; instead, they may make decisions based on personal preferences. Such random selection behavior can lead to imbalanced load distribution across the power grid and increase the overall charging cost. Consequently, the degree of user participation directly determines whether a charging strategy can effectively optimize grid load distribution while minimizing charging costs. To evaluate this impact, Figures 15–18 illustrate the effects of different user participation levels on system performance, considering a total of 1000 charging requests. These figures provide an intuitive understanding of how varying degrees of user participation influence load balancing and charging costs, thereby enhancing the comprehension of system optimization strategies.

From Figure 15, it can be observed that as the EV user participation rate increases, the SLVTPR values for all strategies exhibit an upward trend, indicating that users following the LBMS-based charging station recommendation strategy can significantly improve load balancing. Among all strategies, DRL_Space and DRL_Both demonstrate the best performance. This is because DRL_Space focuses specifically on load balancing, ensuring a more even distribution of load among charging stations. Meanwhile, DRL_Both considers both load balancing and charging costs, optimizing load distribution while maintaining economic efficiency. As a result, DRL_Both can still achieve effective load balancing; however, since it simultaneously accounts for both user cost and load balancing, its spatial load balancing performance is slightly inferior to that of DRL_Space.

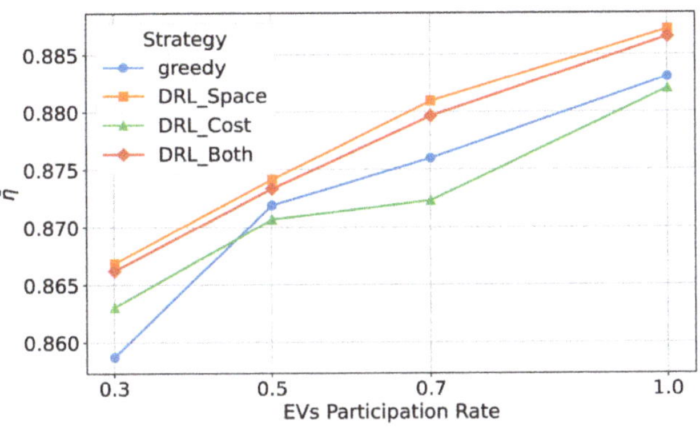

Figure 15. Comparison of space load valley–peak ratio under different EV participation levels.

In contrast, at low user participation levels, the greedy strategy exhibits poor load balancing performance, even lower than that of DRL_Cost. This is because the greedy strategy selects charging stations solely based on the lowest current load, without considering future load variations, which can lead to station overloading. However, as user participation increases, the load balancing performance of the greedy strategy gradually surpasses that of DRL_Cost, indicating that when a greater number of users opt for charging stations with lower loads, the effectiveness of the greedy strategy in achieving load balancing improves. Nevertheless, since DRL_Cost primarily optimizes charging costs, its load balancing performance is slightly inferior to other methods at higher user participation levels.

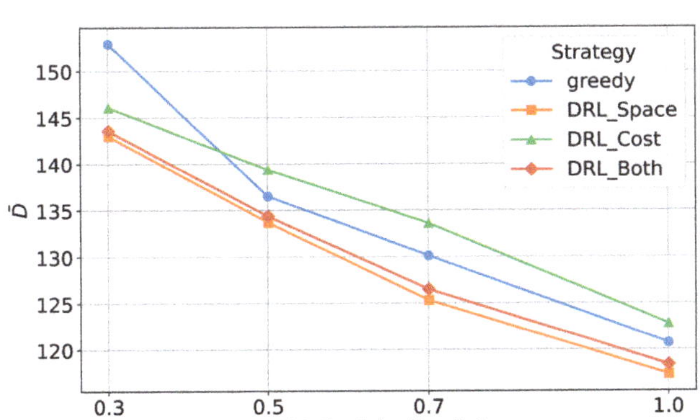

Figure 16. Comparison of space load standard deviation under different EV participation levels.

Figure 16 illustrates the spatial load standard deviation under different levels of user participation. From the figure, it can be observed that as user participation increases, the spatial load standard deviation decreases across all strategies, indicating that higher user participation contributes to improved load balancing. Among all strategies, DRL_Space and DRL_Both exhibit the lowest standard deviation, demonstrating their superior ability to balance loads effectively, particularly at high user participation levels. The greedy strategy shows the highest standard deviation at low user participation levels, indicating that it causes significant fluctuations in load distribution. Although DRL_Cost achieves better load balancing than the greedy strategy at lower user participation levels, its load balancing performance remains inferior to that of the other strategies as user participation further increases.

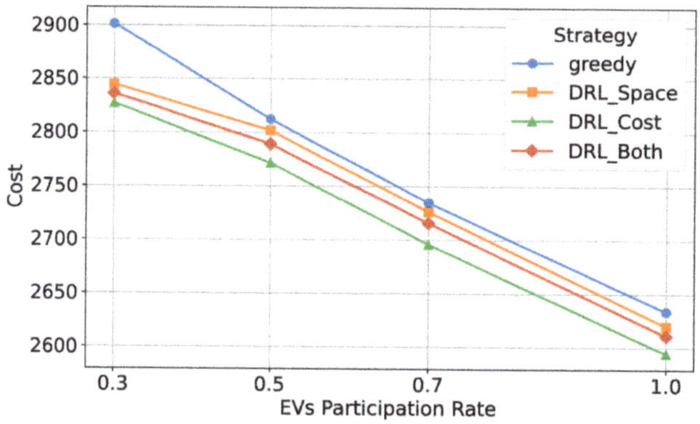

Figure 17. Comparison of total charging costs under different EV participation levels.

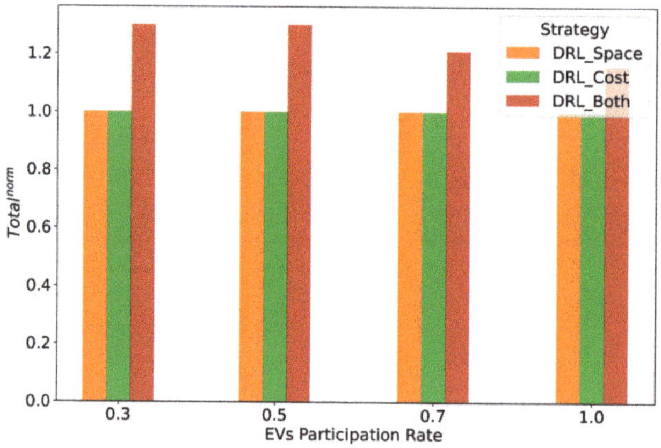

Figure 18. Comparison of normalized results under different EV participation levels.

Figure 17 illustrates the total charging cost under different user participation levels. As user participation increases, the overall charging cost decreases, indicating that the LBMS-based charging station recommendation strategy generally enables users to access lower electricity prices, thereby effectively reducing charging costs. Among all strategies, DRL_Cost consistently achieves the lowest charging cost, as its primary optimization objective is to minimize user costs. Even at low participation levels, this strategy effectively reduces costs, demonstrating strong economic optimization capabilities. DRL_Both, as a comprehensive optimization strategy, performs slightly worse than DRL_Cost in cost reduction but outperforms DRL_Space, indicating that it maintains a balance between cost reduction and load balancing, ensuring overall system stability. In contrast, DRL_Space focuses more on load balancing and gives less consideration to price optimization, resulting in higher costs than DRL_Both and DRL_Cost, though it still significantly outperforms the greedy strategy. The greedy strategy incurs the highest charging cost, especially at low user participation levels, where costs are significantly higher than those of other strategies. This is because the greedy approach selects the charging station with the lowest current load without considering electricity prices or future load variations, leading to station overload and increased charging costs. As a result, DRL_Cost provides the best economic performance, DRL_Both achieves a trade-off between cost control and load balancing, while DRL_Space and the greedy strategy result in relatively higher charging costs.

Figure 18 further illustrates the overall system performance of different optimization strategies. Across all user participation levels, DRL_Both achieves the highest overall performance score, indicating that this strategy effectively balances load distribution and charging costs, ensuring maximum system benefits. In comparison, DRL_Space and DRL_Cost perform well in their respective optimization objectives but fall slightly short of DRL_Both in overall performance. DRL_Space excels in load balancing, but since it does not explicitly optimize charging costs, its overall system performance is slightly lower than that of DRL_Both. On the other hand, DRL_Cost demonstrates the best performance in controlling charging costs but does not sufficiently address load balancing. Therefore, its overall system performance is also inferior to that of DRL_Both.

5. Conclusions and Future Work

The travel patterns and charging demands of EV users are characterized by significant uncertainty, and without proper coordination, this can lead to increased user charging costs and jeopardize the safe operation of the distribution grids. In contrast, if the selection of charging stations and charging behaviors within the stations is properly coordinated, it is possible to effectively control user charging costs and reduce the load imbalance across the distribution grids. In this paper, DRL-based optimization strategies for EV charging are proposed, aiming to minimize user charging costs while achieving load balancing in the distribution grids. By employing a load balancing matching strategy to guide users to the charging stations with the lowest load and integrating the DDQN algorithm to optimize the orderly charging process within the stations, this strategy effectively addresses the uncertainty in user travel and charging demands, significantly reducing charging costs and enhancing the safety of distribution grids.

Simulation results indicate that, compared to the traditional LBMS algorithm, the proposed DRL approaches excel in controlling charging costs and optimizing load balancing across the grids. Specifically, when the reward function focuses on minimizing user charging costs, the strategy effectively reduces user costs; when the reward function aims to minimize the load imbalance in the distribution grids, it significantly improves load balancing. Furthermore, when the reward function considers both user charging costs and grids' load balancing, it achieves an optimal balance between the two. Additionally, the strategy demonstrates robust performance across various levels of user participation, effectively minimizing user costs and optimizing grids' load balance regardless of the participation rate, illustrating the strategy's ability to handle dynamic and uncertain environments effectively.

The model proposed in this paper can be deployed in a large-scale urban environment, in which a centralized dispatching unit can be established to supervise the user's choice of charging stations and the dispatching of charging stations. When a user needs to charge his EV, the dispatching center can adopt the proposed charging station allocation strategy to recommend a suitable charging station for the EV. When the user arrives at the charging station, the dispatching center will send the distribution network information to the station. The charging station can combine this information with its own charging price and use the DRL model proposed in this study to dynamically adjust the weight ratio according to the needs of the charging station and the power grid to achieve the charging power allocation for the EV. In addition, this strategy is not only applicable to EV charging networks, but can also be applied to other complex scenarios, such as energy management (such as virtual power plants, distributed energy scheduling) and resource allocation (such as logistics optimization, fleet scheduling), providing theoretical support and practical guidance for solving dynamic optimization problems in various fields.

In future work, the proposed strategy can be further enhanced in the following three aspects:

1. V2G technology could be incorporated into future research, potentially further reducing users' charging costs and enhancing load balancing across distribution networks.
2. Future research could explore the impact of highly dynamic conditions, such as fluctuations in traffic flow or surges in user demand, on the performance of the system.
3. The time-of-use electricity prices used in this paper are fixed. Future research can incorporate dynamic pricing to further explore the impact of price uncertainty on system performance.

Author Contributions: Conceptualization, Y.X., Z.C., J.Z., and X.C.; methodology, Y.X., Z.C., J.Z., and X.C.; software, Z.C.; validation, Y.X. and X.C.; formal analysis, Y.X. and Z.C.; investigation, Z.C.; resources, Y.X. and X.C.; data curation, Z.C. and X.C.; writing—original draft preparation, Z.C.; writing—review and editing, Y.X. and X.C.; visualization, Z.C.; supervision, Y.X. and X.C.; project administration, Y.X.; All authors have read and agreed to the published version of the manuscript.

Funding: This research received no external funding.

Data Availability Statement: The data presented in this study are not publicly available due to privacy concerns.

Conflicts of Interest: The authors declare no conflicts of interest.

Abbreviations

The following abbreviations are used in this manuscript:

EVs	Electric Vehicles
DRL	Deep Reinforcement Learning
LBMS	Load Balancing Matching Strategy
ADMM	Alternating Direction Method of Multipliers
MPC	Model Predictive Control
MDP	Markov Decision Process
DDQN	Double Deep Q Network
SOC	State of Charge
SLVTPR	Spatial Load Valley-to-Peak Ratio
LLF	Least Laxity First
DQN	Deep Q Network
V2G	Vehicle-to-Grid
DERs	Distributed Energy Resource
ESS	Energy Storage System
MADRL	Multi-Agent Deep Reinforcement Learning

Nomenclature

ΔT	The time interval between the two decision times
(x_i, y_i)	The position of EV i
$\bar{\eta}$	The average spatial load valley-to-peak ratio of distribution networks
$\bar{\theta}$	The parameters of the target neural network
$\bar{D}$	The average standard deviation of load among distribution networks
β	The weight factor in the reward function for balancing the load among distribution networks
$\Delta T_{i,t}^{emr}$	The charging urgency of EV i at time t
$\eta(t)$	The spatial load valley-to-peak ratio of distribution networks at time t
γ	The discount factor of the MDP formulation
μ	The learning rate of the DDQN algorithm
$\mathbf{Cand}_i$	The set of candidate charging stations for EV i
$\boldsymbol{R}$	The distances from EV users to each charging station
θ	The parameters of the training neural network
A	The actions taken by the agent
$a_{j,t}$	The action selected by charging station j at time t
a_t	The action selected at time t
B	The update frequency of the target neural network
C^{max}	The maximum charging time for EVs
$c_{i,t}$	The charging time slots required by the user in the continuous charging condition of EV i at time t
$Cost$	The total charging cost for EV users
$Cost^{norm}$	The normalized value of the charging costs
$D(t)$	The standard deviation of load among distribution networks at time t
D^{norm}	The normalized value of spatial load standard deviations
d_i	The user's charging demand of EV i
E^{max}	The maximum parking time for EVs
E^{tot}	The total number of training episodes
E_p	The episode count
$e_{i,t}$	The estimated parking time slots of EV i at time t
I_t	The set of EVs arriving at the charging station at time t
J_t	The set of EVs that arrived before time t and have not finished their charging
K_t	The set of EVs that need to be charged at time t after I_t arrive
$l_{j,t}$	The difference between the proportion of the baseline load of distribution network j at time t in the total baseline load and the proportion of the ideal load for the distribution network
$l_{j,t}^{nor}$	The normalized value of $l_{j,t}$ at time t for distribution network j
M	The total number of distribution networks
m	The target network update counter
N	The total number of EV users
$N_t^{c,e}$	The total number of EVs with remaining charging time slots c and expected parking time slots e at time t
$N_{j,t}^{c}$	The number of charging EVs at charging station j at time t
$N_{j,t}^{tot}$	The total number of EVs at charging station j at time t
P	The transition probabilities of the environment
p	The charging power of the charging station
p_i	The estimated parking time of EV i
P_t^{tot}	The total baseline load of all distribution networks at time t when no EV users are charging
$P_{j,t}$	The sum of the loads in the distribution network where charging station j is located at time t

$p_{j,t}$	The charging price of charging station j at time t
$P_{j,t}^{ini}$	The initial load of the distribution network j at time t
$p_{j,t}^{nor}$	The normalized charging price of charging station j at time t
$P_{peak}(t)$	The peak power load of distribution networks at time t
$P_{valley}(t)$	The valley power load of distribution networks at time t
$Q_j^{\pi}(s,a)$	The state-action function of charging station j
R	The rewards obtained by the agent
R^{max}	The maximum driving range of the EV when fully charged
$R_{i,j}$	The distance from EV i to charging station j
R_i	The distance that EV i can travel based on its initial SOC S_i^{ini}
$r_{j,t}$	The reward when the optimization objective is to minimize the load disparity among multiple distribution networks and charging cost for users of charging station j at time t
$r_{j,t}^{cost}$	The reward when the optimization objective is to minimize the charging cost for users of charging station j at time t
$r_{j,t}^{disp}$	The reward when the optimization objective is to minimize the load disparity among multiple distribution networks of charging station j at time t
S	The state of the environment
S_i^{ini}	The initial state of charge of EV i
S_{dim}	The dimensionality of the state space
T^{tot}	The number of total time windows
$T_{i,j}^{tra}$	The travel time for EV i to reach charging station j
T_i^{str}	The charging start time for EV i at charging station j
$Total^{norm}$	The normalized sum value that takes into account the charging costs and the spatial load standard deviations
v	The EV's travel speed
V_j^G	The set of charging stations j that EVs can reach

References

1. Jeong, S.; Jang, Y.J.; Kum, D.; Lee, M.S. Charging automation for electric vehicles: Is a smaller battery good for the wireless charging electric vehicles? *IEEE Trans. Autom. Sci. Eng.* **2018**, *16*, 486–497. [CrossRef]
2. International Energy Agency. Global EV Outlook 2024. License: CC BY 4.0. Available online: https://www.iea.org/reports/global-ev-outlook-2024 (accessed on 26 April 2024).
3. Van Vliet, O.; Brouwer, A.S.; Kuramochi, T.; van Den Broek, M.; Faaij, A. Energy use, cost and CO_2 emissions of electric cars. *J. Power Sources* **2011**, *196*, 2298–2310. [CrossRef]
4. Chi, K.T.; Huang, M.; Zhang, X.; Liu, D.; Li, X.L. Circuits and systems issues in power electronics penetrated power grid. *IEEE Open J. Circuits Syst.* **2020**, *1*, 140–156. [CrossRef]
5. Xu, Y.; Çolak, S.; Kara, E.C.; Moura, S.J.; González, M.C. Planning for electric vehicle needs by coupling charging profiles with urban mobility. *Nat. Energy* **2018**, *3*, 484–493. [CrossRef]
6. Chen, L.; Pan, Z.; Yu, T.; Wang, K. Real-time optimal dispatch for large-scale electric vehicles based on dynamic non-cooperative game theory. *Autom. Electr. Power Syst* **2019**, *43*, 32–40. [CrossRef]
7. Mahmud, K.; Hossain, M.J.; Ravishankar, J. Peak-load management in commercial systems with electric vehicles. *IEEE Syst. J.* **2018**, *13*, 1872–1882. [CrossRef]
8. Said, D.; Mouftah, H.T. A Novel Electric Vehicles Charging/Discharging Management Protocol Based on Queuing Model. *IEEE Trans. Intell. Veh.* **2020**, *5*, 100–111. [CrossRef]
9. Hu, Q.; Bu, S.; Terzija, V. A Distributed P and Q Provision-Based Voltage Regulation Scheme by Incentivized EV Fleet Charging for Resistive Distribution Networks. *IEEE Trans. Transp. Electrif.* **2021**, *7*, 2376–2389. [CrossRef]
10. Liu, W.; Wang, X.; Xu, Y. Bilevel Planning of Wireless Charging Lanes in Coupled Transportation and Power Distribution Networks. *IEEE Trans. Transp. Electrif.* **2024**, *10*, 2499–2510. [CrossRef]
11. Bourass, A.; Cherkaoui, S.; Khoukhi, L. Secure optimal itinerary planning for electric vehicles in the smart grid. *IEEE Trans. Ind. Inform.* **2017**, *13*, 3236–3245. [CrossRef]

12. Yang, H.; Deng, Y.; Qiu, J.; Li, M.; Lai, M.; Dong, Z.Y. Electric vehicle route selection and charging navigation strategy based on crowd sensing. *IEEE Trans. Ind. Inform.* **2017**, *13*, 2214–2226. [CrossRef]
13. Sortomme, E.; Hindi, M.M.; MacPherson, S.J.; Venkata, S. Coordinated charging of plug-in hybrid electric vehicles to minimize distribution system losses. *IEEE Trans. Smart Grid* **2010**, *2*, 198–205. [CrossRef]
14. Sun, C.; Wen, X.; Lu, Z.; Zhang, J.; Chen, X. A graphical game approach to electrical vehicle charging scheduling: Correlated equilibrium and latency minimization. *IEEE Trans. Intell. Transp. Syst.* **2020**, *22*, 505–517. [CrossRef]
15. Chen, X.; Wang, H.; Wu, F.; Wu, Y.; González, M.C.; Zhang, J. Multimicrogrid load balancing through EV charging networks. *IEEE Internet Things J.* **2021**, *9*, 5019–5026. [CrossRef]
16. Xia, Y.; Gu, X.; Fu, D.; Chen, X. Large Scale EV Charging Allocation in Unbalanced Multi-Microgrid System. *IEEE J. Emerg. Sel. Top. Circuits Syst.* **2023**, *13*, 797–805. [CrossRef]
17. Liu, J.; Lin, G.; Huang, S.; Zhou, Y.; Li, Y.; Rehtanz, C. Optimal EV charging scheduling by considering the limited number of chargers. *IEEE Trans. Transp. Electrif.* **2020**, *7*, 1112–1122. [CrossRef]
18. Sun, B.; Huang, Z.; Tan, X.; Tsang, D.H. Optimal scheduling for electric vehicle charging with discrete charging levels in distribution grid. *IEEE Trans. Smart Grid* **2016**, *9*, 624–634. [CrossRef]
19. Alinia, B.; Hajiesmaili, M.H.; Crespi, N. Online EV charging scheduling with on-arrival commitment. *IEEE Trans. Intell. Transp. Syst.* **2019**, *20*, 4524–4537. [CrossRef]
20. Gang, X.; Bingxu, Z.; Guangchao, Z. Distributed and robust optimal scheduling model for large-scale electric vehicles connected to grid. *Trans. China Electrotech. Soc.* **2021**, *36*, 565–578. [CrossRef]
21. Shi, Y.; Tuan, H.D.; Savkin, A.V.; Duong, T.Q.; Poor, H.V. Model predictive control for smart grids with multiple electric-vehicle charging stations. *IEEE Trans. Smart Grid* **2018**, *10*, 2127–2136. [CrossRef]
22. Zheng, Y.; Song, Y.; Hill, D.J.; Meng, K. Online distributed MPC-based optimal scheduling for EV charging stations in distribution systems. *IEEE Trans. Ind. Inform.* **2018**, *15*, 638–649. [CrossRef]
23. Wu, W.; Lin, Y.; Liu, R.; Li, Y.; Zhang, Y.; Ma, C. Online EV charge scheduling based on time-of-use pricing and peak load minimization: Properties and efficient algorithms. *IEEE Trans. Intell. Transp. Syst.* **2020**, *23*, 572–586. [CrossRef]
24. Fernandez, V.; Pérez, V. Optimization of electric vehicle charging control in a demand-side management context: A model predictive control approach. *Appl. Sci.* **2024**, *14*, 8736. [CrossRef]
25. Shang, Y.; Li, D.; Li, Y.; Li, S. Explainable spatiotemporal multi-task learning for electric vehicle charging demand prediction. *Appl. Energy* **2025**, *384*, 125460. [CrossRef]
26. Liu, X.K.; He, H.; Wang, Y.W.; Xu, Q.; Guo, F. Distributed Hybrid Secondary Control for a DC Microgrid via Discrete-Time Interaction. *IEEE Trans. Energy Convers.* **2018**, *33*, 1865–1875. [CrossRef]
27. Wan, Z.; Li, H.; He, H.; Prokhorov, D. Model-free real-time EV charging scheduling based on deep reinforcement learning. *IEEE Trans. Smart Grid* **2018**, *10*, 5246–5257. [CrossRef]
28. Wang, S.; Bi, S.; Zhang, Y.A. Reinforcement learning for real-time pricing and scheduling control in EV charging stations. *IEEE Trans. Ind. Inform.* **2019**, *17*, 849–859. [CrossRef]
29. Sadeghianpourhamami, N.; Deleu, J.; Develder, C. Definition and evaluation of model-free coordination of electrical vehicle charging with reinforcement learning. *IEEE Trans. Smart Grid* **2019**, *11*, 203–214. [CrossRef]
30. Chen, G.; Wang, X.; Yuan, C.; Shuai, X.; Zhou, Q. Coordinated Charging Strategy Applicable to Large-scale Charging Stations Based on Deep Reinforcement Learning. *Autom. Electr. Power Syst.* **2023**, *47*, 88–95. [CrossRef]
31. Gao, H.; Jiang, S.; Li, Z.; Wang, R.; Liu, Y.; Liu, J. A Two-Stage Multi-Agent Deep Reinforcement Learning Method for Urban Distribution Network Reconfiguration Considering Switch Contribution. *IEEE Trans. Power Syst.* **2024**, *39*, 7064–7076. [CrossRef]
32. Mehta, R.; Srinivasan, D.; Khambadkone, A.M.; Yang, J.; Trivedi, A. Smart charging strategies for optimal integration of plug-in electric vehicles within existing distribution system infrastructure. *IEEE Trans. Smart Grid* **2016**, *9*, 299–312. [CrossRef]
33. California ISO. California ISO Open Access Same-Time Information System. 2020. Available online: https://www.caiso.com/ (accessed on 10 February 2021).

World Electric Vehicle Journal

MDPI

Article

Unlocking Grid Flexibility: Leveraging Mobility Patterns for Electric Vehicle Integration in Ancillary Services

Corrado Maria Caminiti *, Luca Giovanni Brigatti, Matteo Spiller, Giuliano Rancilio and Marco Merlo

Department of Energy, Politecnico di Milano, 20156 Milano, Italy; lucagiovanni.brigatti@mail.polimi.it (L.G.B.); matteo.spiller@polimit.it (M.S.); giuliano.rancilio@polimi.it (G.R.); marco.merlo@polimi.it (M.M.)

* Correspondence: corradomaria.caminiti@polimi.it; Tel.: +39-02-2399-8505

Abstract: The electrification of mobility has introduced considerable challenges to distribution networks due to varying demand patterns in both time and location. This underscores the need for adaptable tools to support strategic investments, grid reinforcement, and infrastructure deployment. In this context, the present study employs real-world datasets to propose a comprehensive spatial–temporal energy model that integrates a traffic model and geo-referenced data to realistically evaluate the flexibility potential embedded in the light-duty transportation sector for a given study region. The methodology involves assessing traffic patterns, evaluating the grid impact of EV charging processes, and extending the analysis to flexibility services, particularly in providing primary and tertiary reserves. The analysis is geographically confined to the Lombardy region in Italy, relying on a national survey of 8.2 million trips on a typical day. Given a target EV penetration equal to 2.5%, corresponding to approximately 200,000 EVs in the region, flexibility bands for both services are calculated and economically evaluated. Within the modeled framework, power-intensive services demonstrated significant economic value, constituting over 80% of the entire potential revenues. Considering European markets, the average marginal benefit for each EV owner is in the order of 10 € per year, but revenues could be higher for sub-classes of users better fitting the network needs.

Keywords: electric vehicles; flexibility; smart charging; geospatial planning; open-source data

1. Introduction

Driven by ambitious emission reduction targets, the energy sector is moving towards the challenge of decarbonization [1]. The decreasing utilization of thermal plants in favor of non-programmable renewable energy sources is increasing the need for alternative providers of flexibility for the grid [2]. Simultaneously, the automobile industry is undergoing a dramatic transformation toward electrification. Based on current energy, climate, and industrial policy settings, every other car sold globally in 2035 is set to be electric. By 2030, almost one in three cars on the roads in China is projected to be electric, and almost one in five in both the United States and the European Union [3]. The proliferation of Electric Vehicles (EVs), without a proper strategy for managing the charging process, could increase the stress on the grid rather than presenting an opportunity for the development of decentralized energy storage [4]. Mobility habits and the associated incremental electricity demand from charging processes tend to overlap with existing demand peaks. However, if managed wisely, EV technology may represent a valuable asset for the network. Despite their impacts on power systems, EVs also offer great flexibility potential on the demand side, defined as a controlled power adjustment sustained for a required duration [5].

Flexibility from Demand Side Management (DSM) is crucial, with EVs fitting well into the spinning reserve and demand response categories due to their rapid response capabilities [6]. User mobility needs impose strict constraints that limit the potential of EVs to provide flexibility services for periods longer than a number of hours. To implement DSM techniques through EVs, it is necessary to alter the charging process. Authors in [7] describe the different charging schemes currently studied: while passive charging is performed at the maximum available power until the battery is fully charged, not allowing for any flexibility, smart charging involves power modulation within the range allowed by the charging point based on instructions from a controller. Demand shifting allows for smoothing demand peaks while maintaining a more uniform number of EVs performing charging processes throughout the day. The time flexibility of EV demand is numerically evaluated in [8] for a real-life study case: defined as the difference between the charging duration and the connection duration during a transaction, demand flexibility is evaluated for charging stations in the Lombok district of Utrecht, the Netherlands. The findings reveal that 59% of current demand is deferrable for over 8 h and 16% for more than 24 h, indicating a potential for aligning charging with next-day PV generation peaks. The effectiveness of a smart charging algorithm was tested in [9]: involving the largest household-based pilot project, smart charging use cases directly contribute to reducing charging costs and increasing the utilization of renewable energy. However, the impact of potential changes in the charging schedule on both infrastructure and market levels is not evaluated.

To determine the potential for grid services, it is essential to model mobility patterns, and consequently, EV users' needs. The top-down approach proposed in [10] offers a comprehensive analysis of the impacts and benefits of electric vehicle diffusion. While this method is extremely effective for drawing conclusions at a national level in terms of policy framework, its ability to isolate impacts at a sub-regional level from the general conclusions is limited. For this reason, the work in [11] exploits an open-source framework, MATSim [12], to model all trips that take place on a typical day. Synthetic drivers are generated starting from a population sample. From socio-demographic attributes, daily driving profiles are assigned to each individual to simulate traffic for the entire study area. Such an approach has great potential, since it leads to a synthetic database with detailed information on energy expenditures, state of charge, and time scheduling of each vehicle. Authors in [13] employ the same tool for a national-scale analysis of EV impact. Specifically, the emission reduction from electrifying the transportation sector in Croatia is assessed, mapping the mobility needs of the region and evaluating the impact of deeper RES penetration. However, relevant limitations include the large amount of specific data and calibration effort required to adapt the model to a specific case study. An alternative approach is proposed in [14], relying on mapping combustion engine vehicles onto electric vehicle equivalents and national traffic studies in Germany. Additionally, the embedded uncertainty was quantified using the Monte Carlo method. The same method is employed in [15] to deal with the uncertainty embedded in the EV charging load on each busbar of a test system. Specifically, a spatial–temporal model was developed to evaluate the impact of large-scale deployment of plug-in electric vehicles on urban distribution networks. The approach uses Origin–Destination analysis to reduce the uncertainties caused by EV mobility. The EV technical and market information provided by a European project were analyzed to obtain generalized EV characteristics, e.g., the correlation between battery capacity and the maximum travel range. Moreover, a peak shaving charging logic is implemented, constrained by the satisfaction of the energy demand of each user at the end of the modeled day. A similar survey-based approach is used in [7]. To represent the availability of vehicles most realistically, the authors directly integrate the results of a travel survey performed in Denmark into the model developed. This approach, on the one hand, guarantees an optimal representation of the current mobility context in an area, but on the other hand, it does not allow for any scenario evaluation. The study limits its evaluation to domestic charging, neglecting the impacts and potentials of daily charging in workplaces or other publicly available charging stations. Finally, the approach in [16] models a district distribution network to study the potential provision of stability from the growing presence of EVs while PV generation rises as well. The model adopts an optimization methodology to minimize the net load variance of the grid, obtaining a meaningful stabilizing effect as the EV share grows. A similar objective is pursued by authors in [17], who focused on the entire country of Sweden. They assessed the impact of EVs on the LV network, highlighting the urgency of reinforcement in larger urban centers to prevent voltage violations. Although the adopted methodologies can accurately simulate the interaction of EVs with a district network, they lack the accurate upstream modeling of mobility habits and the downstream quantification of the available flexibility band potentially offered by EVs throughout the day. These two elements are crucial to broadening the analysis to larger areas and estimating the economic value of such services in the ancillary services market.

As a result of these considerations, the present work introduces two key innovations. First, from a systemic perspective, a transferable methodology that leverages a physical, bottom-up traffic model, extensively detailed in [18], was developed. This methodology is novel in its ability to quantify the incremental impact of EV charging processes while simultaneously evaluating real-life, independent information. By integrating a self-calibrated, survey-based mobility model into the case study, an incremental demand profile that accounts for both temporal and spatial dynamics was generated, offering a comprehensive assessment of the impacts and needs. Second, flexibility-wise, an advanced logic that optimally designs EV participation in the ancillary services market by considering both user behavior and infrastructural constraints was introduced. This innovation results in the precise calculation of flexibility bands that can be offered in the market, considering users and infrastructural constraints. Additionally, these services were embedded within the current European market framework to evaluate their economic viability, ensuring compliance with existing regulations. The potential revenues from the Frequency Containment Reserve (FCR) and manual Frequency Restoration Reserve (mFRR) services are rigorously evaluated and compared, highlighting the financial benefits of our approach.

2. Materials and Method

The objective of this analysis is to evaluate the flexibility potential of a given region in offering grid services. The methodology is broken down into four hierarchical steps: first, the mobility needs of a given study area, imposed by the socio-economic framework, public transportation capillarity, and morphology are gathered and filtered (Section 2.1). Secondly, a spatial–temporal traffic model is built to evaluate the patterns and habits of the inhabitants (Section 2.2). Subsequently, the incremental demand and flexibility potential of EVs can be assessed (Section 2.3) and ancillary services economically evaluated, including the present market framework (Section 2.4).

2.1. Data Gathering and Preprocessing

The present methodology aims to build a transferable and extendable framework relying on open-source data. A pivotal input is the Origin–Destination (OD) matrix, a shared methodology used to quantify and illustrate the mobility needs of a given region. It consists of mobility records that typically include information on the origin, destination, and reason for each recorded trip. To fully exploit the relevance of the information contained in the OD matrix, the methodology incorporates a series of independent geo-informed datasets. Local administrative cells provide a suitable discretization of the study area, including data on the population and workers recorded in each cell. Additionally, relevant amenities such as parking spaces, schools, and universities are collected and aggregated based on administrative cells. In parallel, given the bottom-up approach, the road transportation network is imported. This involves utilizing a large-scale, real-world transportation network modeled as a graph using open-source data. The data includes geometric information representing road segments with coordinates. The roads are filtered and categorized into seven main groups, each defined by its nominal flow speed and capacity. Moreover, primary substations' conventional areas are included to assess the impact on the electric network.

2.2. Traffic Model

The bottom-up GIS-based approach, extensively presented in [18] and in this work, referred to as Traffic Model (TM), is employed in the present work to trace mobility patterns of a given region and estimate the impact of electrification of the light-duty transportation sector. First, a weighted graph, using the region's traffic network layout and traffic flow patterns, is simulated by considering factors such as traffic congestion and road network design. Dijkstra's algorithm is employed to find the fastest paths, given the current traffic condition, for these trips within the road graph.

To enhance the spatial resolution of the trips' origin and destination, a further sectionalization of the region is conducted, incorporating a gravity model using GIS-empowered probability density functions. Specifically, each section is characterized by detailed attributes such as population, number of workers, university presence, and amenities, which are used to build probability functions that estimate the start and end points of trips within the zones. It employs various influential factors for different trip types, such as work, study, leisure, and return home trips. For instance, sections with higher worker populations are more likely to be selected as destinations for work trips. The model calculates the probability of each section being the origin or destination by using discrete probability density functions, with parameters weighted according to their relevance to the trip purpose.

The OD matrix does not provide details about the fuel type of passenger cars used for traveling. Therefore, a method to differentiate between EV and thermal engine vehicles is required. To resolve this issue, a static penetration rate is defined, indicating that a certain percentage of trips are made with EVs. As a result, a corresponding proportion of trips from the travel survey is chosen to create an electric trip, a subset of the trips simulated in the traffic model set that reflects the defined penetration rate. Moreover, three main outputs from previous stages—arrival time, destination, and distance—are used to estimate the vehicle's energy demand. Each EV trip energy requirement is based on an average consumption rate, ACR (Equation (1)). The requested energy is aggregated into the power profile of the corresponding substation.

$$\Delta E^{consumed} = d \cdot ACR \tag{1}$$

Once the energy spent on each single trip is calculated, it is relevant to state that not every EV owner is willing to charge at the end of each travel. A probabilistic approach is employed here, assuming that the higher the energy consumption for the trip, the higher the probability of needing a charge. This means EVs completing longer trips are more likely to require charging compared to those that have only traveled short distances. Moreover, in longer trips, the energy required for charging is likely to match the energy used, while if a vehicle is selected to charge after a shorter trip, it is reasonable that the charging process will last longer than the energy expenditure for the last travel. This is summarized in Equation (2) through a scaling factor sf representing the relation between the consumed energy in the last travel, $\Delta E^{consumed}$, and the energy to be recharged ΔE^{EV}. Specifically, in the first case, the sf is set to 1, while it reaches the value of 10 for very short urban trips.

$$\Delta E^{EV} = sf \cdot \Delta E^{consumed} \tag{2}$$

As a result of all these steps, the output of the TM is a subset of electric trips in the study area and the corresponding energy demand, resulting in a spatial–temporal incremental demand profile.

2.3. Flexibility Evaluation Model

At this stage, each trip is routed on the road graph and associated with an origin, a destination, an arrival time, and the attribute of being electric, including any need to recharge a specified amount of energy. The incremental EV-related electric demand is now an attribute of each subsection of the study area and exhibits a temporal trend.

2.3.1. Temporal Margin for Flexibility

A pivotal consideration in developing the following section is that charging power modulation, shifting away from passive charging, should not impact the needs of the EV owner. In line with this consideration, each trip, associated with a purpose r, is connected to a characteristic stay time st^{EV} derived from a Gaussian distribution in Equation (3). The typical duration is modeled as a function of the travel typology (e.g., for working purposes μ_r is set to 8 h, while for trips returning home it is set to 11 h) while variability is set constant (0.5 h) for all the different travels.

$$st^{EV} \sim \mathcal{N}(\mu_r, \sigma^2) \tag{3}$$

For each travel, it is now possible to define three notable temporal quantities: the arrival time at^{EV}, derived by the traffic model, the end of charge eoc^{EV}, resulting from Equation (5) with a passive charging paradigm, and a departure time dt^{EV} obtained using Equation (6). A graphical description of the described quantities is reported in Figure 1.

$$ct^{EV} = \frac{\Delta E^{EV}}{p_{nom}} \tag{4}$$

$$eoc_{tvl} = at^{EV} + ct^{EV} \tag{5}$$

$$dt^{EV} = at^{EV} + st^{EV} \tag{6}$$

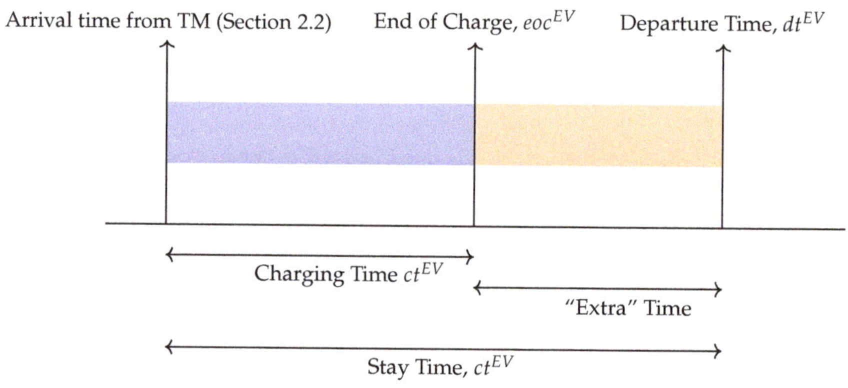

Figure 1. Graphical representation and temporal characterization of the charging process.

2.3.2. Charging Infrastructure Modelling

The model, capable of tracking traffic trends in a region, needs to be integrated with the charging infrastructure network. Despite defining a temporal flexibility margin for each vehicle based on user needs, the availability of a free charging point to connect the EV represents a necessary condition in the evaluation of ancillary service provision. Charging points availability may represent the bottleneck for a given penetration of EVs in terms of offerable bands, the range of power or energy that can be adjusted or controlled within a system to maintain grid stability, balance supply and demand, or optimize energy usage.

From a systemic perspective, the charging infrastructure is adapted to the local reality of the study area, considering the local charging needs. These points are then distributed at a greater granularity using local EV penetration, based on the concept that infrastructure maturity is driven by the charging demand of the population. Nominal characterization is completed with the nominal capacity of each charging point, specifying the nominal power $P_{nom,cs}$. Fast charging has been overlooked due to its still-limited adoption, and public infrastructure capabilities are split between 7.4 kW and 22 kW according to diffusion rate ($p_{7.4kW}$ and p_{22kW}). From an operational point of view, each charging point has attributes related to a time-varying state and power level. Specifically, two possible states have been defined (free and busy) for the attribute of availability S_t, while a numeric value, upper bounded by the nominal power of the charging point, is given for the attribute of charging power $P_{ch,t}$ at each time step.

The infrastructural layer of the charging points can be merged with the physical layer of traffic. EVs requesting charge are labeled and provided with their extensive energy needs. Given the finite number of charging points, a new attribute for each EV requiring charging is defined. Considering the entire stay time, a vehicle's state can range between waiting to be connected, connected, charging, and charged. At the end of each trip, an EV can be directly connected to the charging station, or must wait if all the area's spots are occupied. Once connected, the charging process can start immediately or, as extensively described in Section 2.3.3, be shifted within its stay time. For homeward trips, the states are limited to the last three, as it is assumed that vehicles requiring charging when returning home will always have infrastructure availability. The presented framework is entirely reported in Figure 2.

Systemic attributes		**Operational attributes**	**EV attributes**	
- Location - Nominal Power P^{nom}		- Charging Power P_t^{ch} - State S_t^{cp}	- Stay time st^{EV} - State S_t^{EV}	- Reason of the trip r^{EV} - Energy Request ΔE^{EV}

Figure 2. Charging infrastructure and EVs operational framework.

2.3.3. Uncontrolled Charging Power Demand

Given the operational framework described in the previous paragraph, the incremental eMobility-related demand profile can be constructed, resulting from the interaction of the TM and infrastructural layers. To mitigate inaccuracies introduced by the absence of an updated, geoinformed database of public charging points, the capability limit of the infrastructure is evaluated at a province level. In other words, the arrivals of cars, monitored at a high granularity deriving from the TM, are collected and continuously compared with the occupancy level of the charging points in a broader area. Charging will start as soon as the travel ends, employing Algorithm 1 to simulate the process.

Algorithm 1 Pseudo-code for charging process modelization

```
Assign the charging power of the given charging point;
Compute the charging time needed to fulfill the energy request;
Compute the charging end time frame;
Apply a mask to identify time frames where the trip is charging;
Populate the time-varying vector P_t^ch.
```

According to the purpose of the travel, as discussed in Section 2.3.2, the charging power is either stochastically determined with a given probability or is a fixed value considering domestic charging infrastructure. The charging power is held constant throughout the process, overlooking the dependence between the power set point and the SoC managed by the battery management system.

To address the issue of a trip ending in a region where charging capacity is already saturated, a logical solution that aligns with real-world practices was implemented. The proposed solution relies on the idea that if a charging point is occupied upon the vehicle's arrival, the charging process does not commence immediately. Most Mobility Service Providers (MSPs) offer applications allowing users to book charging points. These apps provide information on when a charging spot will be available, enabling users to plan their charging accordingly. The chosen approach delays the start of the charging process until a charging spot in the same area becomes free. This ensures that the vehicle can begin charging once a spot is available without unnecessary waiting. It is crucial to ensure that the delay in starting the charging process does not exceed the departure time dt^{EV}, calculated as in Equation (6).

Subsequently, once the charging infrastructure has been characterized, the end of charge eoc^{EV} can be easily computed and compared with dt^{EV}. Since these are independent processes, the charging process may be not completed due to a short stay time and long driving distance. These marginal instances will be labeled as originally incomplete and, with an a posteriori procedure, the two time limits are forced to coincide. Employing the described procedure, the operational attribute, charging power, of each charging point is characterized at each time step. The time-varying passive load profile of a given region, resulting from the subset of charging station C for a given instant t, can be calculated as in Equation (7).

$$P_t^{uncontrolled} = \sum_{c=0}^{C} P_t^{ch} \tag{7}$$

2.3.4. Enhanced Band Development Procedure

EVs can contribute to provisioning grid services, typically high-power, characterized by relevant power interaction for a limited amount of time [19]. To significantly impact grid stability, it is essential to offer a uniform, non-negligible flexibility service throughout the day. To achieve this objective, a novel procedure was implemented to enhance the band available to the aggregator for market bidding. The procedure focuses on the number of available vehicles for providing flexibility services during the most critical hours of the day. The core idea is to delay charging sessions that are scheduled to start during periods when the load profile exceeds its average, thereby filling the valleys and aiming for a more uniform load profile. A more uniform demand profile directly impacts the flexibility provision, reducing the offerable band volatility.

Each time a modification to the charging start time is implemented, two pivotal constraints must be considered based on the type of travel being processed. First, to keep the load-shifting procedure neutral on the user side, energy requirements deriving from the TM must be respected. In other words, the charging, despite being delayed, must terminate within the departure time dt^{EV}. Secondly, flexibility services must not impact the profitability of charging spots limiting the charged energy. In other words, postponing a charging process is acceptable as long as no other vehicles are waiting to charge in the area. Practically, for "Return Home" travels, there are no restrictions related to the availability of the charging spot. While, for the other types of travel, it is necessary to check for any reservations at the charging point in the time windows immediately following the end of the charging process. The entire procedure is summarized in Algorithm 2.

Algorithm 2 FCR–Band Enhancement Procedure

```
for iteration in Iterations do
    Find and select all time windows when the load profile is above its average
    for time window in Time Windows do
        for travel in Travels do
            if travel reason r is Return Home then
                Check if the travel can fulfill its energy need even starting its charging
                process one-time window later (charging end does not coincide with charging limit).
                If yes, update the power list P_t^ch postponing the charging.
            else
                Check the same energy constraint and that there is at least one free time
                frame after charging ends in the availability list of the associated charging point. If yes,
                update the power list P_t^ch postponing the charging.
            end if
        end for
    end for
end for
```

Concerning Figure 1, the temporal configuration of the charging session has become more complex. The arrival time, denoted as at^{EV}, no longer directly corresponds to the charging start time, cs^{EV}, as presented in Section 2.3.3. Even when a vehicle is connected to the charging infrastructure, the start of the charging process can be delayed to smooth the incremental power demand. Extreme conditions of these charging delays are illustrated in Figure 3. Essentially, a charging process can be postponed between two static user-related temporal boundaries, at^{EV} and dt^{EV}, as long as the charging can still be completed within this window. Additionally, the intermittency of an ongoing charging profile and partial load charging has not been considered in the current charging logic.

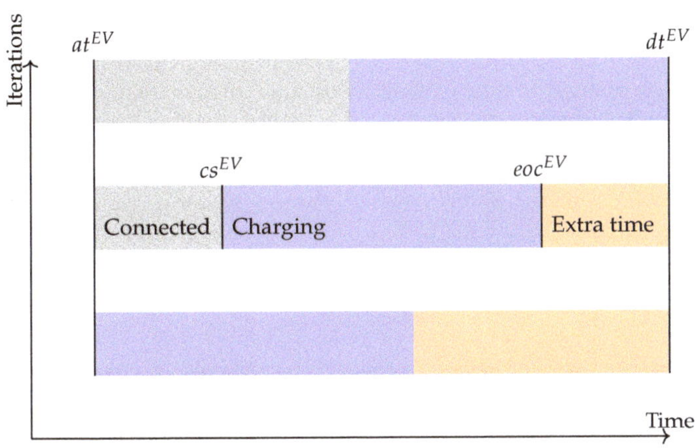

Figure 3. Explanatory application of load shifting algorithm.

2.4. Ancillary Services Modelling

To effectively harness EVs for grid integration, a comprehensive grasp of the regulatory landscape governing their participation in the ancillary services market (ASM) is essential. ASM is the venue where Terna S.p.A., the Italian TSO, procures the resources that it requires for managing and monitoring the system relief of intra-zonal congestions, creating energy reserve and real-time balancing [20]. EVs' aggregations are designated as "Unità Virtuali Abilitate Miste" (UVAMs). Terna defines UVAMs as entities comprising various components such as production units, storage systems, and consumption units. Once qualified, UVAMs must navigate specific procedures for ASM participation. Responsibilities include executing dispatch orders and submitting a daily baseline, with allowances for modification as per guidelines stipulated in the Grid Code [21]. Furthermore, the UVAM manager must communicate power distribution among UVAM points on a quarterly basis and provide structured offers for reserve allocation. These offers encompass quantity–price pairs tailored for the ASM programming phase and the balancing market, ensuring each offer meets or exceeds a threshold of 1 MW in absolute value.

2.4.1. Frequency Containment Reserve

FCR services help to slow and stabilize frequency changes through rapid, automatic adjustments in generator output. These services can be provided by various types of generators, potentially including an aggregate of EVs. Primary reserve is always allocated proactively (well in advance of real-time) through specific auctions in which operators bid for their availability to provide primary regulation reserve during a certain period in exchange for payment [20]. Resources are assumed to provide frequency response based on a droop curve with a deadband, as described in Equation (8) and referenced in [22].

$$K_p = \frac{\frac{\Delta f}{f_{nom}}}{\frac{\Delta P_e}{P_{eff}}} \quad (8)$$

where ΔP_e is the power set point of the frequency regulation, $\frac{\Delta f}{f_{nom}}$ the frequency deviation in p.u., and P_{eff} is the BESS nominal power.

According to their resource capabilities and technical characteristics, ancillary services can offer flexibility in one or both directions. Specifically, within the framework of FCR, a resource can increase its power output if the measured grid frequency exceeds the upper bound or reduce it if the frequency falls below a threshold. Services that operate in only one direction are referred to as one-direction services, while those that operate in both directions are called both-direction services. Given that charging is modeled as occurring at the nominal power of the charging point, a one-direction service that decreases charging power is most suitable for EV charging. However, for this study, a both-direction service is designed. To ensure the capability limits of charging points are not exceeded, the undisturbed charging power was fixed to 90% of P^{nom}. The presented framework regarding charging point capability dependence on the case study is reported in Figure 4.

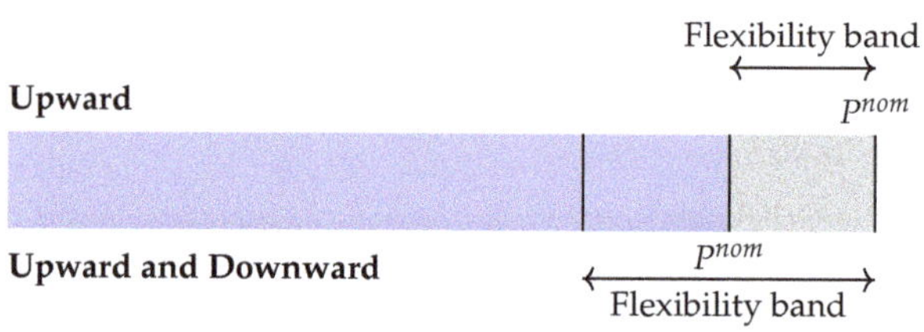

Figure 4. Flexibility margins of charging points.

2.4.2. Restoration Reserve

FCR services act to slow and arrest changes in frequency and are not limited to primary frequency reserve. Specifically, to restore the margin used, TSO can call single producers. This, non-automatic restoration service falls under the name of mFRR. In Italy, Terna allows each qualified unit to participate in the market, offering couples energy–price restorations for each hour of those between each session of ASM. The energy bid on the market corresponds to a reduction or increase in the power output of the plant. In the case of the study, EVs' aggregate may be selected for mFRR in certain hours, the energy uptake must be reduced or increased to a certain value and the new set point reached must be kept for the whole hour duration. The longer time scale of the required variations for this service makes it more challenging to keep the constraints regarding the full charge of single vehicles satisfied. mFRR modelization can be broken down into three subsequent, hierarchically ordered steps, illustrated in Figure 5.

Firstly, as in FCR, mobility needs represent the bottleneck concerning the amount of energy the aggregator can offer on the market. In line with Section 2.4.1, the principles of neutrality on the user side in terms of charging duration and infrastructure availability also guide the modelization of this service. The flexibility margin calculation is case-sensitive, depending on the direction, upward or downward, of the service. In the first case, given an expected charging schedule for each charging vehicle, the quantity Δ, the maximum admitted charging power necessary to supply the required energy within the temporal boundaries, is calculated. As anticipated, a necessary condition for each vehicle to contribute to the offerable band is the infrastructure availability, i.e., the absence of vehicles waiting for the charging point to be free. If this condition is satisfied, the vehicle that has a sufficiently long stay time to complete the charging at a lower charging power, will offer a band bnd^{EV} resulting from Equation (9). On the other hand, when increasing the charging power for stability purposes, the bottleneck is represented by the battery capacity: Δ is in this case the maximum increment that can be imposed on the charging power given the energy demand of the vehicle. The offerable band is again calculated, employing Equation (9). The two calculation mechanisms and the original charging profile are reported in Figure 6. Aggregating charging stations, the available bands are calculated as in Equation (10).

$$bnd_t^{EV} = \min(\Delta, 0.1 \cdot P^{nom}) \quad (9)$$

$$bnd_t^{tertiary} = \sum_{c=0}^{C} bnd_t^{EV} \quad (10)$$

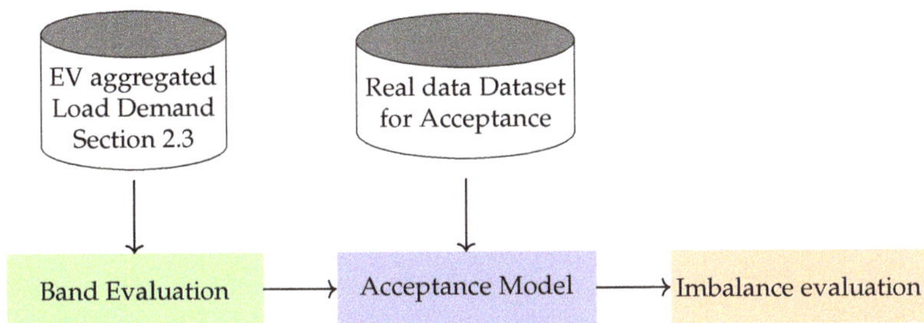

Figure 5. mFRR evaluation procedure.

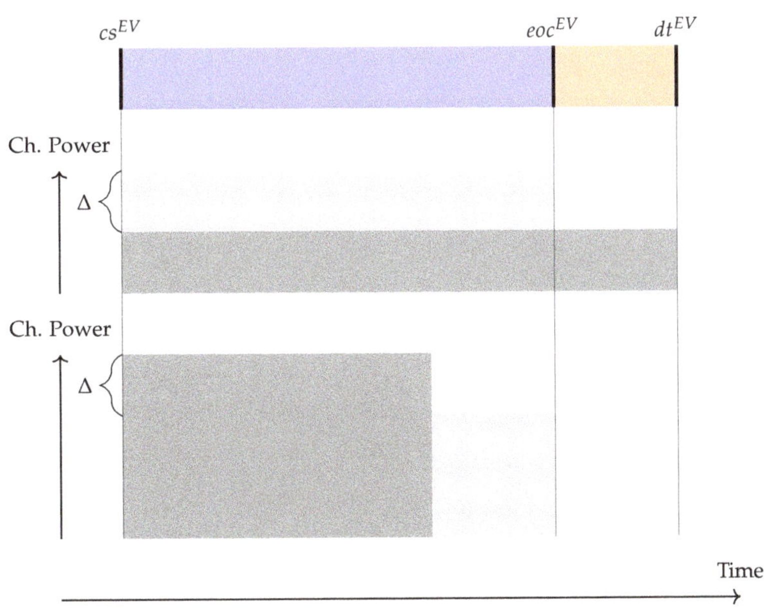

Figure 6. Alternative charging processes for tertiary reserve. The original charging schedule is indicated in light gray, with dark gray for the resulting charging strategy.

As anticipated, while in FCR, set-point variations are automatically triggered by frequency deviation; for mFRR, services are required from the TSO. Understanding recurrences and the criteria for acceptance employed is pivotal to model this service. For this reason, a data-based acceptance model was developed to evaluate the effectiveness of participating in the market as an EV aggregator. This acceptance model relies on an acceptance probability matrix $AP^{d}_{p,b,h}$ with $d \in D$, where D is the set containing the two possible directions of the energy flows (i.e., injecting or absorbing energy). Moreover, as evident in the matrix definition, the acceptance probability is modeled as a function of the offered power (i.e., the energy offered over a time window of one hour can be translated in power terms) and the price [23,24]. The index h represents the temporal coordinate: a different matrix is available for each time step of the simulation. Note that, concerning the temporal granularity set by the traffic model, the same matrix is repeatedly employed over the same hour. Applying a random selection process, according to the case study definition, given the hour of the day and the flexibility band for tertiary reserve deriving from Equation (10), it is possible to derive the acceptance probability, ap, from the TSO to provide the service.

The last step of the procedure involves the load profile and the imbalance evaluation. To satisfy the request from the TSO, the aggregate needs to generate a deviation from the set point equal in power, keeping this new condition for the whole hour in which it is called to provide the service. On the other hand, to complete the charging process, energy deviation must be counterbalanced in the following time steps. At this point, additional complexity is added to the methodology: as with any other participant in the market, aggregators must submit their net consumption in advance, and, although not always detrimental, the impact of imbalances must be included in the economic assessment. The impact is strongly dependent on the "zonal imbalance" which corresponds to the total deviation computed as the sum of all the imbalances of the single production and consumption units belonging to a specific zone. If the imbalance of the single unit has the same sign of the zonal one, it means the plant is exacerbating the existing deviation. Conversely, if the sign of the imbalance of the unit is opposed to the zonal one, the unit is relieving the local imbalance. The TSO can adopt two different approaches when dealing with imbalances: the single pricing (SP) or the dual pricing (DP) algorithm, presented in Table 1.

Table 1. SP and DP remuneration logics.

		Unit Imbalance [+]	Unit Imbalance [−]
Zonal Imbalance [+]	SP	Receives: min (avg ASM upward; DAM)	Pays: min (avg ASM upward; DAM)
	DP	Receives: min (min ASM upward; DAM)	Pays: DAM
Zonal Imbalance [−]	SP	Receives: max (avg ASM downward; DAM)	Pays: max(avg ASM downward; DAM)
	DP	Receives: DAM	Pays: max(max ASM downward; DAM)

2.5. Economic Feasibility Analysis

To conduct a feasibility assessment of the economic value of the flexibility service potential for an area, an economic analysis is performed to estimate the potential revenues generated by the proposed services. Economical values should be assigned to each evaluated service to provide an overall estimation of the possible revenues for the study area. In the current Italian regulatory framework, no remuneration is provided to plants that contribute to primary frequency control, as it is conceived as a mandatory service for relevant plants. Despite this condition, the capacity market in Central Europe offers the possibility to perform a preliminary economic assessment. The market model in place remunerates band availability with a temporal frequency of four hours. Moreover, the existing regulatory framework mandates the service to be offered in both directions. Thus, numerically, the offerable band must be computed as the minimum band that operators can safely offer on the market:

$$bnd_{mw} = \min(\min(\text{Upward Band}), \min(\text{Downward Band})) \tag{11}$$

where mw is one of the six four-hour market windows.

In the remuneration of mFRR, the price deviation from DAM plays a crucial role, since it not only influences the margin for the aggregator but also has an impact on the acceptance probability.

3. Case Study and Input Data

Numerical assumptions, input profiles, and scenario characterization are presented in this section. Firstly, the OD matrix, which maps the mobility needs of a specific area, is pivotal. Data for this case study are provided by Regione Lombardia in [25]. The heatmap in Figure 7 captures the large-scale pattern on a regional scale. The number of trips is represented on a logarithmic scale to highlight non-diagonal elements. Similarly, Figure 8 illustrates the temporal distribution of recorded events, highlighting the varying patterns associated with different travel purposes. Work-related travels are predominantly concentrated in the early morning hours, reflecting typical commuting times. Similarly, study-related travels follow a similar early morning peak, aligning with school and university schedules. In contrast, leisure and return home travels exhibit a different pattern, with a pronounced peak in the afternoon and more temporal dispersion. This suggests that these types of travel are more flexible and spread out across the day, unlike the more regimented schedules of work and study travel. Numerical assumptions for μ_r in Equation (3), which pertains to the average, reason-dependent stay time for each travel instance, are detailed in Table 2.

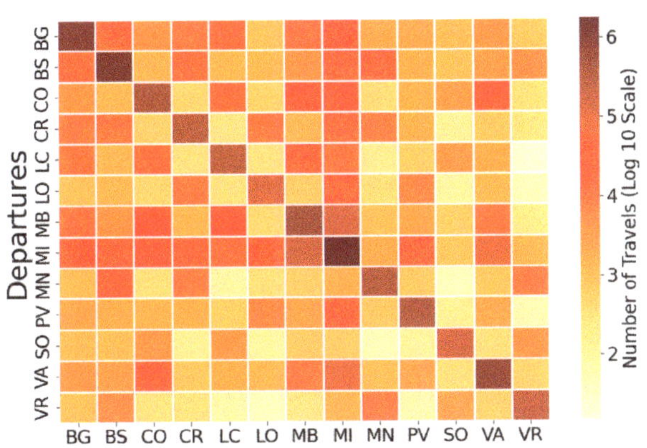

Figure 7. Mobility flows representation on a province level.

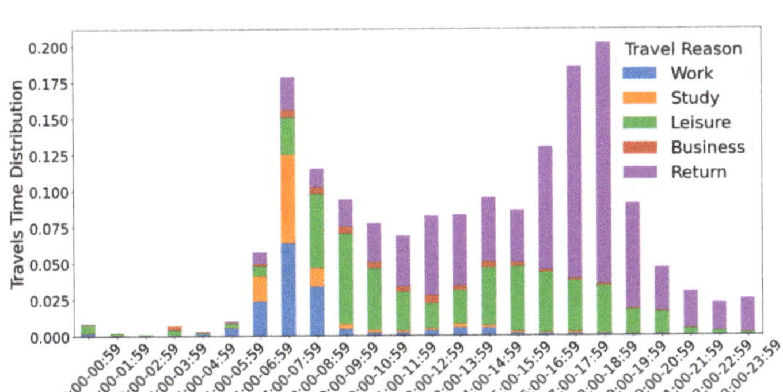

Figure 8. Temporal distribution of the OD travels during a typical working day.

Table 2. Average stay time as a function of travel's purpose.

Staytimes μ_r			
Work	Study	Return Home	Leisure
8 h	6 h	11 h	4 h

Secondly, information on the current electrification level of the light-duty transportation sector is taken from [26]. Although the current level of EV penetration in Lombardy is below 1%, a 2.5% penetration scenario is assumed to be a reliable depiction of a likely near-future condition. Additionally, EVs are geographically distributed, as shown in Table 3, which reports the current distribution in the study region. This distribution is assumed to be the result of the combined effects of morphology, economic factors, and local mobility habits, and is therefore considered invariant over time. Infrastructure-wise, the number of public charging points was set to 15,000; possible nominal power outputs are 7.4 kW and 22 kW, distributed, respectively, with a share of 80/20% [27]. Private charging was assumed to be performed uniquely at 7.4 kW nominal power. A travel-independent ACR equal to 0.2 kWh/km was selected.

In parallel, two relevant profiles are needed to evaluate the ancillary service potential of the study area numerically: first of all, a real-life frequency record is imported into the framework. To fit within the temporal granularity of the mobility model, the worst measurement (i.e., widest deviation) measured over a time window of five minutes, the temporal discretization time step of the TM, was considered. TSO's intervention was assumed to be triggered whenever the frequency value deviates more than 20 mHz from the nominal frequency of 50 Hz. A graphical representation of the considered profile is offered in Figure 9. Economically-wise, as anticipated in Section 2.5, the capacity market in central Europe, the TSOs of those areas of 50 Hertz for Germany, and Tennet for the Netherlands, is taken as a reference [28]. Prices are reported in €/MW and are constant for a given time window.

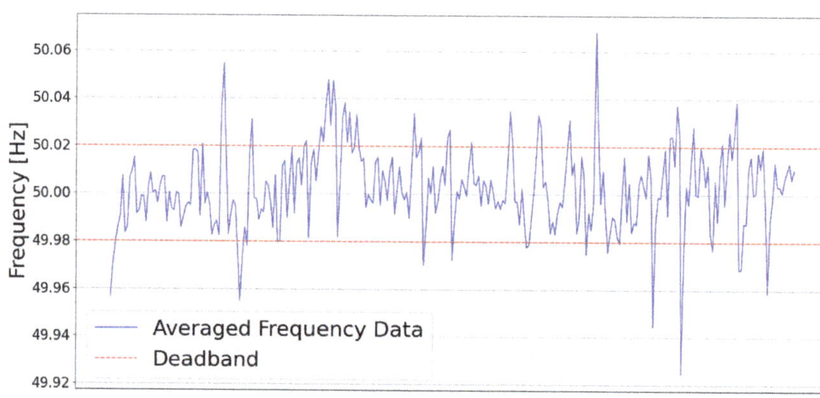

Figure 9. Daily averaged frequency profile.

On the other hand, concerning mFRR, the database of calls by Terna is included in the analysis. Probabilities, derived from historical records and employed in this work, are represented in Figure 10. Finally, the imbalances' economic evaluation relies also on the record for DAM and ASM prices from [29].

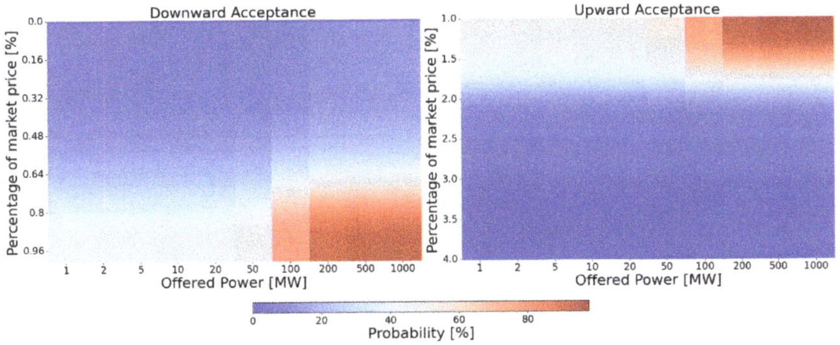

Figure 10. Acceptance probability for mFRR.

Table 3. Probability of assigning an electric travel to each province in Lombardy.

BG	BS	CO	CR	LC	LO	MN	MI	MB	PV	SO	VA
11.4%	14.4%	7.4%	3.1%	3.8%	1.6%	3.2%	38%	8.9%	3%	1.7%	3.4%

4. Results and Discussion

The objective of this procedure is to numerically evaluate the ancillary service potential for a given EV penetration scenario in a specified study area. The chosen case study is the Lombardy region, the most populated and economically significant area in northern Italy. In this section, two different use cases are discussed and compared. Specifically, the additional demand in Simulation 1 is constructed using uncontrolled charging, while Simulation 2 employs the band enhancement algorithm.

4.1. Traffic Model Output and Uncontrolled Charging Profiles

The most relevant output of the TM is the spatial–temporal mapping of each recorded travel of the OD matrix. The two coordinates are depicted in Figure 11, illustrating the additional demand imposed by EV charging on equivalent primary substation areas. The incremental load is not uniform over time due to drivers' habits. Uncontrolled charging reflects this new class of load, representing the mobility needs of EV users. Specifically, peak demand occurs in the late afternoon, induced by return home travels and coinciding with high stress already in place on the electric network. The impact is not equally distributed among the areas. Highly urbanized locations experience the greatest impact due to the combination of population density and economic factors. However, a single conclusion cannot be drawn for semi-urban or rural areas. For certain areas, local mobility, characterized by higher average travel ranges due to nearby commercial or industrial areas, is more impacted compared to localized communities with lower mobility indices.

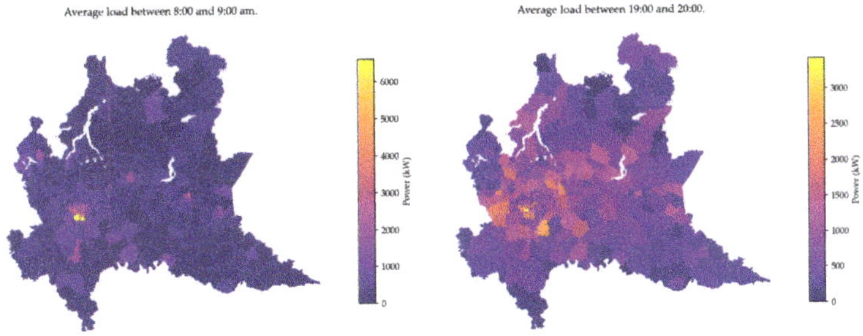

Figure 11. Incremental EV-related demand aggregated on a primary substation equivalent area.

A similar conclusion can be partially derived from Figure 12, which represents the power delivered by public chargers in each province. Although this figure primarily captures work-related demand, as home charging typically occurs on private wallboxes, it is evident that local characteristics significantly influence power demand. Furthermore, given the independent modeling of charging infrastructure capacity based on EV penetration and the mobility needs of each area, the infrastructure's capacity becomes saturated, as illustrated by the province of Brescia (BS) in orange.

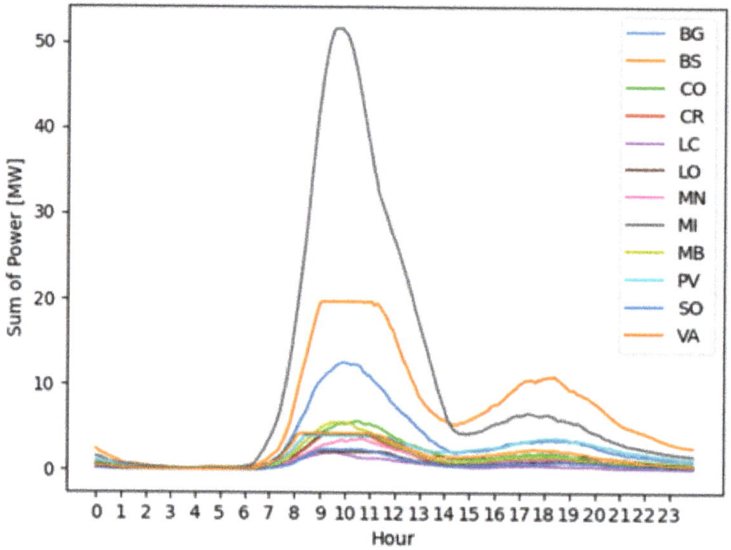

Figure 12. Incremental EV-related load due to public charging processes on a province scale.

An aggregated result for the total demand is reported in Figure 13. Aggregating on a spatial level, the incremental demand of the entire study area is reported in red, split into public charging and home charging. Guided by population habits, the profile shows two peaks: one in the morning between 9 a.m. and 12 a.m. that is mostly attributable to morning travels for work reasons that start charging once arrived at the workplace, and a second, which is located around 8 p.m. and is instead related mostly to return home trips that exploit domestic charging to fulfill their energy needs.

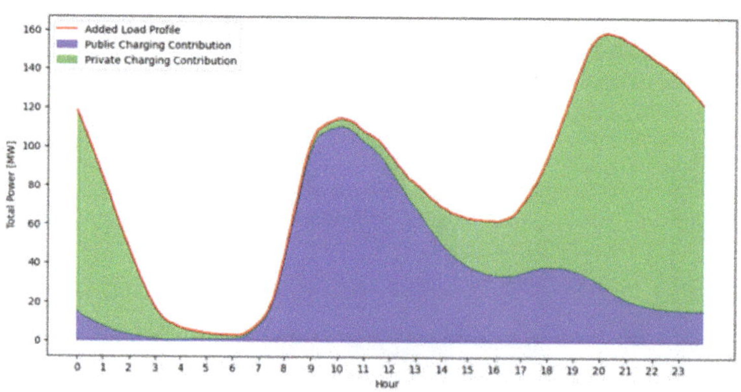

Figure 13. Incremental EV-related load profile with contributions from public and domestic charging.

The additional demand, combined with the stay time of each vehicle, allows for the computation of ancillary service bands using the routines described in Section 2.4. From a flexibility perspective, the available band is a time-varying attribute, dependent on the number of connected EVs and the drivers' needs. For each time interval, the power deviation that the aggregator can handle while satisfying all constraints is reported in Figure 14. Considering the factors mentioned above, it can be concluded that the difference between the blue and red curves is due to the significance of the second constraint over the first. In other words, it can be observed that the available band in the "Increase Charge Direction" directly follows the corresponding load profile trend, as the number of connected EVs is the primary bottleneck, and higher charging powers simply reduce the charging duration. Conversely, delaying full recharges beyond a certain threshold can be unacceptable from both user and infrastructural perspectives. Consequently, the band difference is most pronounced in the middle of the day, when public charging is predominant. The additional limitation, imposed in energy terms, present for tertiary reserve service is evident when comparing the blue lines of Figures 14 and 15: being more energy intensive, the offerable upward band is greatly restricted due to the lower stay time and finite infrastructural capacity.

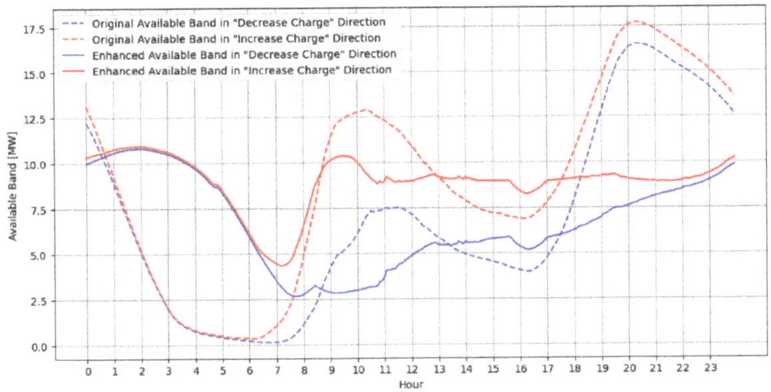

Figure 14. Available bands for FCR.

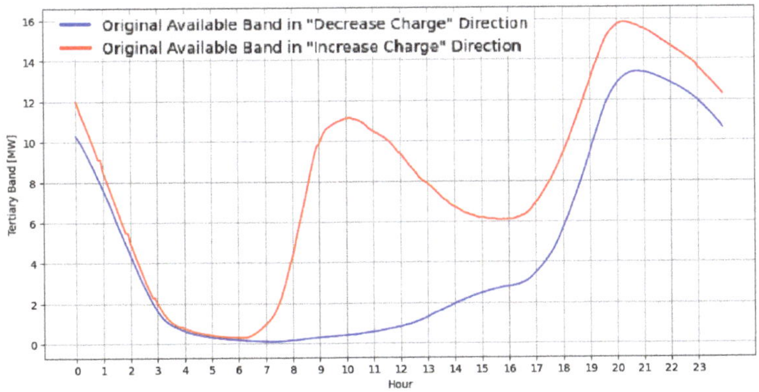

Figure 15. Available bands for mFRR.

4.2. Band Enhancement Impact

Dealing with a single class of consumers in a power network can significantly hinder flexibility and operational efficiency. This is primarily because a homogeneous consumer tends to exhibit similar consumption patterns, leading to periods of synchronized high demand.

The application of Algorithm 2 on the power demand derived from uncontrolled charging logic in Figure 13 results in a more regular profile with reduced valley-to-peak difference. In Figure 16, the evening peak is easily shifted over the night, exploiting the wider stay time of return home travels. Numerically, the global peak is reduced by 37.9%. On the other hand, the work-related peak demand is only partially mitigated due to the intrinsic limitation of public charging infrastructure. The benefits are not limited to mitigation of grid congestion, as decoupling the power provision from the user habits allows us to enlarge the effectiveness of ancillary services. Specifically, bands for primary frequency control are more regular, showing less variability and better fitting in the four-hour window market framework. Employing Equation (11), the offerable band on the market is highly affected by the demand shifting, as is visible in Table 4. Tertiary bands are also enlarged, though to a limited extent due to stricter constraints. The impact of this approach is also evident from a spatial perspective: visually, the absence (Figure 17) and the very localized band availability (Figure 18) are not only spread over time but also diffuse across space, enabling a more homogeneous service.

Table 4. Flexibility bands in MW considering the two strategies.

Time Slot [h]	0–4	4–8	8–12	12–16	16–20	20–24
Uncontrolled charging	0.77	0.13	1.08	4.07	3.95	12.67
Band Enhancement	9.78	2.66	2.75	4.84	5.15	7.58

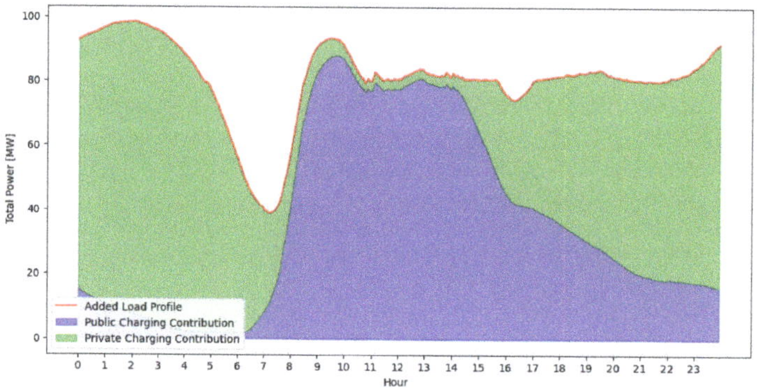

Figure 16. Aggregated demand after the application of band enhancement algorithm.

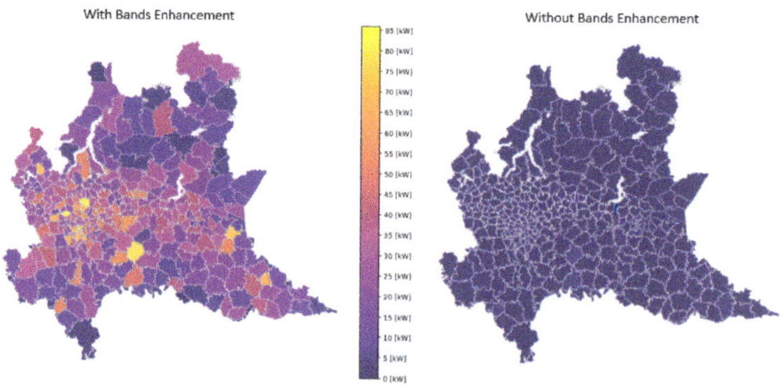

Figure 17. Spatial distribution comparison at 5 AM of FCR band.

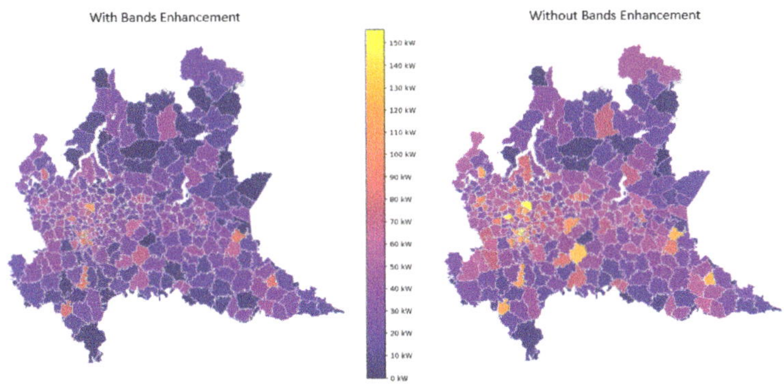

Figure 18. Spatial distribution comparison at 8 PM of FCR band.

4.3. Economic Evaluation

To assess the potential revenues generated by the modeled services, an economic analysis is performed. Economical values are associated with each of the proposed services to gain an estimation of possible revenues, determining technology maturity and competitiveness.

Combining the input dataset presented in Section 3, the economic comparison of the possible alternatives is reported in Figure 19.

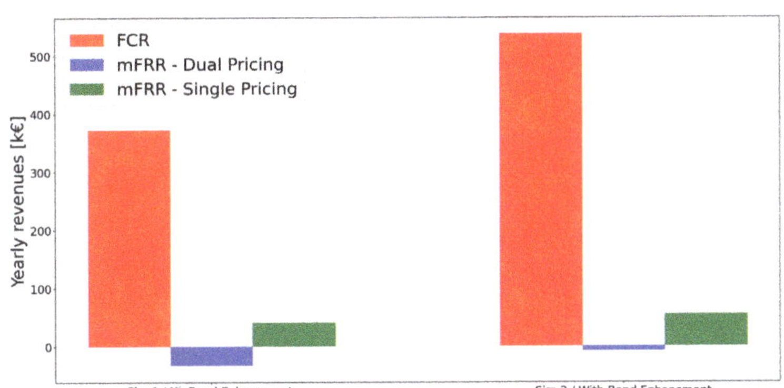

Figure 19. Aggregated demand after the application of band enhancement algorithm.

Multiple considerations are possible: firstly, primary frequency regulation, a power-intensive service, represents the most relevant source of value imposing fewer constraints on the charging processes. In contrast, given the simulated level of penetration and the modeling of the charging infrastructure, energy-related services such as tertiary reserves are less suitable. Secondly, the effectiveness of the band enhancement algorithm is testified by a better economic result: the offerable band in the six market window is greater, and the same improvements are testified by better results in tertiary revenues. Finally, the dual pricing paradigm resulted in far more of a penalization for the evaluation of imbalances, entirely counterbalancing the revenues deriving from the mFRR. Regardless of the simulation number, and considering the total number of vehicles contributing to the economics in Figure 19, the direct marginal economic benefit for each EV owner is limited to below EUR 10/year, which, alone, is insufficient to justify massive participation. However, it is worth noticing that this represents an average value, shared among all the different resources (i.e., EV owners) regardless of the flexibility offered. In other words, the yearly revenues may increase for a subclass of users (i.e., domestic users charging during the evening) better fitting their charging profile and the network infrastructure necessities.

4.4. Policy Implications

Policy-wise, ensuring consistent bandwidth availability throughout the year for flexibility, whether on a 24-h basis or for specific 4-h periods, is essential for effective remuneration. Additionally, while providing greater bandwidth during peak periods (e.g., 17–20) may yield higher ASM prices, it is important to prevent strategic behaviors that could disrupt system stability. Therefore, it is recommended to promote or mandate Time-of-Use (ToU) tariffs for Charging Point Operators (CPOs). Such measures would incentivize better load management and naturally smooth out peak demands, particularly during critical system hours, ensuring a more stable and efficient energy distribution. As an example, the energy cost in two possible electricity bill schemes, namely fixed vs. ToU, and ASM prices for mFRR, as per German 2023 data, is reported in Table 5 [28].

Table 5. Capacity prices for mFRR in Germany in 2023, the data source comes from Ref [28].

Time Slot [h]	0–4	4–8	8–12	12–16	16–20	20–24
Average capacity prices [€/MW/h]	0.46	1.14	3.48	1.52	5.00	3.41

The proposed opportunity value in EUR/MWh is the net value of withdrawing energy and then selling upward flexibility on ASM, considering a fixed award rate for ASM bids (i.e., 20%) and the actual mileage for the mFRR (the activated MWh per awarded MW) in each time slot. Only variable costs are considered, since other costs (e.g., grid costs) are invariant in the proposed analysis. Assuming fixed the total consumption throughout a day, a positive opportunity value in an hour indicates that it is worth increasing the consumption in that hour, to increase available upward flexibility. As can be seen in Figure 20, in the case of a fixed consumer's bill cost, the opportunity value is positive during the evening peak (i.e., 17–21), highlighting a push towards consuming in this period. In contrast, with a well-designed ToU tariff, there is significant convenience in consuming during daylight (i.e., 13–16) and during the night (i.e., 0–4), shaving the peak during the evening.

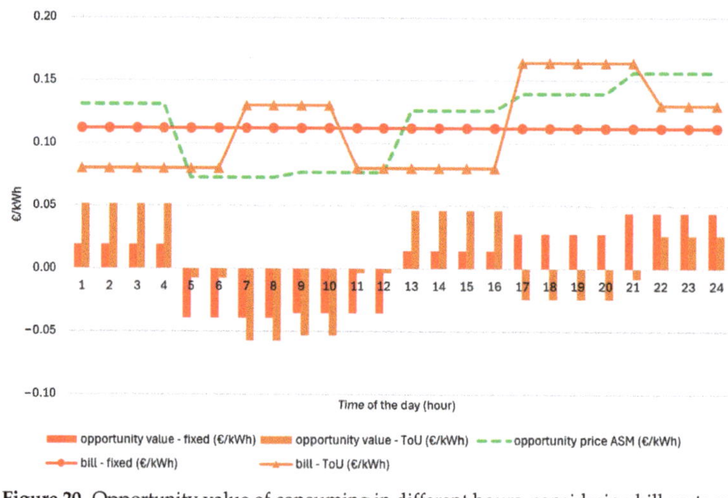

Figure 20. Opportunity value of consuming in different hours, considering bill costs and provision of upward service on ASM.

Secondly, on the market side, single price will definitively be the method used in the future, and is already in use at present in the EU. This approach, elected as the most cost-reflective, also proves itself effective for flexibility provision, as evidenced by the fact that valuable tertiary resources yield a positive net value only under the SP model. In contrast, under the DP model, it is not economically viable to make these resources available. Therefore, adopting and maintaining a SP model is crucial to ensure the efficient utilization and availability of tertiary resources, leading to a more stable and profitable energy market.

Moreover, the band enhancement strategy offers significant benefits to the system by reducing peak demand and peak withdrawal. This can be seen when comparing Figures 13 and 16, roughly halving withdrawal during the evening peak. Nonetheless, if the downward service is provided during peak hours, the withdrawal could increase despite the efforts of the resource. Therefore, it could be beneficial to apply network costs exclusively to charging withdrawals, exempting any dispatching services provided. This is already applied or under discussion in many regulatory frameworks. Additionally, implementing non-firm connections—where withdrawal power is not guaranteed in all time slots—can be a cost-effective solution. These connections should be priced lower, with a specific valuation for any reduction in power during critical hours, to further incentivize peak demand management and enhance overall system efficiency.

As a result of these considerations, indirect economic benefits may contribute to the limited direct profitability deriving from the service provision, increasing the marginal benefit for the EV owner.

5. Conclusions

In an era where the integration of renewable energy sources is accelerating, the increased demand for ancillary services highlights the strategic importance of efficiently managing EV charging processes. The proposed methodology seamlessly combines the accuracy and realism of bottom-up models with the estimated ancillary services potential for wider areas. Additionally, a GIS-based bottom-up methodology has proven effective in assessing the combined impact of mobility and the electric network using sparse information, resulting in outputs that are easy to interpret and apply. Uncontrolled charging limitations from a network perspective are identified through a direct comparison with a novel enhanced charging algorithm. This algorithm mitigates network stresses, by 37% in terms of global peak, while respecting user mobility needs and charging station availability.

In this framework, FCR and mFRR are configured in simulations considering the updated regulatory framework. Starting from the expected power demand, offerable bands are numerically calculated for each time step of the simulations and economically evaluated. FCR was more competitive with respect to mFRR due to the less stringent constraints, higher offerable bands, and a more suitable market framework. This study highlights that the current selection logic for mFRR disproportionately penalizes the resources with limited power capabilities, such as aggregates of EVs. It suggests establishing a framework that facilitates distributed storage resources when technically feasible, as they can offer economically competitive prices with no marginal cost for providing this service. From an MSP's perspective, the ancillary service potential of a specific subarea can serve as a strategic indicator for planning mobility operators' investments. This geo-informed approach can identify areas with high mobility flows but limited charging capacity, helping to prioritize investments based on spatial needs.

Market-wise, results on remuneration appear to be too modest to solely motivate an EV owner's participation in ancillary services, especially given the range anxiety of users and battery degradation. However, for a certain subclass of users, capable of offering flexibility in specific hours, the economic result may be different. In general, to harness the EVs potential as a distributed storage resource, a dedicated regulatory framework should be implemented to norm EVs' participation in ancillary service provision.

Author Contributions: Conceptualization, C.M.C., L.G.B., M.S., G.R. and M.M.; methodology, C.M.C., L.G.B. and M.S.; software, C.M.C., L.G.B. and M.S.; validation, C.M.C., L.G.B. and M.S.; resources, C.M.C., L.G.B., M.S., G.R. and M.M.; data curation, C.M.C., L.G.B. and M.S.; writing—original draft preparation, C.M.C., L.G.B. and G.R.; writing—review and editing, C.M.C., L.G.B., M.S., G.R. and M.M.; visualization, C.M.C.; supervision, G.R. and M.M.; project administration, G.R. and M.M. All authors have read and agreed to the published version of the manuscript.

Funding: This research received no external funding.

Data Availability Statement: The original contributions presented in the study are included in the article, further inquiries can be directed to the corresponding author.

Conflicts of Interest: The authors declare no conflicts of interest.

Abbreviations

The following abbreviations are used in this manuscript:

ASM	Ancillary Service Market
DAM	Day-Ahead Market
DP	Dual Pricing
DSM	Demand Side Management
EV	Electric Vehicle
FCR	Frequency Containment Reserve
mFRR	manual Frequency Restoration Reserve
MSP	Mobility Service Provider
OD	Origin–Destination
PV	Photovoltaic
SOC	State Of Charge
SP	Single Pricing
TM	Traffic Model
TOU	Time Of Use
TSO	Transmission System Operator
UVAM	Unità Virtuali Abilitate Miste
V2G	Vehicle-to-Grid
V2H	Vehicle-to-Home

References

1. International Energy Agency. *Net Zero by 2050*; Technical report; IEA: Paris, France, 2021.
2. Impram, S.; Varbak Nese, S.; Oral, B. Challenges of renewable energy penetration on power system flexibility: A survey. *Energy Strategy Rev.* **2020**, *31*, 100539. [CrossRef]
3. International Energy Agency. *Global EV Outlook 2024*; IEA: Paris, France, 2024.
4. Caminiti, C.M.; Merlo, M.; Fotouhi Ghazvini, M.A.; Edvinsson, J. optimHome: A Shrinking Horizon Control Architecture for Bidirectional Smart Charging in Home Energy Management Systems. *Energies* **2024**, *17*, 1963. [CrossRef]
5. Gonzalez Venegas, F.; Petit, M.; Perez, Y. Active integration of electric vehicles into distribution grids: Barriers and frameworks for flexibility services. *Renew. Sustain. Energy Rev.* **2021**, *145*, 111060. [CrossRef]
6. Hussain, S.; Lai, C.; Eicker, U. Flexibility: Literature review on concepts, modeling, and provision method in smart grid. *Sustain. Energy Grids Netw.* **2023**, *35*, 101113. [CrossRef]
7. Gunkel, P.A.; Bergaentzlé, C.; Græsted Jensen, I.; Scheller, F. From passive to active: Flexibility from electric vehicles in the context of transmission system development. *Appl. Energy* **2020**, *277*, 115526. [CrossRef]
8. Gerritsma, M.K.; AlSkaif, T.A.; Fidder, H.A.; van Sark, W.G.J.H.M. Flexibility of Electric Vehicle Demand: Analysis of Measured Charging Data and Simulation for the Future. *World Electr. Veh. J.* **2019**, *10*, 14. [CrossRef]
9. Spencer, S.I.; Fu, Z.; Apostolaki-Iosifidou, E.; Lipman, T.E. Evaluating smart charging strategies using real-world data from optimized plugin electric vehicles. *Transp. Res. Part D Transp. Environ.* **2021**, *100*, 103023. [CrossRef]
10. Rancilio, G.; Cortazzi, A.; Viganò, G.; Bovera, F. Assessing the Nationwide Benefits of Vehicle–Grid Integration during Distribution Network Planning and Power System Dispatching. *World Electr. Veh. J.* **2024**, *15*, 134. [CrossRef]
11. Yi, Z.; Chen, B.; Liu, X.C.; Wei, R.; Chen, J.; Chen, Z. An agent-based modeling approach for public charging demand estimation and charging station location optimization at urban scale. *Comput. Environ. Urban Syst.* **2023**, *101*, 101949. [CrossRef]
12. Horni, A.; Nagel, K.; Axhausen, K. (Eds.) *Multi-Agent Transport Simulation MATSim*; Ubiquity Press: London, UK, 2016; p. 618. [CrossRef]
13. Novosel, T.; Perković, L.; Ban, M.; Keko, H.; Pukšec, T.; Krajačić, G.; Duić, N. Agent based modelling and energy planning—Utilization of MATSim for transport energy demand modelling. *Energy* **2015**, *92*, 466–475. [CrossRef]
14. Hanemann, P.; Graupner, S.; Grunert, P.; Koepke, M.; Bruckner, T. Grid Load Contributions Through Electric Vehicles and Their Uncertainties. *Z. Für Energiewirtschaft* **2017**, *41*, 269–281. [CrossRef]
15. Mu, Y.; Wu, J.; Jenkins, N.; Jia, H.; Wang, C. A Spatial–Temporal model for grid impact analysis of plug-in electric vehicles. *Appl. Energy* **2014**, *114*, 456–465. [CrossRef]

16. Secchi, M.; Barchi, G.; Macii, D.; Petri, D. Smart electric vehicles charging with centralised vehicle-to-grid capability for net-load variance minimisation under increasing EV and PV penetration levels. *Sustain. Energy Grids Netw.* **2023**, *35*, 101120. [CrossRef]
17. Hartvigsson, E.; Taljegard, M.; Odenberger, M.; Chen, P. A large-scale high-resolution geographic analysis of impacts of electric vehicle charging on low-voltage grids. *Energy* **2022**, *261*, 125180. [CrossRef]
18. Yousefi, G.; Dimovski, A.; Radaelli, L.; Merlo, M. Estimating the impact of electric mobility on distribution networks through GIS techniques. *Sustain. Energy Grids Netw.* **2024**, *38*, 101379. [CrossRef]
19. Jaffal, H.; Guanetti, L.; Rancilio, G.; Spiller, M.; Bovera, F.; Merlo, M. Battery Energy Storage System Performance in Providing Various Electricity Market Services. *Batteries* **2024**, *10*, 69. [CrossRef]
20. Gestore Mercati Elettrici. *Italian Electricity Market Description*; Technical report; GME: Rome, Italy, 2023.
21. Terna. *Capter 4 to Grid Code*; Technical report; Terna: Rome, Italy, 2023.
22. Rancilio, G.; Vicario, A.; Merlo, M.; Berizzi, A. Battery energy storage contribution to system frequency for grids with high renewable energy sources penetration. In *Power System Frequency Control—Modeling and Advances*; Academic Press: Cambridge, MA, USA, 2023; pp. 133–150. [CrossRef]
23. Bovera, F.; Rancilio, G.; Falabretti, D.; Merlo, M. Data-Driven Evaluation of Secondary- and Tertiary-Reserve Needs with High Renewables Penetration: The Italian Case. *Energies* **2021**, *14*, 2157. [CrossRef]
24. Spiller, M.; Rancilio, G.; Bovera, F.; Merlo, M. Opportunities for DERs in Ancillary Services Markets: Analysis and deployment strategies in a national framework. In Proceedings of the 2022 18th International Conference on the European Energy Market (EEM), Ljubljana, Slovenia, 13–15 September 2022; pp. 1–6. [CrossRef]
25. Regione Lombardia. *Matrice Origine Destinazione*. Available online: https://www.dati.lombardia.it/Mobilit-e-trasporti/Matrice-OD2014-DISAGGREGATA-file-compresso-per-dow/rwsg-m4kj/about_data (accessed on 3 September 2024).
26. ANFIA. *Distribution of EVs*; Technical report; ANFIA: Torino, Italy, 2023.
27. Motus-E. *Le Infrastrutture di Ricarica a Uso Pubblico in Italia—Quinta Edizione*; Technical report; Motus-E: Rome, Italy, 2023.
28. REGELLEISTUNG.NET. *Prices for Primary Frequency Control*. Available online: https://www.aemc.gov.au/sites/default/files/2018-03/Advice%20from%20Nick%20Miller%20-%20Costs%20of%20primary%20frequency%20regulation.PDF (accessed on 3 September 2024).
29. GME. *Prices of MGP*; Technical report; GME: Rome, Italy, 2023.

World Electric Vehicle Journal

MDPI

Article

Solar–Hydrogen-Storage Integrated Electric Vehicle Charging Stations with Demand-Side Management and Social Welfare Maximization

Lijia Duan, Gareth Taylor and Chun Sing Lai *

Department of Electronic and Electrical Engineering Brunel University London, London UB8 3PH, UK; lijia.duan@brunel.ac.uk (L.D.); gareth.taylor@brunel.ac.uk (G.T.)
* Correspondence: chunsing.lai@brunel.ac.uk

Abstract: The reliable operation of a power system requires a real-time balance between supply and demand. However, it is difficult to achieve this balance solely by relying on supply-side regulation. Therefore, it is necessary to cooperate with effective demand-side management, which is a key strategy within smart grid systems, encouraging end-users to actively engage and optimize their electricity usage. This paper proposes a novel bi-level optimization model for integrating solar, hydrogen, and battery storage systems with charging stations (SHS-EVCSs) to maximize social welfare. The first level employs a non-cooperative game theory model for each individual EVCS to minimize capital and operational costs. The second level uses a cooperative game framework with an internal management system to optimize energy transactions among multiple EVCSs while considering EV owners' economic interests. A Markov decision process models uncertainties in EV charging times, and Monte Carlo simulations predict charging demand. Real-time electricity pricing based on the dual theory enables demand-side management strategies like peak shaving and valley filling. Case studies demonstrate the model's effectiveness in reducing peak loads, balancing energy utilization, and enhancing overall system efficiency and sustainability through optimized renewable integration, energy storage, EV charging coordination, social welfare maximization, and cost minimization. The proposed approach offers a promising pathway toward sustainable energy infrastructure by harmonizing renewable sources, storage technologies, EV charging demands, and societal benefits.

Keywords: renewable energy; electric vehicle charging station; electric vehicle charging time uncertainty; social welfare maximization

1. Introduction

The continued expansion of the world's economies has resulted in escalating energy shortages and worsening air quality. In response, low-carbon and renewable energy (RE) sources have garnered substantial interest as potential avenues to help alleviate both energy scarcity and environmental degradation. Numerous nations have enacted carbon emission regulations with the goals of curbing energy usage, boosting the efficiency of electricity production, minimizing greenhouse gas emissions, and fostering the advancement of clean energy alternatives. While conventional fossil fuel-based power generation technologies offer ample capacity and reliable operations, their overall energy utilization is inherently inefficient and inflexible in meeting fluctuations in supply and demand. Clean energy can be categorized as either renewable or non-renewable. Renewable sources encompass wind, hydroelectric, solar, geothermal, tidal, and other types. Non-renewable clean energy includes nuclear power and biomass. Renewable options provide competitively priced electricity with minimal carbon footprints; however, the intrinsic variability and intermittency of these sources inject uncertainty and volatility into energy availability.

The origins of Electric Vehicle (EV) development can be traced back to 1830 when Robert Anderson created a prototype carriage powered by electricity. However, it was not until 1884 that the first mass-produced EV was introduced by British inventor Thomas Parker [1]. Subsequently, in 1897, an electric taxi company commenced operations in New York City. Nonetheless, the EVs of that era were plagued by a multitude of issues that hindered their widespread adoption in the automotive market. A notable challenge was the inability to reliably recharge the car batteries, necessitating frequent replacement with new ones. Even in instances where recharging was possible, the limited driving range and scarcity of charging stations severely restricted the utility of EVs [2]. According to [1,2], the development of electric ignition systems for combustion engines in 1834, coupled with the availability of inexpensive fuel and gradual improvements in the reliability of gasoline-powered cars, further impeded the advancement of EVs.

Since the early 2000s, there has been a growing sense of urgency to address the greenhouse gas emissions and environmental pollution stemming from excessive reliance on fossil fuels [3]. Consequently, RE sources have increasingly become a major focal point and are now viewed as crucial alternatives in numerous countries worldwide. Notably, the electricity generation and transportation sectors are significant contributors to carbon dioxide (CO_2) emissions, accounting for approximately 64% of total emissions [3]. This situation has raised substantial public concerns regarding the potential irreversible environmental damage caused by these emissions. According to sources [4], the adoption of RE is essential for meeting carbon dioxide reduction targets in these sectors. EVs are recognized as a promising solution to control CO_2 emissions. Comprehensive research reviews [5,6] on EVs indicate that their use not only mitigates environmental impacts but also offers economic benefits to owners through lower operating costs. However, the expansion of EV charging infrastructure presents a major obstacle to increasing EV adoption [7]. Slow progress in developing charging facilities hinders potential buyers' willingness to transition to EVs [8]. Furthermore, high investment costs and uncertainties surrounding EV demand further impede infrastructure development. In London, for example, projections suggest that over 500,000 charging points will be needed by 2040, with nearly 50,000 required in public areas [9].

Figure 1 shows the Solar–Hydrogen-Storage Integrated Electric Vehicle Charging Station (SHS-EVCS), which harnesses PV, a hydrogen storage system, and battery storage to charge EVs. The station includes a solar array that converts solar power into electrical energy. This energy can be used immediately for charging EVs, fed into the grid, or stored in batteries for later use. The hydrogen storage system employs an electrolyser to split water into hydrogen and oxygen. The hydrogen is then compressed and stored for future needs. During high-demand periods, the stored hydrogen is converted back into electricity through a fuel cell to charge EVs. Additionally, a battery storage unit stores any excess energy from the solar array or the fuel cell, ensuring a steady energy supply for EV charging even when the primary renewable sources are not available.

In this paper, Section 2 provides a review of RE usage in EV charging stations (EVCSs) and the application of demand-side management (DSM) in electricity trading. Section 3 presents the problem formulation for the SHS-EVCSs. Section 4 includes a case study with forecasts for three different scenarios. Section 5 analyses the results, and Section 6 concludes by summarizing the significant findings of this paper.

The goal of this paper is to maximize social welfare, ensuring that both the SHS-EVCS and EV drivers receive the maximum possible benefits. This involves optimizing the interactions between the EVCS and EV users, aiming for a balanced and mutually beneficial outcome. By achieving this, the study seeks to demonstrate how the management of resources can lead to enhanced efficiency and satisfaction for all stakeholders involved.

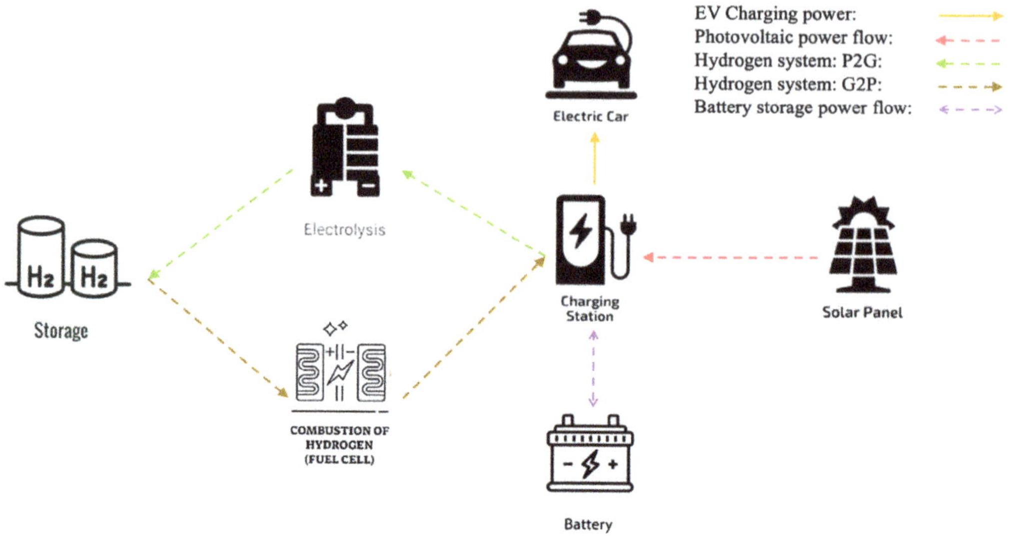

Figure 1. Single Solar–Hydrogen-Storage Integrated Electric Vehicle Charging Station topology.

2. Literature Review

To establish an efficient, resilient, and sustainable infrastructure, EVCSs can integrate RE sources [10]. Incorporating hydrogen systems, battery storage, and solar energy will play a critical role, presenting both opportunities and significant technological challenges [11]. This approach is particularly appealing for charging EVs, as battery storage systems powered by solar energy can ensure reliable charging even on cloudy days, thus improving the reliability of the charging capacity. Hydrogen tanks can store excess RE, providing high-density energy for backup power and fuel cell vehicles, thus creating versatile energy storage solutions [12]. However, integrating these diverse energy sources into a cohesive charging infrastructure poses technical challenges, with energy management being a primary concern. Ensuring efficient use, conversion, and storage of energy requires complex systems. Balancing battery solutions involves data monitoring and analysis [13]. Despite the high costs, limited lifecycle, and high-capacity features of hydrogen systems, optimizing infrastructural compatibility remains challenging, especially with the complex needs of RE installations [14]. EVCSs must accommodate large battery packs, electrolysers for clean energy production, and smart grid capabilities for dynamic energy management [14]. Interoperability issues also arise, requiring standardized interfaces and protocols to enable the seamless operation of different power generation and storage technologies. Scalability is essential to meet the increasing demand for charging stations and EVs [15]. The performance and integration of RE systems can be enhanced through advanced EV charging stations. Machine learning and artificial intelligence can improve energy management systems, optimizing the integration of supply and demand [16]. Innovations in battery technology, including solid-state batteries, will enable fast charging times and high energy densities, potentially transforming battery storage options at EV charging stations [17,18]. Additionally, hydrogen production efficiency driven by RE through green electrolysis can reduce costs, enhancing scalable energy storage solutions in EV charging infrastructure [19]. Paper [20] proposes a strategy to optimize small-scale PV systems for homes, reducing energy costs by up to 40% and allowing for extra loads without expanding the network, while addressing EV charging and unsynchronized demand challenges. A review [21] examined advancements in energy storage from 1850 to 2022, highlighting their evolution, principles, and role in addressing renewable energy intermittencies to help decarbonize the environment.

Demand-Side Management (DSM) is a key strategy within smart grid systems, encouraging end-users to actively engage and optimize their electricity usage [22,23]. DSM implements various measures to help consumers plan and use electricity more rationally and efficiently, thereby improving the efficiency of terminal electricity consumption. Additionally, DSM is crucial in resource allocation by adapting to demand-side participation in power market management, which helps prevent large-scale power outages due to peak loads or the need for additional power generation infrastructure, ultimately reducing costs within the smart grid system [24]. In power market transactions, electricity's unique characteristics—being instantly consumable and hard to store—require rapid data exchange, swift decision-making, and real-time adjustments from both supply and demand sides. This level of responsiveness is difficult to achieve in traditional power grids. However, in smart grid systems, advanced technologies like smart meters, smart homes, and cloud computing significantly enhance DSM capabilities [25,26].

As the energy Internet evolves, it enables the integration of multiple energy sources and dynamic management practices. This transformation allows for more sustainable and efficient energy consumption patterns, aligning with the shift towards comprehensive, interconnected energy networks. Demand response (DR) is a vital mechanism where electricity consumers adjust their power usage in response to market signals or incentives from power suppliers [27]. DR is a critical component of DSM, and studies have shown that it significantly improves end-use electricity consumption efficiency, reduces resource waste due to supply–demand imbalances, and contributes to overall grid stability [28,29]. The rapid development of smart grids, incorporating technologies like smart homes and EVs, provides substantial support for DR initiatives. Furthermore, DR benefits consumers by reducing electricity bills through altering traditional consumption patterns and strategically shifting usage away from peak periods, helping to prevent large-scale power outages [30]. As distributed energy resource integration progresses, the synergy between regional energy systems—including various forms of energy generation and storage—and power grid systems has transitioned DR from a conventional approach to a holistic strategy of managing diverse load types, such as wind, solar, electrical, and storage. This evolution has normalized DR integration into power system dispatch strategies, highlighting its growing importance in achieving a more resilient and efficient energy landscape [31,32]. Integrating DR into energy management enhances operational flexibility and optimizes the economic operation of power systems by maximizing resource utilization. The increasing presence of RE sources and their variability in the power grid further emphasize the need for sophisticated DR strategies.

Maintaining stable power system operations necessitates a real-time balance between supply and demand. However, achieving this balance solely through supply-side regulation is challenging. Therefore, effective DSM measures are crucial [33]. Real-time pricing, as the most direct and effective DSM method, employs electricity price signals to encourage consumers to voluntarily shift their electricity usage to off-peak periods, thereby reducing peak loads and achieving the objectives of peak shaving and valley filling [33,34]. Study [35] pioneered the application of the social welfare maximization model to real-time pricing in smart grids. The goal was to maximize user utility while minimizing the power provider's costs, with the objective function being the user's utility minus the provider's cost. This study established a social welfare maximization model for smart grid real-time pricing, subject to the constraint of power supply and demand balance. It utilized a dual optimization method to determine supply, electricity consumption, and real-time electricity prices [35].

Societal welfare has long been a fundamental preoccupation, and the interplay between individual needs and collective interests forms the crux of modern social welfare discourse. "Social welfare" emerges as the outcome of public choices regarding the distribution of societal benefits within specific institutional frameworks, representing a collective or group interest [36]. On the other hand, "utility" denotes the psychological satisfaction derived by consumers during the consumption process. Economics employs these two fundamental notions of welfare and utility to elucidate and illustrate the potential balance between personal motivations and social choices [34]. Table 1 shows the relevant work.

Table 1. Literature review and benchmarking of relevant work.

Reference	Electricity Sources	P2P and Game Theory Applied	EV/RE Uncertainty Prediction	Include Social Welfare	Multi-EVCSs Energy Exchange
[37]	Wind, PV, battery energy storage, grid	No	Yes	No	No
[38]	Wind, PV, battery energy storage, grid	Yes	Yes	Yes	No
[39]	Grid	Yes	Yes	No	No
[40]	PV, wind	No	Yes	Yes	No
[41]	Energy management system	No	Yes	No	No
[42]	Battery storage system	No	Yes	No	No
[43]	Wind, PV, battery energy storage, grid	No	No	No	No
[44]	Grid, PV, wind	No	Yes	No	No
[45]	PV, battery energy storage, hydrogen storage, grid	No	No	No	Yes
[46]	PV, battery energy storage, hydrogen storage, grid	Yes	No	No	Yes
This work	PV, battery energy storage, hydrogen storage, grid	Yes	Yes	Yes	Yes

This paper examines the practical feasibility of EVCSs through two primary models. The first is a non-cooperative game model for a single charging station aimed at minimizing construction, operation, and maintenance costs. The second model is the internal EVCS energy transaction cooperation model, expanded from [46], which integrates operational load and an internal dispatch center for demand response. Additionally, a model focuses on maximizing social welfare by considering the economic interests of EV owners, including acceptable charging prices, essentially minimizing the cost of electricity supply. In the general social welfare model, another significant factor is the capital cost of the electricity provider (grid, RE, storage systems, etc.). However, in the SHS-EVCS model, this capital cost is calculated in the non-cooperative game model for a single charging station and is, therefore, excluded from the social welfare maximization model.

This paper makes three main contributions:

- Develops an EV charging time model using a Markov decision process to address uncertainties in charging times and a Monte Carlo simulation to assess and predict EV charging demand, which is expected to fluctuate based on probabilities associated with different charging times.
- Creates a bi-level optimization model that combines non-cooperative and cooperative game theory to minimize capital costs and maximize social welfare.
- Improves the social welfare maximization real-time pricing model based on the dual theory, considering the benefits to drivers, the SHS-EVCS, and the grid side, according to EV charging demand.

3. Problem Formulation

This paper organizes the multi-SHS-EVCS system into two distinct levels. The first level focuses on an individual SHS-EVCS, employing a non-cooperative game theory model to reschedule energy dispatch with the aim of minimizing capital costs. The second level, similar to the approach in paper [46], but introduces an Internal Management System

(IMS), also known as a demand response centre, into the SHS-EVCS coalition to manage the sequencing of charging and discharging activities. At this level, the goal is to maximize the coalition's profit. The ultimate objective of this bi-level model is to maximize social welfare.

3.1. First-Level SHS Model

This section combines solar energy, hydrogen storage, battery storage (SHS), and the grid into a non-cooperative game theory model for EVCS. In this model, hydrogen storage provides long-term energy buffering, batteries offer quick responses to demand changes, solar energy adds renewable generation, and the grid ensures a stable supply. The goal is to minimize construction, operation, and maintenance costs while optimizing energy distribution. Figure 2 illustrates the specific interaction model.

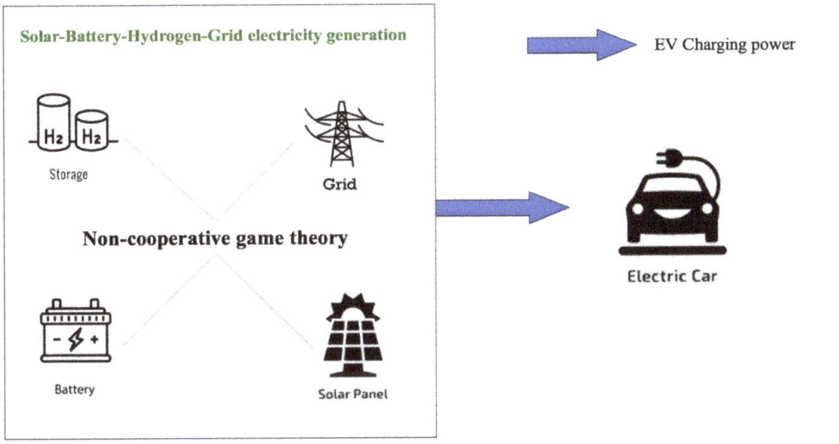

Figure 2. Electricity dispatch depends on non-cooperative game theory within SHS system.

The problem formulation of the first level's objective function is given in reference [45] in Equations (1) and (3)–(10).

3.1.1. Non-Cooperative Game Theory

The model discussed in this section is a multi-agent framework where each piece of equipment aims to minimize costs independently. The strategic decisions of each participant are made without considering others, creating a scenario of mutual interaction and fair competition, typical of a non-cooperative game model. The interactions among all energy sources can be described as follows:

$$G = \{P; N; E\} \tag{1}$$

where P represents the game players, N is the strategy set, and E is the payoff function.

- Game player: In this context, each decision-making entity is referred to as a game player. The participants in this study include solar energy (S), hydrogen energy storage (H), battery energy storage (B), and the power grid (G). These participants are collectively represented as

$$P = \{S,\ H,\ B,\ G\} \tag{2}$$

- Strategy Set: Each participant selects a set of strategies during the game to maximize their benefits, which in this context means minimizing costs. The strategies for each energy source are solar energy (N_S), hydrogen energy storage (N_H), battery energy storage (N_B), and the power grid (N_G). The strategy set is

$$N = \{N_S;\ N_H;\ N_B;\ N_G\} \tag{3}$$

- Payoff: The payoff function assesses each participant's cost and provides feedback for strategy adjustments in subsequent rounds. The payoff for each participant depends on their strategy and the strategies of others, represented as

$$E_{SHS}(N_S;\ N_H;\ N_B;\ N_G) \tag{4}$$

where E_{SHS} is the minimum cost for the SHS system.

Non-cooperative games aim to find a Nash equilibrium, achieving the optimal objective function. To meet the Nash equilibrium condition, minimizing the grid's cost means reducing power purchases from the grid. Therefore, the optimal solution for the grid P_g^t requires fixing the power outputs of P_{pv}^t, P_{hs}^t, and P_{es}^t. Similarly, for the optimal solutions of solar power P_{pv}^t, battery storage P_{es}^t, and hydrogen storage P_{hs}^t, the power outputs of the other sources must be fixed.

Step to solve Nash Equilibrium:

(1) Initialization: set initial strategies N_S, N_H, N_B, N_G; replace by P_g^t, P_{pv}^t, P_{hs}^t, and P_{es}^t.

(2) Iterative optimization:

- Grid:
    - Fix P_{pv}^t, P_{hs}^t, and P_{es}^t.
    - Solve the optimal P_g^t.
- Solar power:
    - Fix P_g^t, P_{hs}^t, and P_{es}^t.
    - Solve the optimal P_{pv}^t.
- Battery storage:
    - Fix P_g^t, P_{hs}^t, and P_{pv}^t.
    - Solve the optimal P_{es}^t.
- Hydrogen storage system:
    - Fix P_g^t, P_{es}^t, and P_{pv}^t
    - Solve the optimal P_{hs}^t.

(3) Check convergence:

- Verify if all participants' strategies converge to a fixed point. If not, return to Step 2 for further iteration.

4. Verify Nash Equilibrium:

- Ensure all strategy combinations meet the Nash equilibrium condition, where no participant can further reduce its cost by changing its strategy while others' strategies remain unchanged.

3.1.2. EV Uncertainty Model

To capture the variability in EV charging times, this paper uses a Markov chain probability model. This model is ideal for describing random processes with discrete time intervals and states, where the future state depends solely on the current state, not on the sequence of previous states. The use of transition probabilities between adjacent states simplifies the complexity of the random process. Each battery power level of an EV can be considered a discrete state, with the random fluctuations in battery power represented as a non-stationary Markov chain. This chain effectively models the likelihood of an event occurring based solely on the present state [47].

$$P\left\{x_f^{dis}\middle|x_1^{dis}=s_1^{dis},\ x_2^{dis}=s_2^{dis},\ldots,x_{f-1}^{dis}=s_{f-1}^{dis}\right\}=P\left\{x_f^{dis}\middle|x_{f-1}^{dis}=s_{f-1}^{dis}\right\} \tag{5}$$

$$s_f^{dis}\in S^{dis},\ 1\le f\le f_{max}$$

where x_f^{dis} is the $f'th$ discrete variable; s_f^{dis} is the $f'th$ discrete state; S^{dis} is the discrete state space; f_{max} is the maximum number of discrete variables.

Based on (5), it is also needs to satisfy

$$P\left\{x_f^{dis}\middle|x_{f-1}^{dis}=s_{f-1}^{dis}\right\}=P\left\{x_{f+1}^{dis}\middle|x_f^{dis}=s_f^{dis}\right\} \tag{6}$$

To capture the randomness and reversibility of EV charging and driving power consumption, the Markov sampling process incorporates the detailed balance condition. This condition ensures that transitions between any two states are reversible:

$$p_{ij}P\left\{x_{f+1}^{dis}=s_j^{dis}\middle|x_f^{dis}=s_i^{dis}\right\}=p_{ji}P\left\{x_{f+1}^{dis}=s_i^{dis}\middle|x_f^{dis}=s_j^{dis}\right\} \tag{7}$$

where p_{ij} is the transition probability of state i to j; p_{ji} is the transition probability of state j to i.

EVs can be classified into two operational states: charging and non-charging. These correspond to two modes: charging mode and waiting mode [47–50]. In the charging mode, the EV is connected to the SHS-EVCS and is actively charging. Conversely, in the waiting mode, the EV is at the SHS-EVCS but not charging, meaning it is in a parked state. The variable c_f represents the charging probability, capturing the randomness of EV charging, while d_f denotes the probability of interrupted charging, reflecting the variability within the charging process.

The state transition probabilities are illustrated in Figure 3. In this context, EVC_f and EVD_f represent the states of the charging mode and waiting mode, respectively. The transition probabilities between these states are denoted by d_f and c_f, while a_f signifies the waiting probability. ΔT_1 and ΔT_2 correspond to the charging time and waiting time, respectively. All the arrows represent the probability to change the charging or discharging state.

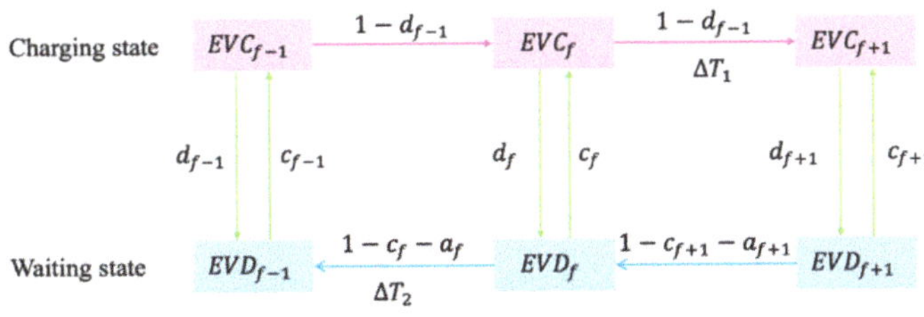

Figure 3. State transition diagram.

If the EV drivers are anxious of battery level, in this paper, the level is $B_{min}^{arrive}\le 35\%$, which means if the EV battery's SOC is less than or equal to 35%, the drivers will charge their EV. When charging the car, if $B_{min}^{leave}\ge 95\%$, the EV drivers will choose to discharge their cars. N is the full battery level which is 100. The probability for c_f and d_f can be demonstrated as [47–49]:

$$c_f=\begin{cases}\frac{1}{1+e^{\left(f-0.5\left(B_{min}^{arrive}+N\right)\right)^{0.14}}} & f\ge B_{min}^{arrive}\\ 0 & f<B_{min}^{arrive}\end{cases} \tag{8}$$

$$d_f = \begin{cases} \frac{1}{1+e^{(0.5(B_{min}^{leave}+N)-f)^{0.32}}} & f \geq B_{min}^{leave} \\ 1 & f < B_{min}^{leave} \end{cases} \quad (9)$$

The state probability EVC_f and EVD_f of EV can be denoted as [45–47]

$$P_s\left(EVC_f\right) = \begin{cases} \left(1 - d_{f-1}\right) P_s\left(EVC_{f-1}\right) + c_f P_s\left(EVD_f\right) \\ 1 \leq f \leq N \\ c_0 P_s(EVD_0) \\ f = 0 \end{cases} \quad (10)$$

$$P_s\left(EVD_f\right) = \begin{cases} d_f P_s\left(EVC_f\right) + a_f P_s\left(EVD_f\right) + \left(1 - c_{f+1} - a_{f+1}\right) P_s\left(EVD_{f+1}\right) \\ 0 \leq f \leq N-1 \\ d_N P_s(EVC_N) + a_N P_s(EVD_N) \\ f = N \end{cases} \quad (11)$$

where $Ps(EVC_f)$ is the state probability of EVC_f, and $Ps(EVD_f)$ is the state probability of EVD_f.

These variables are subject to

$$\sum_{f=0}^{100} \left\{P_s\left(EVC_f\right) + P_s\left(EVD_f\right)\right\} = 1 \quad (12)$$

$$d_0 = 0,\ c_0 = 1,\ d_{100} = 1,\ c_{100} = 0 \quad (13)$$

where (12) indicates that the total sum of all state probabilities of the EV at any given time is equal to 1; (13) denotes that when the battery level is equal to 0, it must be in the EVC_f mode, and when the battery is in a full state, it must be in the EVD_f mode.

ΔT_1 and ΔT_2 are [47]

$$\Delta T_1 = min\left\{\frac{B_{cap}\Delta(+f)}{N\eta_c P_c},\ T_{leave} - T_{arrive}\right. \quad (14)$$

$$\begin{cases} \Delta T_2 = \frac{B_{cap}\Delta(-f)}{NE_{km}v} \\ 1 \leq \Delta(-f) \leq N \end{cases} \quad (15)$$

where

- η_c is the EV battery efficiency, which is 95;
- P_c is the fixed charging power, which is 7 kW;
- B_{cap} is the battery capacity, which is 60 kWh (Tesla model 6 battery capacity);
- $\Delta(-f)$ is the electricity loss when driving the EV;
- E_{km} is energy consumption, which is 16.25 kWh/100 km;
- v is the average driving speed, which is 32 km/h.

To predict the electricity demand for all SHS-EVCSs catering to numerous EV drivers, it is crucial to first establish a prediction model for the charging demand of a single EV. Following this, the overall electricity demand of EVs can be calculated by aggregating the individual EV charging demands using the Monte Carlo random simulation method. The total charging power demand at any specific time can be represented as follows:

$$P_{cha,sum} = \sum_{i}^{N_{ev}} P_{cha,i} \quad (16)$$

where $P_{cha,sum}$ is the total charging power; $P_{cha,i}$ is the i'th EV charging power; N_{ev} is the total EV number.

For each EV's SOC, it should follow

$$SOC_{start} = \frac{1}{SOC_{start,max} - SOC_{start,min}} \tag{17}$$

where SoC_{start} is the initial SoC of the EV; $SoC_{start,max}$ and $SoC_{start,min}$ are the maximum and minimum SoC of the EV.

For the EV's battery SoC_{arrive} when arriving at SHS-EVCS, it is denoted as

$$SOC_{arrive} = SOC_{start} - \frac{D_{EV}E_{km}}{B_{cap}} \tag{18}$$

where D_{EV} is the EV's driving distance and subject to

$$0 \leq D_{EV} < \frac{B_{cap}}{E_{km}} \tag{19}$$

Meanwhile, for all the EV drivers, they have z times charging opportunities [51]:

$$1 \leq z \leq 3 \tag{20}$$

While the Monte Carlo method is known for its high prediction accuracy, it requires an understanding of the probability distribution of the input samples. Predicting EV charging demand involves many uncertain factors that often do not adhere to a standard probability distribution, making it challenging to determine the corresponding probability density function [47]. For example, the probability of EV owners choosing a simple travel chain for commuting and the likelihood of EVs charging from the grid usually fall under discrete probability distributions [47–49]. Hence, it is necessary to first establish the probability distribution of the uncertain factors before applying the Monte Carlo method to predict EV charging demand. The algorithm for forecasting EV charging is as follows (Algorithm 1):

Algorithm 1: EV Charging Demand Forecast Algorithm

Input: the EV data: B_{cap}, η_c, P_c, E_{km}, v, N_{ev}...

(a) Repeat 1: calculate the EV's charging demand
(b) Get initial EV data SoC_{start}, $P_{cha,sum}$ through Monte Carlo random simulation method
(c) Repeat 2: calculate single EV charging demand
(d) Update the current battery level
(e) Calculate the charging and discharging probability c_f, d_f
(f) Calculate the charging duration ΔT_1
(g) Calculate the waiting duration ΔT_2
(h) Obtain the charging frequency $1 \leq z \leq 3$
(i) Repeat 2 to calculate the single EV charging demand
(j) End repeat 2
(k) Repeat 1 based on (17) to obtain the total EV charging demand
(l) End repeat 1

End: Output the total EV charging demand to obtain the SHS-EVCS charging load

3.2. Second-Level Social Welfare Model

The second level adopts a similar approach to [46], utilizing a game theory-based P2P energy exchange process, while integrating aspects of information transfer and load demand considerations. Figure 4 illustrates the operation of this level: the blue line depicts internal P2P energy trading, demonstrating the exchange of energy among the SHS-EVCSs. The pink line indicates the flow of information, outlining the communication pathways between each EVCSs and the central information hub, as well as between the EVCSs and individual EVs. The yellow line symbolizes the flow of EV charging information, with each EVCS gathering data when EVs are either charging or waiting to charge. These data feed into the Markov decision process to accurately forecast EV charging demand in a real-time scenario.

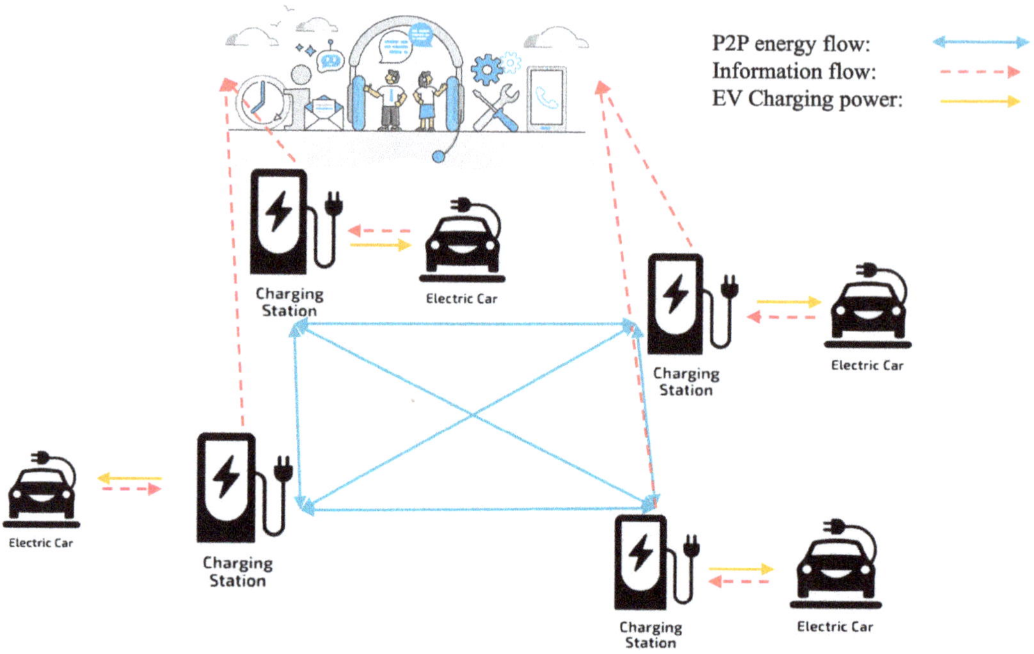

Figure 4. Topology for SHS-EVCSs coalition information and energy flow.

Within the SHS-EVCSs coalition, which includes four EVCSs, D users, and an IMS, each SHS-EVCS produces electricity to satisfy the demand of EV drivers within its locale. They are interconnected through a communication network that facilitates the exchange of real-time data on electricity prices and power requirements. The power consumption cycle is segmented into k time periods. At the start of each period, the IMS determines an electricity price reflecting the current power market conditions. Both EV drivers and SHS-EVCSs gauge their optimal power usage and generation based on this pricing and communicate their data back to the IMS. The IMS then adjusts the electricity price in response to the incoming data on consumption and supply. This cyclical adjustment process persists, with both drivers and SHS-EVCSs tuning their power usage and output to align with the updated pricing information. This exchange continues until equilibrium is reached between supply and demand, thus setting the electricity price for that interval. The established price then serves as the real-time electricity rate for that time.

The problem formulation of the game theory and P2P process is giving in paper [46]; this level considers social welfare maximization, and the formulation is as follows. A novel price-based real-time electricity pricing model [35,52,53] is introduced to optimize the gap between the total utility of the EV driver and the costs incurred by the electricity provider (SHS-EVCS). This sets up the subsequent optimization challenge:

$$max \sum_{k=1}^{K} \left(\sum_{d=1}^{D} V_1\left(x_d^k, \omega_d^k\right) - C_k(G_k) \right) \tag{21}$$

$$s.t. \sum_{d \in D} x_d^k \le G_k,\ k = 1, 2, \ldots, K,$$

$$x_d^k \geq 0,\ k = 1, 2, \ldots, K.$$

The social welfare maximization model (21) can be defined as:

$$max \sum_{k=1}^{K} \left(\sum_{d=1}^{D} V_d\left(x_d^k,\ \omega_d^k\right) - C_k(G_k) \right) \tag{22}$$

$$s.t.\ \sum_{d \in D} x_d^k \leq G_k,\ k = 1, 2, \ldots, K$$

$$m_d^k \leq x_d^k \leq M_d^k,\ d = 1, 2, \ldots, D; k = 1, 2, \ldots, K$$

$$G_k^{min} \leq G_k \leq G_k^{max},\ k = 1, 2, \ldots, K$$

where x_d^k is the electricity consumption of EV driver d at time k; m_d^k and M_d^k are the minimum and maximum electricity consumptions of EV driver d at time k, and they satisfy $m_d^k \leq x_d^k \leq M_d^k$; G_k is SHS-EVCS's electricity supply at time k; G_k^{min} and G_k^{max} are the minimum and maximum electricity supplies of the SHS-EVCS, and $G_k^{min} \leq G_k \leq G_k^{max}$; generally, $G_k^{min} \geq \sum_{d=1}^{D} m_d^k$ and $G_k^{max} \geq \sum_{d=1}^{D} M_d^k$; $C_k(G_k)$ means the cost for the SHS-EVCS to provide electricity for G_k at time k, and it is the convex function of G_k; C_k is the capital cost for SHS-EVCS from first-level output; $V_d\left(x_d^k,\ \omega_d^k\right)$ is the utility of EV driver d using x_d^k at time k, and it is the concave function of x_d^k; ω_d^k is driver d's elasticity of charging his EV at time k, for $\omega_d^k \in [1,\ 3.5]$.

The objective function aimed at maximizing social welfare is concave, and the associated constraints form a convex set. Consequently, this constitutes a convex programming issue that can be efficiently resolved using standard convex programming methods. In Equation (22), the critical decision variables are the electricity consumption of the driver, x_d^k, and the power supply from SHS-EVCS, G_k. However, the real-time electricity price, a central variable in the model, is absent, rendering a direct solution to Equation (22) unfeasible. In convex optimization, the original problem can often be mirrored by a dual problem where the dual decision variables, namely, the Lagrange multipliers from the Lagrange function of the original problem, equate to the shadow prices in economic theory. These shadow prices are deemed theoretically optimal for pricing. Consequently, this work applies the dual theory to effectively determine the real-time electricity pricing [34,35,54–56].

Each interval in (22) is independent of each other, so a distributed algorithm can be used to solve each interval separately. The optimization problem corresponding to the k-th interval is

$$max \sum_{d=1}^{D} V_d\left(x_d^k,\ \omega_d^k\right) - C_k(G_k) \tag{23}$$

$$s.t.\ \sum_{d=1}^{D} x_d^k \leq G_k$$

$$m_d^k \leq x_d^k \leq M_d^k,\ d = 1, 2, \ldots, D$$

$$G_k^{min} \leq G_k \leq G_k^{max}$$

The Lagrange function for (23) is denoted as

$$L\left(x_d^k,\ G_k, \lambda_k\right) = \sum_{d=1}^{D} V_d\left(x_d^k,\ \omega_d^k\right) - C_k(G_k) + \lambda_k\left(G_k - \sum_{d=1}^{D} x_d^k\right) \tag{24}$$

$$= \sum_{d=1}^{D}\left(V_d\left(x_d^k,\ \omega_d^k\right) - \lambda_k x_d^k\right) + \left(\lambda_k G_k - C_k(G_k)\right) \tag{25}$$

where $\lambda_k > 0$ is the Lagrangian multiplier, and according to the dual theory, (23) can be transform to

$$\min \max \ L\left(x_d^k, G_k, \lambda_k\right) \tag{26}$$

where $\lambda_k > 0$, $x_d^k \in \left[m_d^k, M_d^k\right]$, $G_k \in \left[G_k^{min}, G_k^{max}\right]$.

Remark:

$$x_d^{k*} = argmax\left(\left(V_d\left(x_d^k, \omega_d^k\right) - \lambda_k x_d^k\right)\right) \tag{27}$$

$$x_d^k \in \left[m_d^k, M_d^k\right]$$

$$G_k^* = argmax(\lambda_k G_k - C_k(G_k)) \tag{28}$$

$$G_k \in \left[G_k^{min}, G_k^{max}\right]$$

$$A(\lambda_k) = \max \ L\left(x_d^k, G_k, \lambda_k\right) \tag{29}$$

$$x_d^k \in \left[m_d^k, M_d^k\right]$$

$$G_k \in \left[G_k^{min}, G_k^{max}\right]$$

$$= \sum_{d=1}^{D}\left(V_d\left(x_d^{k*}, \omega_d^k\right) - \lambda_k x_d^{k*}\right) + \left(\lambda_k G_k^* - C_k\left(G_k^*\right)\right) \tag{30}$$

If λ_k represents the Time of Use (TOU) or real-time electricity price for the k-th interval, then Equation (27) defines the driver's optimal electricity consumption for charging that maximizes personal welfare (by balancing utility gain against the cost of electricity). Equation (28) shows how the SHS-EVCS optimizes its welfare by maximizing the difference between revenue from electricity sales and the cost of power supply to determine the optimal electricity provision for charging. Therefore, it is logical to use the Lagrangian multiplier λ_k as the real-time electricity price.

To address the dual problem, where the goal is to minimize $A(\lambda_k)$, with $\lambda_k > 0$, the following method can be employed [57,58]:

$$\lambda_k^{h+1} = \lambda_k^h + \gamma_k^h a_k^h \tag{31}$$

where h is the iteration time; γ_k^h is the distance and ≥ 0; a_k^h is the downward direction of function $A(\lambda_k)$; let $a_k^h = \sum_{d=1}^{D} x_d^{k*} - G_k^*$; (31) proves that when $\sum_{d=1}^{D} x_d^{k*} > G_k^*$, $a_k^h > 0$, while $\lambda_k^{h+1} > \lambda_k^h$ due to the demand and supply theory, and the electricity price increases; when $\sum_{d=1}^{D} x_d^{k*} < G_k^*$, $a_k^h < 0$, then $\lambda_k^{h+1} < \lambda_k^h$, and the electricity price decreases; when $\sum_{d=1}^{D} x_d^{k*} = G_k^*$, while $\lambda_k^{h+1} = \lambda_k^h$, the electricity price stays balanced. In paper [30,34,51], γ is a fixed positive number, which is 0.8. Figure 5 shows the power and information interaction between SHS-EVCS, EV drivers, and IMS. Real-time electricity price distributed algorithm shows in Algorithm 2.

Algorithm 2: Real-time electricity price distributed algorithm

Initialization parameters: γ, D, k, m_d^k, M_d^k, and stop error ε (ε is a positive number 0.1), for h = 0. The IMS randomly releases electricity price λ_k^h to the SHS-EVCS and EV driver.

1. According to the random electricity price, use (26) to calculate the optimal electricity consumption $x_d^{k*}\left(\lambda_k^h\right)$ by the EV driver and report the optimal electricity consumption to the ISM.
2. Use $G_k^* = argmax(\lambda_k G_k - C_k(G_k))$ to calculate the optimal electricity supply $G_k^*\left(\lambda_k^h\right)$ by the SHS-EVCS, and report it to the ISM.
3. Accept the optimal $x_d^{k*}\left(\lambda_k^h\right)$ and $G_k^*\left(\lambda_k^h\right)$, using $\lambda_k^{h+1} = \lambda_k^h + \gamma_k^h a_k^h$ to calculate the optimal electricity price.
4. If $0 < G_k^*\left(\lambda_k^h\right) - \sum_{d=1}^{D} x_d^{k*}(\lambda_k^h) \le \varepsilon$, let $\lambda_k = \frac{1}{2}\left(\lambda_k^h + \lambda_k^{h+1}\right)$; if not, input λ_k^{h+1} to 1

End: output the optimal solution for x_d^{k*} and G_k^*

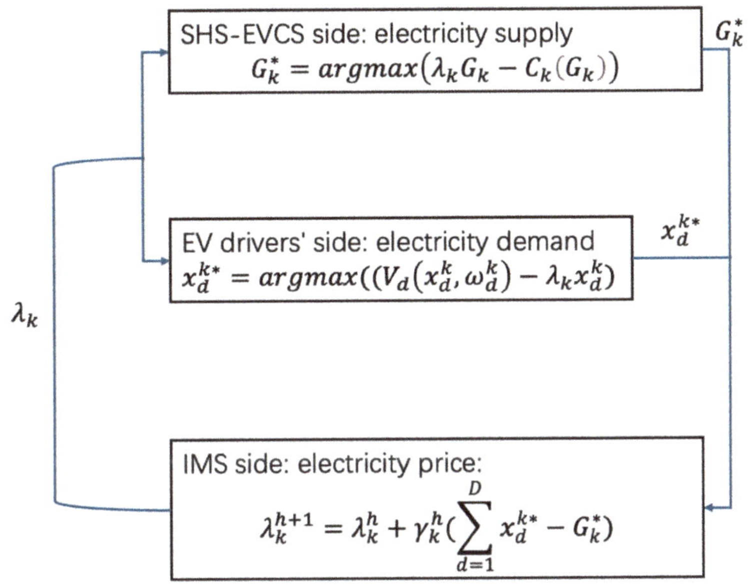

Figure 5. Power and information interaction between SHS-EVCS, EV drivers, and IMS.

4. Case Study

Three scenarios are explored in this analysis: the first involves day-ahead forecasting without the Time-of-Use (TOU) strategy, focusing on uncertainties associated with EVs. The second scenario extends this to include real-time forecasting alongside day-ahead predictions without a TOU strategy. The third introduces both day-ahead and real-time forecasts with an active TOU strategy.

In the TOU scenarios, load management decisions are proactively shaped by anticipated demand and supply data from the transportation department, enabling planning and optimization of energy resources. In contrast, the real-time TOU scenario leverages a Markov decision process to dynamically adjust and optimize the load based on actual data. This approach compares fixed loads, which are immutable and must be met under any circumstances, with flexible loads that can be adjusted based on real-time information. The objective is to assess the impact of these load adjustments on social welfare, with the goal of enhancing efficiency, reducing costs, and improving the synchronization of energy supply and demand.

Table 2 and Figure 6 show the initial scenario that involves fixed loads, where SHS-EVCS parameters are based on the forecasting and scheduling methods based on transport statistics from gov.uk, which is addressed in paper [59]. In Figure 6, the fixed load values for Hammersmith & Fulham and Hounslow are based on weekday averages, which is why the charging peaks occur before and after working hours. The fixed load values for the other two EVCSs are based on weekend averages. Figure 7 displays the SHS-EVCS's load curves for day-ahead and real-time forecasts under a non-TOU framework, showing that variables such as EV charging times, arrival and departure battery capacities, and charging rates remain constant.

Table 2. Four London areas' SHS-EVCS parameters [45,46,60,61].

Parameters	Hammersmith & Fulham	Richmond upon Thames	Hounslow	Ealing
Charger capacity (kW)	360	360	360	360
Number of chargers per station	3	8	14	21
PV installed capacity (kWp)	500	1000	1000	1000
PV installation cost (£/kWp)	1112	1112	1112	1112
Battery capacity (kWh)	1000	800	800	800
Battery installed cost (£/kWh)	331.55	331.55	331.55	331.55
Battery initial state of charge (%)	40	40	40	40
Rated charge and discharge power of battery (kWh)	500	500	500	500
Minimum battery state of charge (%)	25	25	25	25
Maximum battery state of charge (%)	100	100	100	100
Battery charge and discharge efficiency (%)	85	85	85	85
Hydrogen tank capacity (m^3)	1000	1000	1000	1000
Initial capacity of gas tank (%)	30	30	30	30
Hydrogen tank cost (£/m^3)	27.63	27.63	27.63	27.63
Tank storage efficiency (%)	95	95	95	95
Electric to gas efficiency (%)	75	75	75	75
Electricity-to-gas coefficient (kWh/m^3)	0.2	0.2	0.2	0.2
Fuel cell generator capacity (kW)	800	1500	1000	1200
Gas-to-electric efficiency (%)	65	65	65	65
Gas-to-electricity coefficient (m^3/kWh)	0.295	0.295	0.295	0.295
RE feed-in tariff (£/kWh)	0.03	0.03	0.03	0.03

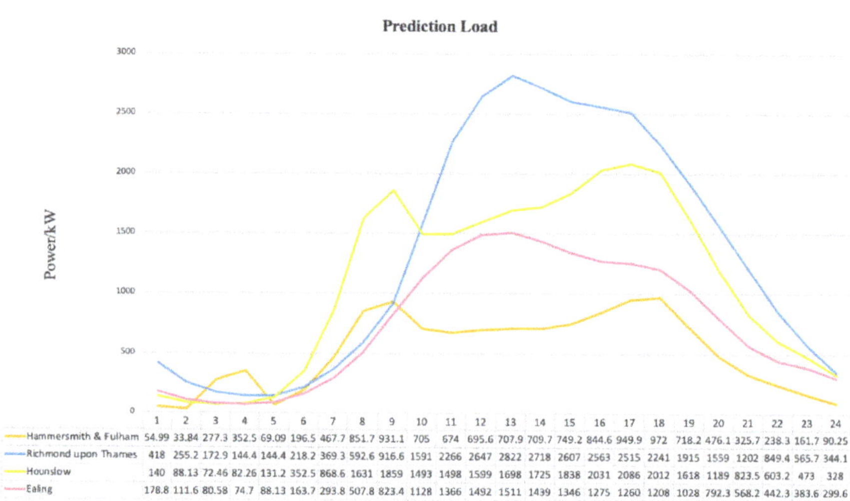

Figure 6. SHS-EVCSs prediction load using [56] method.

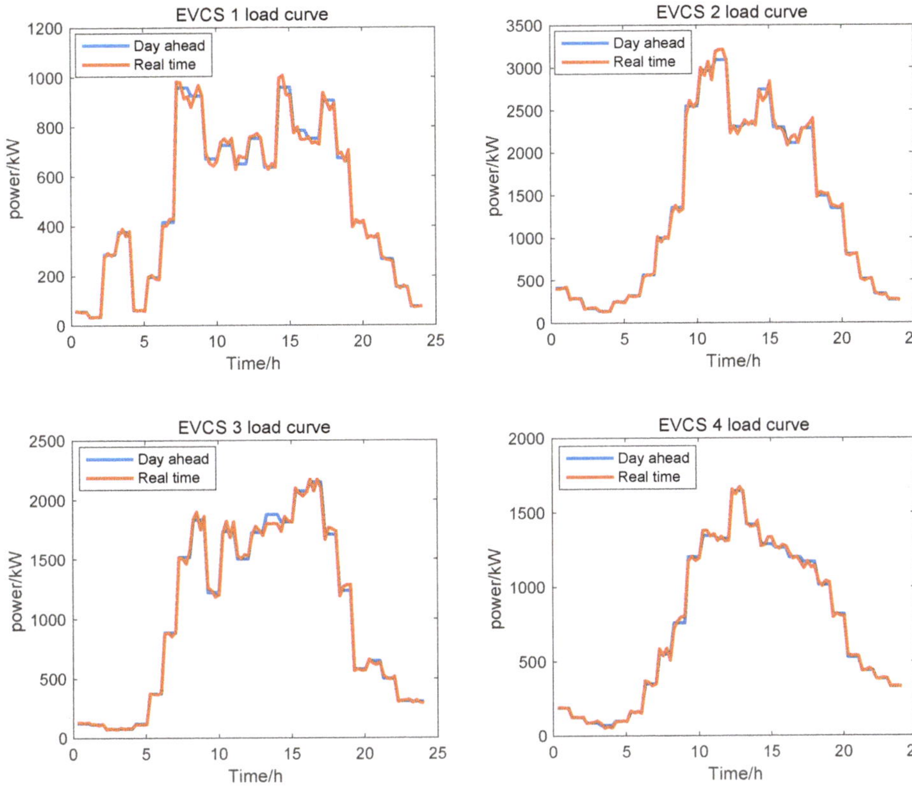

Figure 7. SHS-EVCSs day ahead and real-time load curve without TOU strategy.

Figure 8 shows the anticipated EV load and the impact of implementing a TOU strategy. The blue curve is the standard EV charging pattern without TOU modifications, with pronounced peaks for SHS-EVCSs 1 and 3 around 9 a.m. and 7 p.m.—times that align with typical charging periods. During the early hours (0 to 5 a.m.), power demand is notably lower. In contrast, SHS-EVCSs 2 and 4 experience their peak loads at noon, with lower demand levels mirroring those of SHS-EVCSs 1 and 3. The orange curve, indicating post-TOU implementation, shows a significant smoothing of peaks across all SHS-EVCSs, confirming the strategy's effectiveness in peak shaving. Notably, SHS-EVCSs 2 and 4 exhibit a singular peak at noon, and the mean absolute percentage error (MAPE) for this is 11.37% and 13.56%, further validating the efficiency of the TOU approach. The benefits of the TOU strategy include a substantial decrease in peak loads, visible in the shift from the blue to the orange curve, which alleviates stress on the electrical grid. This could reduce the need for further grid infrastructure enhancements and diminish electricity costs. By more evenly distributing power consumption, the TOU strategy enhances the operational efficiency of power generation and distribution networks, contributing to more consistent grid performance. Furthermore, this strategy can yield financial savings for both utility providers and consumers by minimizing the dependence on more costly and environmentally unfriendly peak power generation.

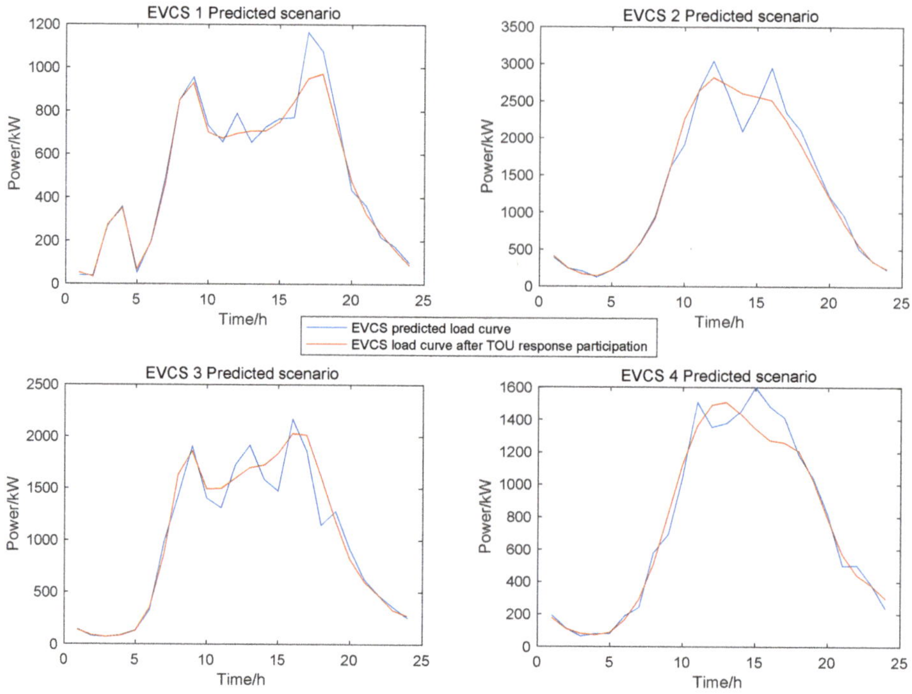

Figure 8. EV prediction load and load after demand side management.

Figure 9 shows the prediction and reduced scenarios for SHS-EVCS prediction. For each SHS-EVCS, a total of 50 scenarios were predicted. In the scenario reduction process, there is a need to identify a subset of the initial scenario set and reassign probabilities to these retained scenarios. This reassignment aims to minimize the probabilistic distance between the probability distribution of the retained scenarios and the initial scenario set.

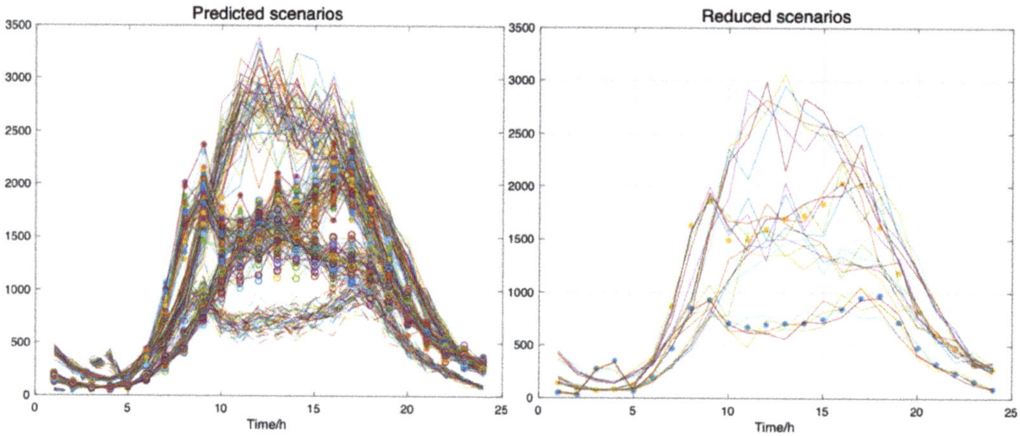

Figure 9. Prediction scenarios and reduced scenarios.

To identify the scenario with the smallest probability to be reduced, one scenario is removed after each iteration. By the end of the entire iterative process, at least the five scenarios with the highest probabilities are retained by the end of the total iteration. This method ensures that while the number of scenarios is reduced, the most representative scenarios are preserved, thereby maintaining the key probabilistic characteristics of the initial scenario set. This approach significantly reduces computational complexity while accurately reflecting the statistical properties of the initial scenario set. Over the span of a day, the prediction scenarios display a high level of variation and complexity, with numerous intersecting lines and extraneous data. In contrast, the reduced scenarios offer a clearer, more focused view, emphasizing the primary trends in electricity demand. Notably, peak demand times at 9 a.m. and 6 p.m. are clear in both scenarios. However, the streamlined scenarios strip away minor variations and outliers, thereby simplifying the information and enhancing its applicability for strategic decision-making.

5. Results and Discussion

Figure 10 shows the EV traffic flow for each SHS-EVCS over a 24 h period, highlighting distinct peaks and valleys in demand. The most significant peak occurs around 10 a.m., with all stations experiencing a marked increase in EV numbers, signalling a morning rush. EVCS 2 sees the highest vehicle count during this period, likely due to its location in a commercial area, making it a popular choice for morning charging. In contrast, EVCS 3, although less busy overall, has notable peaks at 6 a.m. and 9 a.m., indicating specific times of high demand. Another notable surge in charging activity is observed at 8 p.m., aligning with evening charging routines. This pattern is consistent across all stations, with each experiencing an increase in vehicle numbers during this time. EVCS 4 also shows a significant evening peak, indicating steady demand during these hours. The early hours from midnight to 5 a.m. show the lowest vehicle numbers, suggesting minimal charging activity. This low level of activity during the early morning hours might reflect reduced travel and charging demand, potentially due to effective TOU strategies that shift usage away from these times.

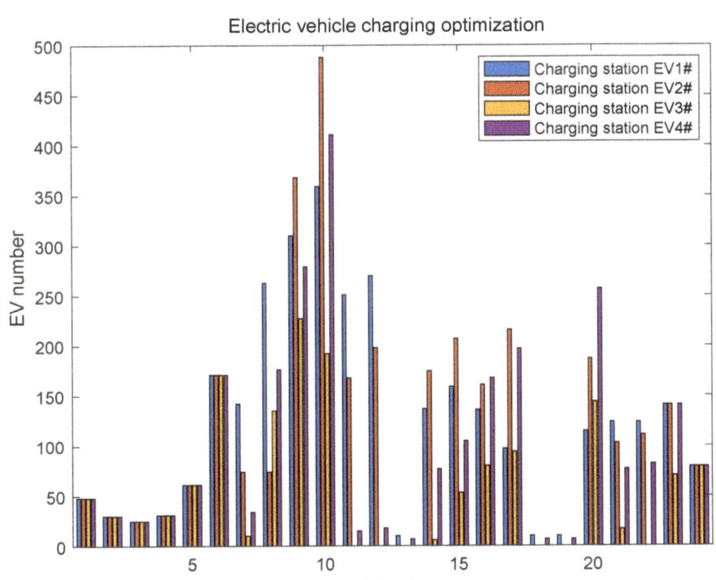

Figure 10. EV charging number optimization.

The charging trends observed are typically linked to daily routines, with morning peaks aligning with pre-work charging and evening peaks with post-work charging. Most EV drivers prefer charging their vehicles before and after work. By adopting strategies to manage these peak times, such as expanding charging station capacity or fine-tuning charging schedules, the efficiency of the entire system and the user experience can be significantly improved.

Figures 11–14 show the energy consumption and electricity load profiles for four SHS-EVCSs. Throughout the daytime, PV systems serve as the main electricity source due to their efficiency and sustainability, making them well-suited for daytime energy production. This reduces reliance on grid-supplied electricity and helps to lower carbon emissions. The use of solar power not only cuts operational costs but also supports broader RE initiatives. With the setting of the sun and the cessation of solar energy production, the SHS-EVCSs transition to using hydrogen fuel cells for power generation. The hydrogen used is typically produced during periods of low demand or when there is surplus solar energy, contributing to a sustainable and reliable energy supply. Additionally, these systems are supplemented by battery storage that accumulates excess energy produced during daylight hours, particularly from the PV systems and the electricity purchased from the grid when its buying price is lower than the P2P purchase price. These batteries then discharge power during peak usage times or when the production from renewable sources is reduced, thereby ensuring a consistent and dependable electricity supply. This capability to balancing the variability of renewable sources underscores the importance of battery integration in maintaining energy stability.

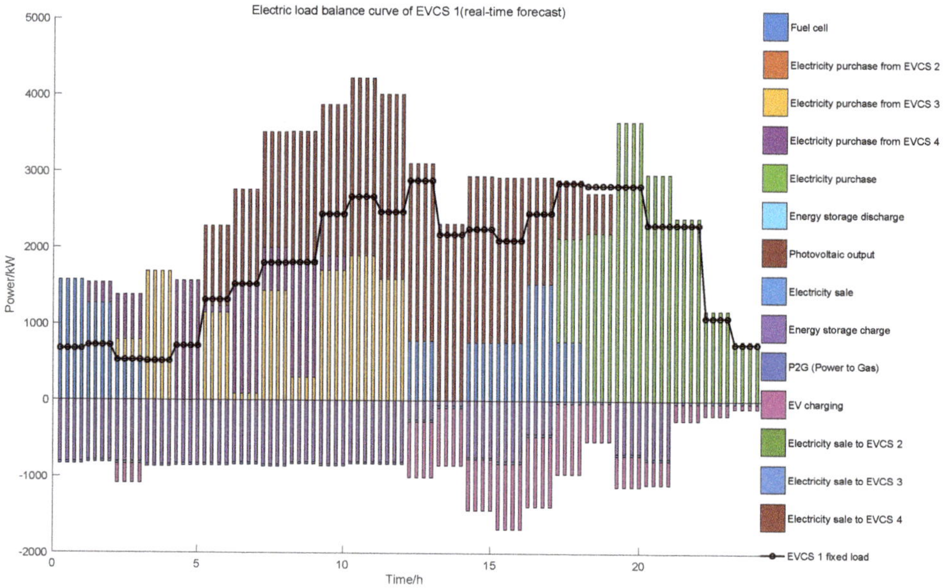

Figure 11. SHS-EVCS1 RE usage and electricity load curve.

Moreover, each SHS-EVCS can engage in electricity trading with other stations in the network, aimed at minimizing costs. This practice of inter-station energy trading enhances flexibility and efficiency in energy management, allowing each station to optimize its operations and reduce expenses. With diverse energy sources and trading options, the SHS-EVCSs are well-equipped to respond to changing energy demands and market conditions, thereby improving the overall stability and sustainability of the electrical grid.

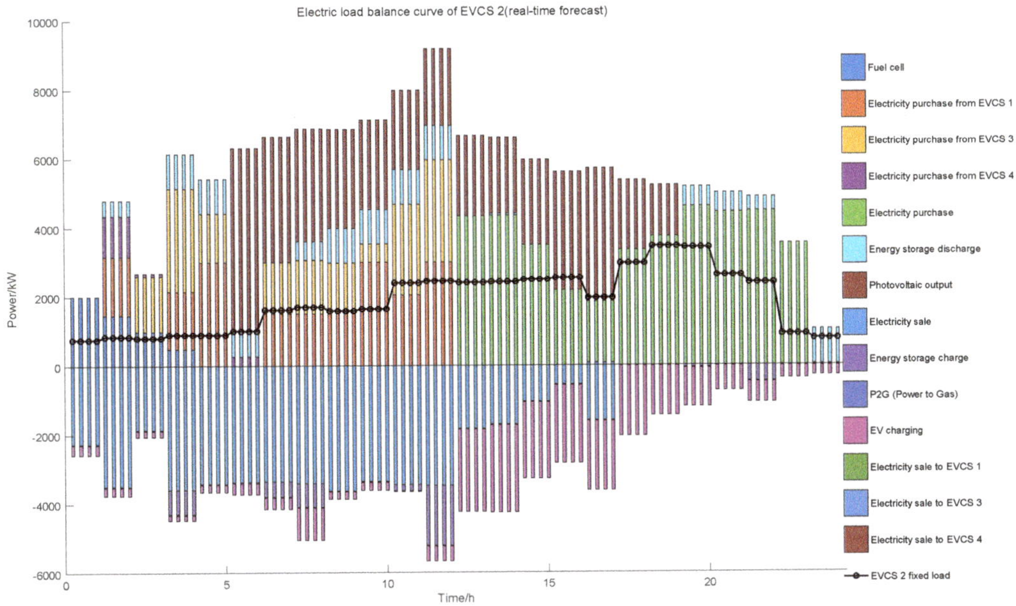

Figure 12. SHS-EVCS2 RE usage and electricity load curve.

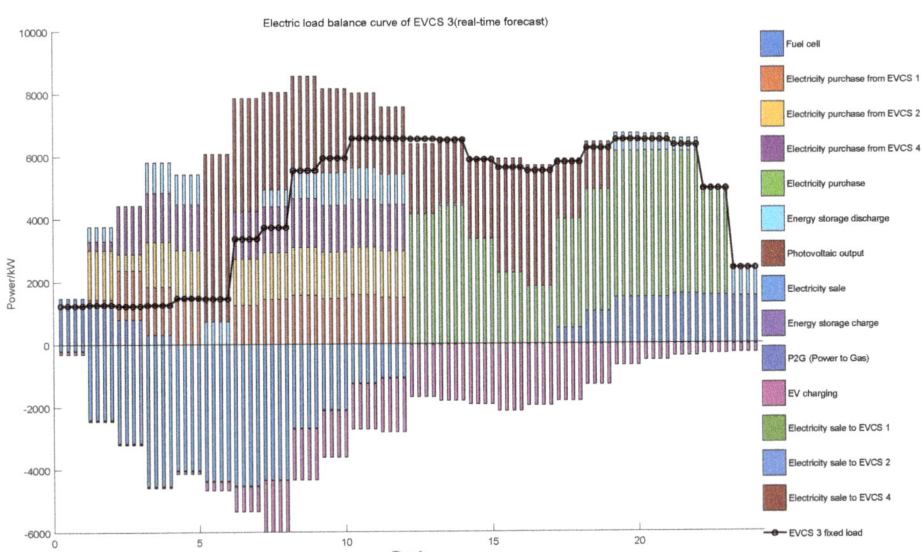

Figure 13. SHS-EVCS3 RE usage and electricity load curve.

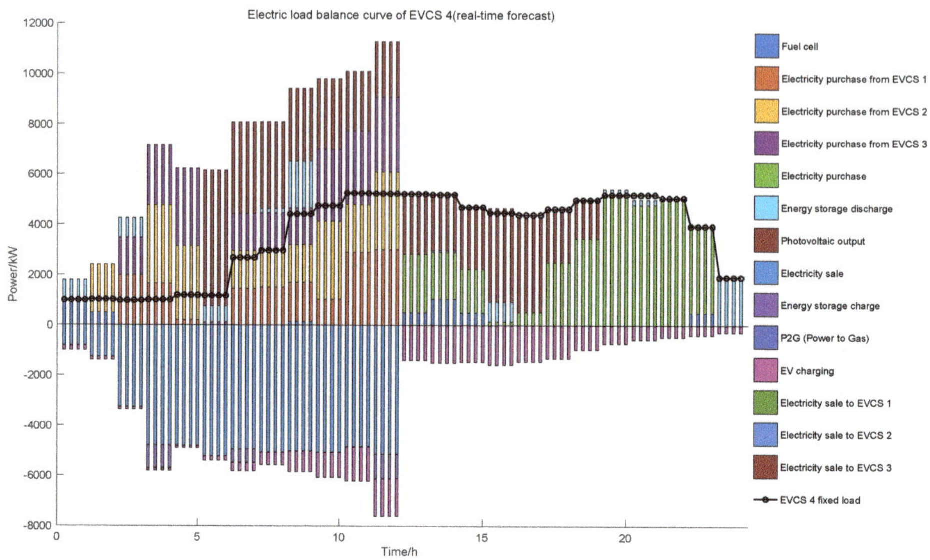

Figure 14. SHS-EVCS4 RE usage and electricity load curve.

Figure 15 shows the relationship between electricity prices and social welfare over 24 h. Initially, the electricity price is relatively high, at around £0.40. The price drops significantly from 11 a.m. to 11 p.m. and remains low, hovering around £0.20. After 11 p.m., the price rises again, peaking at approximately £0.35. Social welfare trends start at a low level when the electricity price is high. As the electricity price decreases, social welfare increases, indicating an inverse relationship between the two variables. This inverse relationship is especially noticeable between 3 p.m. and 11 p.m. when lower electricity prices correspond with higher and more stable levels of social welfare. This relationship suggests that lower electricity prices contribute to higher social welfare, likely due to more affordable energy access benefiting consumers and overall societal well-being. In the economic market, price fluctuations in electricity can be attributed to changes in demand and supply, peak and off-peak hours, or other market dynamics. However, in this paper, the fluctuations in electricity prices are due to the SHS-EVCSs coalition. These fluctuations impact social welfare, but the general trend indicates that as electricity prices decrease, social welfare stabilizes and increases slightly.

Figure 16 shows the relationship between social welfare and time across intervals (day 1, day 2, and day 3). The consistent patterns observed across these scenarios suggest that short-term forecasting can be nearly as effective as long-term forecasting for immediate social welfare planning. This provides insight into the relationship between various time horizon scenarios and social welfare in energy management.

Figure 17 depicts the optimal dynamic power load over 24 h. During the valley load period, from 11 p.m. to 9 a.m., the power load increases from approximately 4000 kW to 13,000 kW. The charging demand for EVs is low during this time, as some drivers charge their EVs at home, and some EVs are already fully charged or have stopped charging. From 6 a.m. to 7 a.m., the sharp increase in load likely indicates that most EV drivers are charging their vehicles as they prepare to leave for work. The SHS-EVCS optimizes the EV charging load through demand response, load forecasting algorithms, and TOU pricing, enhancing power utilization and balancing the internal grid load. It also regulates power distribution to avoid resource wastage during the valley period. At 10 a.m. and from 4 p.m. to 5 p.m., the power load stabilizes around 14,000 kW. During these times, the charging demand for

EVs is relatively stable. Dynamic pricing strategies can further smoothen the load curve and reduce grid pressure fluctuations.

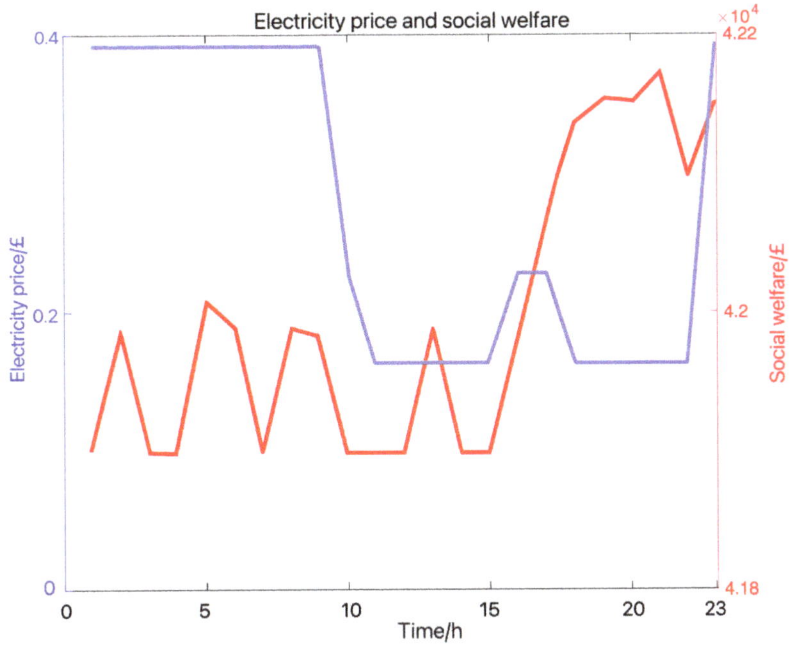

Figure 15. Electricity price and social welfare.

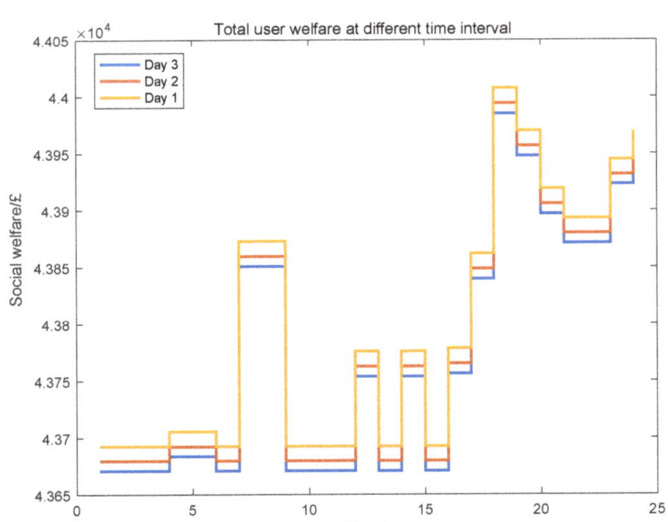

Figure 16. Total user welfare at different time intervals.

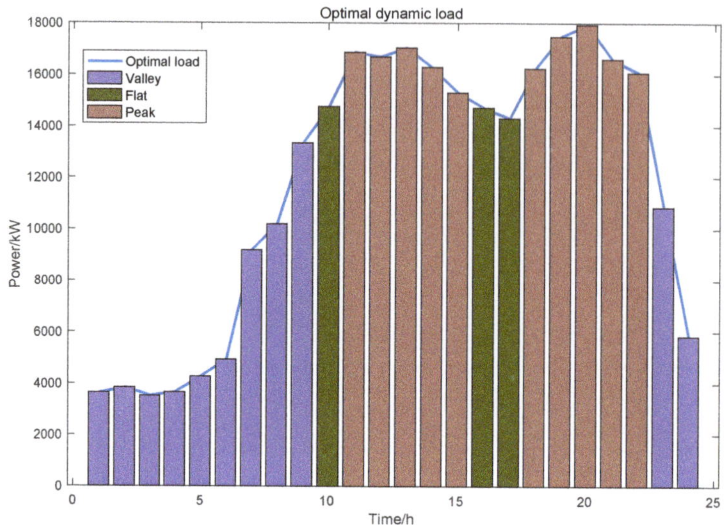

Figure 17. Total optimal dynamic load.

Peak load periods occur from 11 a.m. to 3 p.m. and 6 p.m. to 10 p.m., with loads exceeding 16,000 kW and peaking at 18,000 kW. During these times, the charging demand for EVs rises sharply, especially in the evening when many drivers charge their EVs after returning home from work. Peak shaving strategies such as time-based charging fees, restrictions, and intelligent scheduling systems can be adopted to address this situation. Adjusting the output power of charging stations and coordinating power demand can reduce the grid pressure during peak periods. Additionally, SHS-EVCSs utilize energy storage systems to feed excess power back into the grid during peak times, effectively achieving peak shaving and valley filling. Combined with the IMS, this enables the real-time monitoring and dynamic management of EV charging demand, further enhancing the reliability and stability of both internal and external grids.

Figure 18 shows the profit versus iteration, with iterations ranging from 0 to 25 and profit values from £0 to £18,000. The profit rapidly increases from iteration 1 to 2, reaching approximately £16,800, and then stabilizes around this value for all subsequent iterations. This indicates that this process achieves optimal profit quickly, with minimal changes needed afterward. The initial setup or optimization has an important impact, suggesting that focusing resources on this phase is crucial.

Table 3 presents the energy storage and generation technologies across Hammersmith & Fulham, Richmond upon Thames, Hounslow, and Ealing. Several trends and strategic differences are evident in the deployment of battery storage systems and hydrogen storage systems. The analysis reveals that Ealing has the highest battery storage capacity at 2000 kWh. In contrast, Richmond upon Thames shows a more diversified approach, with significant investments across all technologies. Richmond highlights the highest expenditures in battery systems, with a capacity of 1762.6758 kWh at a cost of £584,415.16, and fuel cell capacity at 1988 kWh costing £997,260.32. All areas have hydrogen storage system capacities of around 1300–1500 kWh. However, Richmond upon Thames again stands out with a slightly higher capacity and cost. Richmond's investment in fuel cells is particularly notable, possibly due to their efficiency and operational flexibility benefits compared to other forms of energy storage and generation. Ealing and Hounslow, while also investing substantially, have lesser capacities and costs associated with their fuel cells.

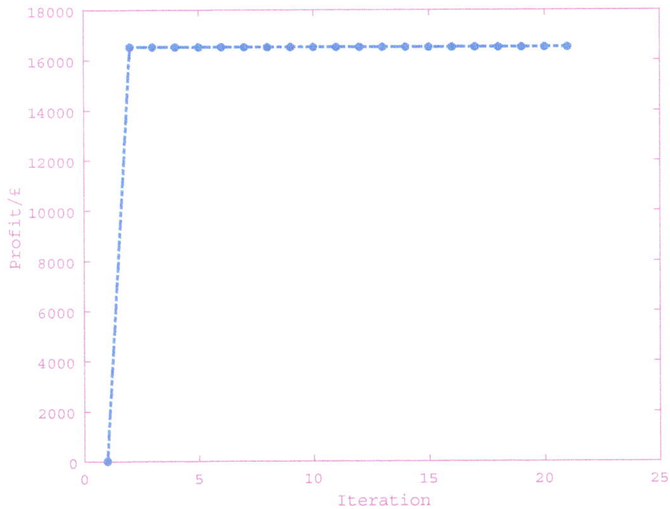

Figure 18. Convergence of real-time electricity price distributed (SHS-EVCS profit) algorithms.

Table 3. Energy optimal parameters and cost.

Parameters	Hammersmith & Fulham	Richmond upon Thames	Hounslow	Ealing
Battery storage system (kWh)	800	1762.68	1000	2000
Battery storage system investment and O/M cost (£/kWh)	265,240	584,415.16	497,330	663,100
Hydrogen storage tank capacity (kWh)	1369.49	1524.98	1379.16	1277.06
Hydrogen storage tank investment and O/M cost (£/kWh)	55,100.09	61,356.08	55,489.2	51,381.16
Fuel cell (kWh)	1582.5	1988	1500	1000
Fuel cell investment and O/M cost (£/kWh)	793,845.3	997,260.32	752,460	501,640

The Table 4 shows the MAPE for four areas, Hammersmith & Fulham (13.04%), Richmond upon Thames (11.37%), Hounslow (11.8%), and Ealing (13.56%). Richmond upon Thames has the lowest MAPE, indicating the highest prediction accuracy, closely followed by Hounslow. Hammersmith & Fulham and Ealing have higher MAPE values, with Ealing having the highest, indicating the least accurate predictions. Overall, while these values are within a reasonable and good range, there is a need for improved forecasting methods in the future.

Table 4. MAPE for 4 SHS-EVCSs.

	Hammersmith & Fulham	Richmond upon Thames	Hounslow	Ealing
MAPE/%	13.04	11.37	11.8	13.56

6. Conclusions

Incorporating a Markov decision process to model EV charging times and using Monte Carlo simulations to predict charging demand has been instrumental in capturing the stochastic nature of EV usage patterns. These methodologies provide a high degree of accuracy in forecasting, enabling the optimization model to adapt to varying conditions and enhancing the reliability of the SHS-EVCS system. Applying the dual theory to determine real-time electricity prices also helps manage demand-side energy consumption. The system can incentivize off-peak charging, reduce peak load stress, and contribute to grid stability by leveraging these theoretically optimal prices. This aligns with broader demand-side management goals, achieving peak shaving and valley filling, reducing overall energy costs, and improving system sustainability. Empirical analysis and case studies demonstrate the effectiveness of TOU strategies in peak load reduction and load balancing. Comparative results between predicted and actual load scenarios highlight the significant impact of TOU in mitigating demand spikes and promoting efficient energy utilization. This benefits the power grid by reducing the need for additional infrastructural investments and offers economic advantages to both EVCS operators and consumers through lower electricity prices and operational costs.

In conclusion, integrating advanced optimization techniques and game theory models in managing SHS-EVCS systems offers a promising pathway toward sustainable and efficient energy infrastructure. The findings of this paper offer valuable insights into the complex interactions between renewable energy sources, storage systems, and EV charging demands. Social welfare could reach up to £4.218×10^4 if the EV charging electricity price is below £0.20. Future research should continue to explore the scalability of these models and their application in diverse geographical and economic contexts, thereby contributing to the global transition toward cleaner and more resilient energy systems.

Author Contributions: Conceptualization, L.D., G.T. and C.S.L.; methodology, L.D. and C.S.L.; software, L.D.; validation, L.D.; formal analysis, L.D.; investigation, L.D.; resources, L.D.; data curation, L.D.; writing—original draft preparation, L.D.; writing—review and editing, L.D., C.S.L. and G.T.; visualization, L.D.; supervision, C.S.L. and G.T.; project administration, C.S.L. and G.T. All authors have read and agreed to the published version of the manuscript.

Funding: This research is supported by Oracle for Research Grant (3146375).

Data Availability Statement: The original contributions presented in the study are included in the article, further inquiries can be directed to the corresponding author.

Conflicts of Interest: The authors declare no conflicts of interest.

Nomenclature

a_f	Waiting probability
a_k^h	The downward direction of function $A(\lambda_k)$
B_{cap}	Battery capacity, 60 kWh (Tesla model 6 battery capacity)
c_f, d_f	The charging probability and the probability of interrupted charging
$C_k(G_k)$	The cost of SHS-EVCS to provide electricity for G_k at time k
C_k	The capital and O&M cost for SHS-EVCS from first-level output
D	Total driver
E_{km}	Energy consumption, 16.25 kWh/100 km
EVC_f, EVD_f	The states of charging mode and wating mode
f_{max}	Maximum number of discrete variables
$\Delta(-f)$	The electricity loss when driving the EV
G_k	SHS-EVCS's electricity supply at time k
G_k^{min}, G_k^{max}	Minimum and maximum electricity supplies of SHS-EVCS
m_d^k, M_d^k	Minimum and maximum electricity consumptions of EV driver d at time k
N_{ev}	Total EV number
p_{ij}, p_{ji}	Transition probability of state i to j, transition probability of state j to i
$P_s(EVC_f)$, $P_s(EVD_f)$	The state probability of EVC_f and EVD_f
P_c	Fixed charging power, 7 kW

$P_{cha,sum}$	Total charging power
$P_{cha,i}$	$i'th$ EV charging power
s_f^{dis}	The $f'th$ discrete state
S^{dis}	Discrete state space
SoC_{start}, $SoC_{start,max}$, $SoC_{start,min}$	Initial, maximum, and minimum SoCs for EV
ΔT_1, ΔT_2	Charging and waiting times
v	Average driving speed, 32 km/h
$V_d\left(x_d^k, \omega_d^k\right)$	Utility of EV driver d using x_d^k at time k
x_f^{dis}	The $f'th$ discrete variable
x_d^k	The electricity consumption of EV driver d at time k
ω_d^k	Driver d's elasticity of charging his EV at time k, $\omega_d^k \in [1, 3.5]$
η_c	EV battery efficiency, 95%
λ_k	The real-time electricity price
γ_k^h	The distance, 0.8

Abbreviations

CSs	Charging stations
DR	Demand response
DSM	Demand-side management
EV	Electric vehicle
G2P	Gas to power
IMS	Internal Management System
MAPE	Mean Absolute Percentage Error
P2G	Power to gas
PV	Photovoltic
RE	Renewable energy
SHS-EVCSs	Solar–Hydrogen-Battery Storage Electric Vehicle Charging Stations

References

1. Liu, C.; Chau, K.T.; Wu, D.; Gao, S. Opportunities and Challenges of Vehicle-to-Home, Vehicle-to-Vehicle, and Vehicle-to-Grid Technologies. *Proc. IEEE* **2013**, *101*, 2409–2427. [CrossRef]
2. Kubański, M. Prospects for the Use of Electric Vehicles in Public Transport on the Example of the City of Czechowice-Dziedzice. *Transp. Res. Procedia* **2020**, *44*, 110–114. [CrossRef]
3. Zeng, J.; Li, M.; Liu, J.F.; Wu, J.; Ngan, H.W. Operational optimization of a stand-alone hybrid renewable energy generation system based on an improved genetic algorithm. In Proceedings of the 2010 IEEE Power and Energy Society General Meeting, Providence, RI, USA, 25–29 July 2010.
4. Tushar, W.; Yuen, C.; Huang, S.; Smith, D.B.; Poor, H.V. Cost Minimization of Charging Stations with Photovoltaics: An Approach With EV Classification. *IEEE Trans. Intell. Transp. Syst.* **2016**, *17*, 156–169. [CrossRef]
5. Keyhani, A. *Design of Smart Power Grid Renewable Energy Systems*; John Wiley & Sons: Hoboken, NJ, USA, 2019; pp. 1–606. [CrossRef]
6. Das, H.S.; Rahman, M.M.; Li, S.; Tan, C.W. Electric vehicles standards, charging infrastructure, and impact on grid integration: A technological review. *Renew. Sustain. Energy Rev.* **2019**, *120*, 109618. [CrossRef]
7. Bohnsack, R.; Pinkse, J.; Kolk, A. Business models for sustainable technologies: Exploring business model evolution in the case of electric vehicles. *Res. Policy* **2014**, *43*, 284–300. [CrossRef]
8. Markkula, J.; Rautiainen, A.; Jarventausta, P. The business case of electric vehicle quick charging–no more chicken or egg problem. *World Electr. Veh. J.* **2013**, *6*, 921–927. [CrossRef]
9. Madina, C.; Zamora, I.; Zabala, E. Methodology for assessing electric vehicle charging infrastructure business models. *Energy Policy* **2015**, *89*, 284–293. [CrossRef]
10. Ehteshami, S.M.M.; Chan, S.H. The role of hydrogen and fuel cells to store renewable energy in the future energy network–potentials and challenges. *Energy Policy* **2014**, *73*, 103–109. [CrossRef]
11. Yue, M.; Lambert, H.; Pahon, E.; Roche, R.; Jemei, S.; Hissel, D. Hydrogen energy systems: A critical review of technologies, applications, trends and challenges. *Renew. Sustain. Energy Rev.* **2021**, *146*, 111180. [CrossRef]
12. Yap, K.Y.; Chin, H.H.; Klemeš, J.J. Solar Energy-Powered Battery Electric Vehicle charging stations: Current development and future prospect review. *Renew. Sustain. Energy Rev.* **2022**, *169*, 112862. [CrossRef]

13. Barman, P.; Dutta, L.; Bordoloi, S.; Kalita, A.; Buragohain, P.; Bharali, S.; Azzopardi, B. Renewable energy integration with electric vehicle technology: A review of the existing smart charging approaches. *Renew. Sustain. Energy Rev.* **2023**, *183*, 113518. [CrossRef]
14. Manousakis, N.M.; Karagiannopoulos, P.S.; Tsekouras, G.J.; Kanellos, F.D. Integration of Renewable Energy and Electric Vehicles in Power Systems: A Review. *Processes* **2023**, *11*, 1544. [CrossRef]
15. Rajendran, G.; Vaithilingam, C.A.; Misron, N.; Naidu, K.; Ahmed, R. A comprehensive review on system architecture and international standards for electric vehicle charging stations. *J. Energy Storage* **2021**, *42*, 103099. [CrossRef]
16. Jauhar, S.K.; Sethi, S.; Kamble, S.S.; Mathew, S.; Belhadi, A. Artificial intelligence and machine learning-based decision support system for forecasting electric vehicles' power requirement. *Technol. Forecast. Soc. Change* **2024**, *204*, 123396. [CrossRef]
17. Thompson, L.; Davidson, P. Future of Battery Storage: Impact of Solid-State Technology on EV Charging Stations. *Energy Storage Trends* **2023**, *7*, 99–115.
18. Ali, A.; Shakoor, R.; Raheem, A.; Abd Muqeet, H.A.; Awais, Q.; Khan, A.A.; Jamil, M. Latest energy storage trends in multi-energy standalone electric vehicle charging stations: A comprehensive study. *Energies* **2022**, *15*, 4727. [CrossRef]
19. Franco, A.; Giovannini, C. Recent and Future Advances in Water Electrolysis for Green Hydrogen Generation: Critical Analysis and Perspectives. *Sustainability* **2023**, *15*, 16917. [CrossRef]
20. Boglou, V.; Karavas, C.-S.; Karlis, A.; Arvanitis, K.G.; Palaiologou, I. An Optimal Distributed RES Sizing Strategy in Hybrid Low Voltage Networks Focused on EVs Integration. *IEEE Access* **2023**, *11*, 16250–16270. [CrossRef]
21. Mitali, J.; Dhinakaran, S.; Mohamad, A.A. Energy Storage Systems: A Review. *Energy Storage Sav.* **2022**, *1*, 166–216. [CrossRef]
22. Huang, H.; Deng, L.; Zhang, L. Review of Real-time Pricing based on Demand Response. *Electr. Technol.* **2015**, *16*, 1–6.
23. Deng, R.; Yang, Z.; Chow, M. A survey on demand response in smart grids: Mathematical models and approaches. *IEEE Trans. Ind. Inform.* **2015**, *11*, 570–582. [CrossRef]
24. Palensky, P.; Dietrich, D. Demand side management: Demand response, intelligent energy systems, and smart loads. *IEEE Trans. Ind. Inform.* **2011**, *7*, 381–388. [CrossRef]
25. Mariano-Hernández, D.; Hernández-Callejo, L.; Zorita-Lamadrid, A.; Duque-Pérez, O.; García, F.S. A review of strategies for building energy management system: Model predictive control, demand side management, optimization, and fault detect & diagnosis. *J. Build. Eng.* **2020**, *33*, 101692. [CrossRef]
26. Onile, A.E.; Machlev, R.; Petlenkov, E.; Levron, Y.; Belikov, J. Uses of the digital twins concept for energy services, intelligent recommendation systems, and demand side management: A review. *Energy Rep.* **2021**, *7*, 997–1015. [CrossRef]
27. Demand Response as Low-Cost Reliability Service: Techno-Economic Benefits and Challenges. Available online: https://smartgrid.ieee.org/newsletters/november-2017/1070-demand-response-as-low-cost-reliability-service-techno-economic-benefits-and-challenges (accessed on 30 November 2017).
28. Paterakis, N.G.; Erdinç, O.; Catalão, J.P.S. An overview of demand response: Key-elements and international experience. *Renew. Sustain. Energy Rev.* **2017**, *69*, 871–891. [CrossRef]
29. Märkle-Huß, J.; Feuerriegel, S.; Neumann, D. Large-scale demand response and its implications for spot prices, load and policies: Insights from the German-Austrian electricity market. *Appl. Energy* **2018**, *210*, 1290–1298. [CrossRef]
30. Zhu, H.; Gao, Y.; Hou, Y. Real-Time Pricing for Demand Response in Smart Grid Based on Alternating Direction Method of Multipliers. *Math. Probl. Eng.* **2018**, *2018*, 1–10. [CrossRef]
31. Pfeifer, A.; Dobravec, V.; Pavlinek, L.; Krajačić, G.; Duić, N. Integration of renewable energy and demand response technologies in interconnected energy systems. *Energy* **2018**, *161*, 447–455. [CrossRef]
32. Mimica, M.; Sinovčić, Z.; Jokić, A.; Krajačić, G. The role of the energy storage and the demand response in the robust reserve and network-constrained joint electricity and reserve market. *Electr. Power Syst. Res.* **2022**, *204*, 107716. [CrossRef]
33. Lu, R.; Hong, S.H.; Zhang, X. A Dynamic pricing demand response algorithm for smart grid: Reinforcement learning approach. *Appl. Energy* **2018**, *220*, 220–230. [CrossRef]
34. Asadi, G.; Gitizadeh, M.; Roosta, A. Welfare maximization under real-time pricing in smart grid using PSO algorithm. In Proceedings of the 2013 21st Iranian Conference on Electrical Engineering (ICEE), Mashhad, Iran, 14–16 May 2013; pp. 1–7. [CrossRef]
35. Samadi, P.; Mohsenian-Rad, H.; Schober, R.; Wong, V.W.S. Advanced demand side management for the future smart grid using mechanism design. *IEEE Trans. Smart Grid* **2012**, *3*, 1170–1180. [CrossRef]
36. Fu, Z.; He, X.; Huang, T.; Abu-Rub, H. A distributed continuous time consensus algorithm for maximize social welfare in micro grid. *J. Frankl. Inst.* **2016**, *353*, 3966–3984. [CrossRef]
37. Shafiei, M.; Ghasemi-Marzbali, A. Electric vehicle fast charging station design by considering probabilistic model of renewable energy source and demand response. *Energy* **2023**, *267*, 126545. [CrossRef]
38. Meng, W.; Song, D.; Huang, L.; Chen, X.; Yang, J.; Dong, M.; Talaat, M.; Elkholy, M.H. Distributed energy management of electric vehicle charging stations based on hierarchical pricing mechanism and aggregate feasible regions. *Energy.* **2024**, *291*, 130332. [CrossRef]
39. Huang, X.; Lin, Y.; Lim, M.K.; Zhou, F.; Liu, F. Electric vehicle charging station diffusion: An agent-based evolutionary game model in complex networks. *Energy* **2022**, *257*, 124700. [CrossRef]
40. Li, G.; Yang, J.; Hu, Z.; Zhu, X.; Xu, J.; Sun, Y.; Zhan, X.; Wu, F. A novel price-driven energy sharing mechanism for charging station operators. *Energy Econ.* **2023**, *118*, 106518. [CrossRef]

41. Wu, Y.; Ravey, A.; Chrenko, D.; Miraoui, A. Demand side energy management of EV charging stations by approximate dynamic programming. *Energy Convers. Manag.* **2019**, *196*, 878–890. [CrossRef]
42. Adetunji, K.E.; Hofsajer, I.W.; Abu-Mahfouz, A.M.; Cheng, L. An optimization planning framework for allocating multiple distributed energy resources and electric vehicle charging stations in distribution networks. *Appl. Energy* **2022**, *322*, 119513. [CrossRef]
43. Abid, S.; Ahshan, R.; Al Abri, R.; Al-Badi, A.; Albadi, M. Techno-economic and environmental assessment of renewable energy sources, virtual synchronous generators, and electric vehicle charging stations in microgrids. *Appl. Energy* **2024**, *35*, 122028. [CrossRef]
44. Güven, A.F. Integrating electric vehicles into hybrid microgrids: A stochastic approach to future-ready renewable energy solutions and management. *Energy* **2024**, *303*, 131968. [CrossRef]
45. Duan, L.; Guo, Z.; Taylor, G.; Lai, C.S. Multi-Objective Optimization for Solar-Hydrogen-Battery-Integrated Electric Vehicle Charging Stations with Energy Exchange. *Electronics* **2023**, *12*, 4149. [CrossRef]
46. Duan, L.; Yuan, Y.; Taylor, G.; Lai, C.S. Game-Theory-Based Design and Analysis of a Peer-to-Peer Energy Exchange System between Multi-Solar-Hydrogen-Battery Storage Electric Vehicle Charging Stations. *Electronics* **2024**, *13*, 2392. [CrossRef]
47. Ma, Q.; Tong, X.; Huang, J.; Wang, P.; Li, J. Nodal charging demand forecast of EVs considering drivers’ psychological bearing ability based on NMC-MCS. *IET Gener. Transm. Distrib.* **2021**, *16*, 467–478. [CrossRef]
48. Fotouhi, Z.; Hashemi, M.R.; Narimani, H.; Bayram, I.S. A General Model for EV Drivers Charging Behavior. *IEEE Trans. Veh. Technol.* **2019**, *68*, 7368–7382. [CrossRef]
49. Tan, J.; Wang, L. Real-Time Charging Navigation of Electric Vehicles to Fast Charging Stations: A Hierarchical Game Approach. *IEEE Trans. Smart Grid* **2015**, *8*, 846–856. [CrossRef]
50. Deilami, S.; Masoum, A.S.; Moses, P.S.; Masoum, M.A.S. Real-Time Coordination of Plug-In Electric Vehicle Charging in Smart Grids to Minimize Power Losses and Improve Voltage Profile. *IEEE Trans. Smart Grid* **2011**, *2*, 456–467. [CrossRef]
51. Electric Vehicle Charging Research Survey with Electric Vehicle Drivers Research Report. Department of Transportation. 2022. Available online: https://assets.publishing.service.gov.uk/media/628f5603d3bf7f037097bd73/dft-ev-driver-survey-summary-report.pdf (accessed on 30 April 2022).
52. Samadi, P.; Mohsenian-Rad, A.-H.; Schober, R.; Wong, V.W.S.; Jatskevich, J. Optimal real-time pricing algorithm based on utility maximization for smart grid. In Proceedings of the 2010 First IEEE International Conference on Smart Grid Communications, Gaithersburg, MD, USA, 4–6 October 2010; pp. 415–420. [CrossRef]
53. Mohsenian-Rad, A.-H.; Wong, V.W.S.; Jatskevich, J.; Schober, R.; Leon-Garcia, A. Autonomous demand-side management based on game-theoretic energy consumption scheduling for the future smart grid. *IEEE Trans. Smart Grid* **2010**, *1*, 320–331. [CrossRef]
54. Chao, H.; Hsieh, P.; Yang, T.; Hsiung, A. Model predictive optimization for distribution management in smart grids. In Proceedings of the IECON 2016-42nd Annual Conference of the IEEE Industrial Electronics Society, Florence, Italy, 23–26 October 2016; pp. 3984–3989.
55. Boyd, S.; Vandenberghe, L. *Convex Optimization*; Cambridge University Press: Cambridge, UK, 2004.
56. Deng, R.; Yang, Z.; Hou, F.; Chow, M.-Y.; Chen, J. Distributed Real–Time Demand Response in Multiseller–Multibuyer Smart Distribution Grid. *IEEE Trans. Power Syst.* **2015**, *30*, 2364–2374. [CrossRef]
57. Xu, D.; Zhong, F.; Bai, Z.; Wu, Z.; Yang, X.; Gao, M. Real-time multi-energy demand response for high-renewable buildings. *Energy Build.* **2023**, *281*, 112764. [CrossRef]
58. Trichakis, N.; Zymnis, A.; Boyd, S. Dynamic network utility maximization with delivery contracts. *IFAC Proc. Vol.* **2008**, *41*, 2907–2912. [CrossRef]
59. Duan, L.; Zhang, X.; Balta-Ozkan, N.; Etminan, S. Design and operation of solar-hydrogen-storage integrated electric vehicle charging station in smart city. In Proceedings of the CIRED 2021 Conference, Geneva, Switzerland, 20–23 September 2021. [CrossRef]
60. Barakat, S.; Samy, M.M. A Hybrid Photovoltaic/Wind Green Energy System for Outpatient Clinic Utilizing Fuel Cells and Different Batteries as a Storage Devices. In Proceedings of the 2022 23rd International Middle East Power Systems Conference (MEPCON), Cairo, Egypt, 13–15 December 2022; pp. 1–6.
61. Allan, M. Hydrogen Cars: How the Fuel Cell Works, Where the UK's Filling Stations Are and How Expensive They Are to Run. The Scotsman. Available online: https://www.scotsman.com/lifestyle/cars/hydrogen-cars-how-the-fuel-cell-works-where-the-uks-filling-stations-are-and-how-expensive-they-are-to-run-1386285# (accessed on 23 November 2023).

World Electric Vehicle Journal

MDPI

Review

Relocation Optimization for Shared Electric Vehicles: A Literature Review

Ye Zou [1], Qian Yu [1,*], Dongming Jiang [2] and Youjun Deng [3]

1 College of Business Administration, Hunan University of Finance and Economics, Changsha 410205, China; zouye_hutb@163.com
2 College of Accounting, Guangzhou College of Technology and Business, Guangzhou 510850, China; jiangdongming@gzgs.edu.cn
3 School of Electrical and Information Engineering, Southwest Petroleum University, Chengdu 610500, China; youjun_deng0910@tju.edu.cn
* Correspondence: yuqian@hufe.edu.cn

Abstract: Car sharing has become an increasingly popular mode of travel. This paper provides a comprehensive literature survey on relocation optimization for shared electric vehicles. The literature is reviewed and categorized based on two types of relocation: static and dynamic relocation. Static relocation is analyzed in terms of operator-based relocation and user-based relocation, while dynamic relocation is analyzed in terms of four methodologies: optimization models, simulations, multi-stage methods, and deep reinforcement learning. The paper finally provides some interesting future research topics, such as considering the nonlinear charging process of electric vehicles in the process of constructing relocation optimization models and designing algorithms for shared electric vehicles.

Keywords: relocation; shared electric vehicles; optimization; review

1. Introduction

Car sharing is a form of Internet vehicle rental for short-distance travel in cities. Affected by the subsidy policy of new energy vehicles and the double integral policy, the vast majority of car sharing platforms involve mainly small and mini electric vehicles. Electric vehicle sharing can alleviate traffic congestion and reduce greenhouse gas emissions, and it has become an important mode of travel. In a one-way electric vehicle sharing system, due to the fact that users return their cars from other places, some stations often have excessive numbers of vehicles, which leads to the inability of users to return their cars, and some stations have a shortage of vehicles, which leads to the inability of users to access cars. Therefore, it is necessary to relocate vehicles to alleviate the imbalance of vehicles in time and space distribution.

Vehicle relocation strategies are divided into user-based relocation strategies and operator-based relocation strategies. The former aims to change the user's leasing behavior through some operational mechanism, thus changing the distribution of vehicles. In the latter, dispatchers are dispatched by the operator to relocate vehicles between stations. At this time, all dispatchers start from the relocation center, reach a pick-up demand point by riding a folding bicycle, put the bicycle into the trunk of the vehicle, and then drive the collected vehicle to a delivery demand point, park, remove the bicycle from the trunk of the vehicle, ride to the next pick-up demand point, and so on. Eventually, the dispatcher rides back to the relocation center. Obviously, the relocation route is composed of an alternating sequence of pick-up and delivery points, and the service time window constraints, the electric vehicle power, and the dispatcher's working time constraints must be satisfied.

This paper aims to review the relocation problem for shared electric vehicles, which is divided into static and dynamic aspects, to propose future research directions for researchers. Especially when designing shared electric vehicle relocation optimization algorithms, deep reinforcement learning algorithms combining the strong feature extraction ability of deep neural networks and the characteristics of reinforcement learning methods that are good at learning beneficial actions can be considered.

2. Literature Review

2.1. Research on Static Relocation for Shared Electric Vehicles

Figure 1 shows the relocation process for shared electric vehicles. Specifically, the static relocation problem for shared electric vehicles assumes that the parameters such as vehicle relocation demand and vehicle travel time are known values given in advance.

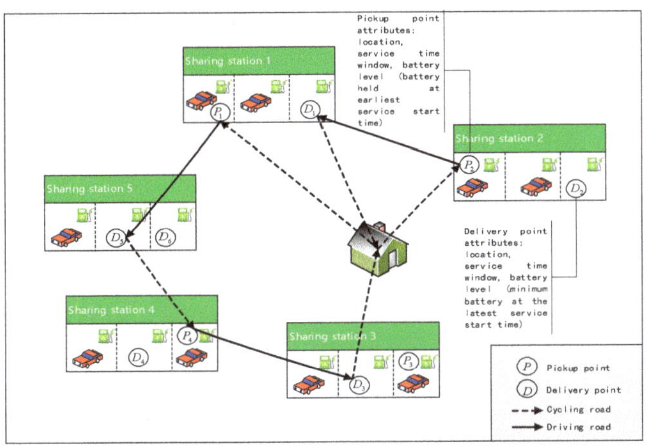

Figure 1. Relocation process for shared electric vehicles.

Many scholars generally assume that by monitoring the real-time demand of users, parking capacity constraints, and the state of shared electric vehicles (e.g., location, remaining battery level, etc.), the operation center of the electric vehicle sharing system can run an appropriate algorithm to obtain the vehicle relocation demand between the sharing stations (i.e., the pick-up demand and the delivery demand), and the demand is considered to be a predetermined value [1–3]. In this section, the research survey of the static relocation problem for shared vehicles is divided into two categories: operator-led and user-led. Table 1 lists the main related research literature and each study's characteristics.

Table 1. Main related research literature and characteristics.

Authors	Publication Year	Systems	Relocation Strategy	Method
C.S. et al. [4]	2024	Free floating	Operator-based	RT exchange-based local search algorithm
Tian et al. [5]	2024	Station-based	Operator-based	Multi-iteration probabilistic greedy algorithm
Wang et al. [6]	2023	Station-based	Operator-based	Gurobi solver and charging priority ranking

Table 1. *Cont.*

Authors	Publication Year	Systems	Relocation Strategy	Method
Qin et al. [1]	2022	Station-based	Operator-based	Branch-and-price-and-cut
Li, et al. [7]	2022	Station-based	Operator-based	Simulation
Zou et al. [2]	2021	Station-based	Operator-based	Large neighborhood search algorithm
Brendel et al. [8]	2017	Free floating	User-based	Design science research
Jorge et al. [9]	2015	Station-based	User-based	Iterated local search
Willing et al. [10]	2017	Free floating	User-based	Spatial decision support
Cepolina et al. [11]	2015	Station-based	User-based	Object-oriented simulation
Brendel et al. [12]	2018	Station-based	User-based	Discrete event simulation
Angela et al. [13]	2019	Station-based	User-based	Discrete event simulation
Boyacı et al. [14]	2017	Station-based	Operator-based	Integer programming and discrete event simulation
Nourinejad et al. [15]	2015	Station-based	Operator-based	Heuristic algorithm
Bruglieri et al. [16]	2017	Station-based	Operator-based	Random heuristics/Destruction and reconstruction metaheuristics
Bruglieri et al. [17]	2019	Station-based	Operator-based	Metaheuristics for adaptive large neighborhood search
Ait-Ouahmed et al. [18]	2018	Station-based	Operator-based	Greedy algorithm and tabu search algorithm
Kek et al. [19]	2009	Station-based	Operator-based	Optimization trend simulation decision support system
Fan et al. [20]	2008	Station-based	Operator-based	Stochastic optimization methods based on Monte Carlo sampling simulation
Alfian et al. [21]	2014	Station-based	Operator-based	
Gambella et al. [22]	2018	Station-based	Operator-based	Heuristic algorithm
Nourinejad & Roorda [23]	2014	Station-based	Operator-based	Discrete event simulation
Weikl & Bogenberger [24]	2015	Free floating	Operator-based	Integer programming
Correia & Antunes [25]	2012	Station-based	User-based/Station location	Mixed integer programming
Deng & Cardin [26]	2018	Station-based	Operator-based/Station location	Discrete event simulation/particle swarm optimization
Boyacı et al. [27]	2015	Station-based	Operator-based/Station location	Integer programming
Li et al. [28]	2016	Station-based	Operator-based/Station location	Continuum approximation
Correia et al. [29]	2014	Station-based	Operator-based/Station location	Mixed integer programming

2.1.1. Operator-Based Static Relocation for Shared Vehicles

Operator-based vehicle relocation strategies are suitable for scenarios where the start and destination of the relocation have been determined. This strategy makes it easy to integrate vehicles and has high reliability, but it also incurs labor costs, cycling costs, driving costs, and costs related to the operation of the relocation center [30]. For the operator-based shared vehicle relocation problem, scholars have studied it from different aspects, which can be roughly divided into the following two categories.

Research on Separate Relocation for Shared Vehicles

Many scholars have devoted themselves to the study of the vehicle relocation problem in shared vehicle systems. They generally adopt an optimization model consisting of a defined objective function and constraints, and most of them aim to maximize the profit of the system operator (the difference between the demand satisfaction revenue of the pick-up and delivery vehicles and the labor cost of the dispatcher) or minimize the total cost. A mixed integer programming model for shared electric vehicle relocation decisions based on folding-bicycle-assisted relocation is constructed, and heuristic algorithms or exact algorithms such as adaptive large neighborhood search algorithms are designed to solve it, and compared with other algorithms such as a tabu search algorithm or a mixed integer linear programming optimal solution [2,16,17]. The latest research aims to establish a set packing model for the shared vehicle relocation problem, propose a branch, cut, and price algorithm, and design a bidirectional label-setting algorithm to deal with the pricing sub-problem [1]. Cai et al. [31] studied the electric vehicle relocation problem when triangle inequality does not hold, and designed a hybrid adaptive large neighborhood search and tabu search algorithm to maximize the total profit under the constraints of time window, relocation duration, and charging demand. In addition, Zou et al. [2] considered the dispatcher's familiarity with the path, constructed a mixed integer linear programming model with the goal of maximizing the quantity of satisfied demands, and designed a large neighborhood search algorithm to solve it. In addition, as the high or low relocation efficiency directly affects the user service experience, some scholars weigh the correlation between the user service level and the vehicle relocation cost from the perspective of users, construct a linear programming vehicle relocation model with this as the optimization objective, and design a heuristic algorithm to solve it [18]. When optimizing vehicle relocation in an electric vehicle sharing system, the particularity of the charging/consumption process of electric vehicles is the key consideration. Some scholars have specifically considered this characteristic, built an electric vehicle relocation model during the operation of the sharing system, and extended the time interval of the model to the shutdown of the system [22].

Different from the above station-based vehicle sharing systems, Weikl & Bogenberger [24] demonstrated the operation process of a multi-stage vehicle sharing relocation management tool in a one-way free-floating vehicle sharing network. Based on the predicted demand, vehicles are relocated between regions and within these regions. These stages begin by strategically defining the area and calculating the expected demand. If there is a mismatch between supply and demand, mixed integer programming (MIP) is applied in macro and micro regions, respectively, to find the relocation strategy that maximizes the profit. Dividing the model into different stages helps to apply the branch and cut algorithm (CPLEX) to obtain the exact solution, but it only uses the known demand to identify the required vehicle. For example, customer backflow into the area is not considered in the actual vehicle inventory level [24]. Li et al. [28] also adopted a similar approach by dividing the operation area into multiple regions.

Research on Joint Relocation for Shared Vehicles, Personnel, and Station Planning

Many vehicle sharing relocation models assume that vehicles can be relocated from one station to another, while ignoring the spatial and temporal distribution of dispatchers.

In fact, dispatchers performing vehicle relocation operations will lead to the problem of an unbalanced spatiotemporal distribution of dispatchers among stations [15]. Therefore, many scholars have devoted themselves to the joint relocation for shared vehicles and dispatchers. Tian et al. [5] studied the joint relocation optimization problem of vehicles and dispatchers in a one-way vehicle sharing system using a combined mode of curbside parking spaces and parking hubs, and proposed a new multi-iteration probabilistic greedy algorithm. Given the user's vehicle reservation data, Nourinejad et al. [15] jointly optimized the respective vehicle and personnel sizes and relocation operations through two integrated multiple traveling salesman models. Wang et al. [32] obtained the relocation demand data of vehicles and personnel by setting the critical value and the vehicle imbalance prevention model of the station, and then constructed an integer linear programming model to optimize the relocation operation under the condition of unknown vehicle reservation data of users.

Vehicle relocation operation not only interacts with personnel relocation, but also is related to decisions such as station location, fleet size, and trip pricing, which can be jointly optimized [27,33]. A typical research approach is to construct a multi-objective mixed integer linear programming model that jointly optimizes vehicle relocation decisions, fleet size, station location, size, and quantity decisions, and set up a virtual relocation center. The vehicles that need to be transferred from each station gather in the virtual center and are transferred from the virtual center to each station with demand for delivery [27]. On this basis, Zou et al. [34] further introduced a user reward and punishment mechanism into the joint optimization model. That is, users will be rewarded when they park their rented vehicles at the operator's own stations. If you park in a public parking lot, they will be penalised.

2.1.2. User-Based Static Relocation for Shared Vehicles

Since a user-based vehicle relocation strategy does not need to hire humans to move vehicles, its cost is lower than that of an operator-based vehicle relocation strategy. However, in a user-based vehicle relocation strategy, it is difficult to control the user's driving behavior, predict the user's acceptance and protect the user's privacy [30]. The user-based vehicle relocation problem for shared vehicles can be roughly divided into two categories as follows.

Research on Shared Vehicle Relocation Considering Trip Pricing

Trip pricing can affect users' travel behavior, so as to balance the supply and demand of vehicles, and has been studied by many scholars. In a station-based one-way vehicle sharing system, a mixed integer nonlinear programming model is constructed to optimize the travel price with the goal of profit maximization, and an iterative local search algorithm matching the problem characteristics is designed to solve it. The results show that a variable travel pricing strategy can improve system profit by balancing vehicle supply and demand [9].

For free-floating vehicle sharing systems, some scholars construct spatial decision support systems to solve the imbalance problem between vehicle supply and user demand, in which the spatiotemporal variation of demand can be used to divide the pricing area in the existing operation area. The results show that such a method can be used to define new operating areas, including high-demand areas and excluding low-demand areas, thereby reducing the risk of vehicle supply and demand imbalance [10].

As in the research on operator-based vehicle relocation strategies, user-based vehicle relocation decisions and station location planning can also be jointly optimized. A mixed integer programming model is constructed to optimize the station location decision, and a variety of trip selection mechanisms are considered to solve the problem of vehicle imbalance. Some of these mechanisms stipulate that the vehicle sharing system operator decides whether to accept a certain travel demand of the user according to the profit maximization goal, some of them stipulate that all user travel demands will be accepted, and some of them stipulate that the trip will be rejected if, and only if, there is no available car at the pick-up point [25].

Research on Shared Vehicle Relocation Considering User Motivation

Through the user incentive mechanism, the user's travel decision is evaluated so as to be beneficial to the balance of vehicle supply and demand. For example, Wang & Liao [35] proposed non-incentive and incentive vehicle relocation strategies for one-way vehicle sharing systems, gave the trigger conditions and travel disutility of response incentive strategies, and proposed a bounded rational dynamic user equilibrium model and a clustering-based column generation algorithm for multimodal transportation systems. Stokkink [36] constructed a system state prediction model based on a Markov chain for a one-way vehicle sharing system, designed a vehicle relocation method based on user incentives, and determined the optimal incentive by weighing the incentive cost and demand loss. At the same time, considering the value of customer time, Stokkink designed a learning algorithm to estimate unknown customer preferences to find the optimal incentive. Schiffer et al. [37] introduced integer programming to optimize the user-based vehicle relocation problem in free-floating vehicle sharing systems, formulated the problem as a k-disjoint shortest path problem, and proposed an exact algorithm. Curtale et al. [38] studied the travel preferences and user-based relocation strategies of one-way electric vehicle sharing services in a framework that integrated the user's current travel choices and contextual variables. The incentive mechanisms for picking up and delivering shared vehicles and sharing rides at distant alternative locations were investigated in two sub-experiments with different travel distances. Brendel et al. [12] constructed a user-driven vehicle relocation framework for station-based one-way vehicle sharing systems. The framework consists of a prediction module, a relocation module, and a user interaction module. Each station has a critical value for the number of vehicles required to meet the users' future travel demands, and relocation will be arranged if the vehicle supply falls below the critical value. The relocation module identified a user who went to the nearest station of another system with unbalanced supply and demand, and the user interaction module was used to calculate the incentive value for a relocation demand and its new destination station. In addition, Angela et al. [13] proposed a user-based shared vehicle relocation mechanism in which users can get a discount on rental fees when they park their vehicles at different destination stations. Based on this, a two-stage optimization model was proposed to optimize the user's destination stations and maximize the profit of the operator. The results show that such a relocation strategy can significantly reduce the number of rejected orders and improve the profit of operators.

2.2. Research on Dynamic Relocation of Shared Electric Vehicles

In the model shown in Figure 2a, the vehicle relocation problem contains two relocation centers ($R_1 - R_2$), seven demand points for vehicle pick-up ($P_1 - P_7$), and seven demand points for vehicle delivery ($D_1 - D_7$). The initial optimal relocation path L1 is $R_2 - P_7 - D_7 - R_2$, and the path completion time is 11. The relocation path L3 is $R_1 - P_3 - D_3 - P_4 - D_4 - R_1$ with a path completion time of 21.

Suppose that during the execution of the initial path, the driving time between P_4 and D_4 changes from 5 to 11 due to weather and accidents, and the driving time between P_7

and D_7 changes from 7 to 20 due to dynamic changes in the environment. In this case, the initial optimal path is no longer optimal, the completion time of path L1 is 24, and the completion time of path L3 is 27. If the optimal path is changed, the pick-up demand point P_4 is served by path L1, and the pick-up point P_7 is served by path L3, a new relocation path L1 $R_2 - P_4 - D_7 - R_2$ is formed, and the path completion time is 19. The new scheduling path L3 is $R_1 - P_3 - D_3 - P_7 - D_4 - R_1$ and the path completion time is 27. Obviously, the improved relocation scheme saves 5 units of time over the static scheme, as shown in Figure 2b.

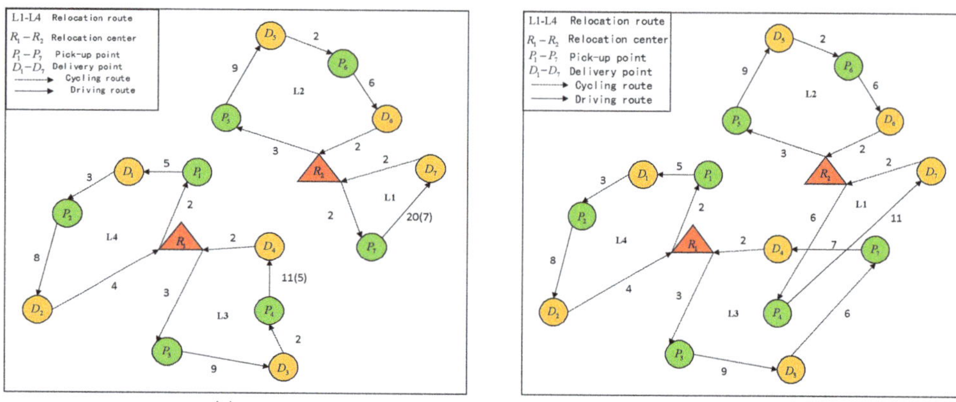

(a) (b)

Figure 2. Relocation of shared electric vehicles. (**a**) is the solution before optimization, and (**b**) is the solution after optimization.

In the existing research, the methods used to solve the dynamic relocation problem of shared vehicles are mainly divided into four categories: optimization models, simulation models, multi-stage methods, and deep reinforcement learning. Therefore, the following section reviews the research on dynamic relocation of shared vehicles based on these four categories.

2.2.1. Research on Dynamic Relocation of Shared Vehicles Based on Optimization Models

In order to optimize the relocation operations of electric free-floating car-sharing system operators, a mixed integer linear programming (MILP) model is established to solve the dynamic relocation, electric vehicle charging, and relocation personnel scheduling problems of FFCS in the case of advance reservation. A new local search algorithm based on relocation tuple (RT) structure has been proposed to solve this problem [4]. Considering uncertain demand, Eilertsen et al. [39] jointly optimized vehicle relocation and vehicle sharing prices between operating areas, and proposed a two-stage integer stochastic programming model and an adaptive large neighborhood search algorithm. For station-based one-way shared vehicle travel systems, Yoshida & Takano [40] used a linear control strategy, calculated a real-time vehicle relocation scheme according to user travel history data, and constructed a linear optimization model based on various scenarios.

Chang et al. [41] proposed a two-stage integer programming model for joint optimization of vehicle and personnel relocation in the context of multiple vehicle sharing operators sharing dispatching personnel, vehicles, and stations, and used a novel deep learning algorithm with an attention-enhanced temporal graph convolutional network to predict the traffic flow of stations. Yang et al. [42] studied the dynamic relocation problem of vehicles operated by temporary workers at night in a one-way vehicle sharing system. An integer programming model was constructed with the objectives of minimizing the relocation cost and considering the hiring relocation of temporary workers, rework restrictions, and carpooling.

Considering the new employment restrictions for each relocation model, the proposed iterative optimization method iteratively solves a set of cluster migration problems and inter-cluster relocation problems. Based on the real-time requirements of users, Yang et al. [43] used two different time granularities to obtain the vehicle relocation task and designed a hybrid solution algorithm combining a rolling horizon algorithm and a custom decomposition algorithm for the joint relocation problem of vehicles and dispatchers. Lin & Kuo [44] studied the vehicle deployment and relocation problem for single-row and free-floating hybrid electric vehicle sharing systems considering demand and parking space randomness. In the first stage, electric vehicles are deployed into a sharing system based on historical user data to maximize profits. In the second stage, with uncertain demand, a dispatcher is used to relocate the electric vehicles in the system to achieve the maximum profit.

Based on random and dynamic user travel demand, Li et al. [28] constructed a continuous approximate model for joint optimization of shared station location and fleet size, with the goal of minimizing the total system cost (including station construction investment, vehicle charging, transportation, and relocation costs). Considering the spatiotemporal variation characteristics of user vehicle reservation data, Zhao et al. [45] constructed a mixed integer linear programming model for joint optimization of shared vehicles and dispatchers, and decomposed the master problem into several sub-problems to be solved by a solution method based on Lagrangian relaxation.

2.2.2. Research on Dynamic Relocation of Shared Vehicles Based on Simulation Models

Li et al. [7] studied the vehicle imbalance problem in a one-way station electric vehicle sharing system. A simulation-based optimization framework was proposed to determine the optimal OSECS operation strategy, including dynamic pricing, vehicle migration, and personnel rebalancing. Researchers created a specific discrete event simulation tool to compare different relocation configurations or management systems, demonstrating the applicability of discrete event simulation methods to a variety of problems [12]. On this basis, the MATSim tool proposed by Balac et al. [46] simulates two car sharing systems with simultaneous operation, where one operator relocates vehicles and the other does not. The simulation results show that the operating revenue of the former may be smaller than that of the latter. Some scholars have specifically proposed a dynamic optimization simulation model for shared vehicle relocation operations; in the discrete time simulation framework, the arrival of each new user is regarded as an event to optimize vehicle relocation, and the model is compared with the prior benchmark relocation model. Case studies have shown that increasing the time between ordering and picking up a car from 0 to 30 min reduces the vehicle fleet size by 86% [23]. Deng & Cardin [26] considered the demand uncertainty and vehicle relocation operation in terms of vehicle distribution and parking space distribution decisions, and designed a solution method based on a discrete event simulator, a particle swarm optimization algorithm, and an optimal computation allocation to solve the model. Finally, a case study of a vehicle sharing system in Singapore was carried out.

In an electric vehicle sharing system based on user reservations, vehicles and personnel can be relocated by constructing an integrated multi-objective mixed integer linear programming optimization and discrete event simulation framework. Such a framework consists of three modules: station clustering, relocation operation optimization, and personnel relocation [14]. Or the relocation problem can be addressed by constructing a multi-stage decision support system to decide a series of near-optimal manpower and operation parameters related to the relocation operation [19].

2.2.3. Research on Dynamic Relocation of Shared Vehicles Based on Multi-Stage Methods

Wang et al. [6] proposed a multi-stage vehicle relocation optimization method for one-way station vehicle sharing systems. In the first stage, a micro-level prediction model of shared travel demand was established to capture the number of orders in the short term. In the second stage, the stations were classified according to the prediction results of user travel demand. In the third stage, a solution method combining a Gurobi solver and charging prioritization was designed to minimize the driving distance and carbon emissions. When a new customer enters the car sharing network, the model is recalculated. In each event, the relocation operation is optimized by solving a binary integer programming model using CPLEX [23]. The main difference between this method and other MIP models is that those models calculate all known requirements only once and have fewer dynamic changes. Another good example is the multi-stage simulation approach, where microscopic traffic simulation, forecasting, and inventory replying policies are solved in sequence, and their results interact [47].

Some scholars use MIP to optimize strategic data, such as fleet size, and to simulate demand and travel trip generation or relocation decisions [48]. Another multi-stage approach proposed by Kek et al. [19] uses MIP in the first step to achieve cost minimization and improves the results by applying additional relocation simulation methods. After the trend filter generates feasible operational solutions based on the optimization results, the time-stepping simulation part is used for efficiency evaluation. The system optimized a problem set consisting of 1236 trips via the simplex algorithm provided in CPLEX.

2.2.4. Research on Dynamic Relocation of Shared Vehicles Based on Deep Reinforcement Learning

Some scholars have applied deep reinforcement learning algorithms to the dynamic vehicle relocation problem of online ride-hailing systems. Based on dynamic ride-hailing demand and supply, Liu et al. [49] proposed a single-agent deep reinforcement learning method to solve the vehicle relocation problem of a ride-hailing platform. Deng et al. [50] studied the vehicle relocation problem for large-scale ride-hailing systems with asymmetric demand. They formulated a semi-Markov decision problem framework with the objective of minimizing passenger waiting time and total empty vehicle miles traveled, and used deep reinforcement learning algorithms to determine near-optimal solutions. Jiao et al. [51] proposed a framework based on deep reinforcement learning and decision time planning to perform real-time vehicle relocation for ride-hailing platforms, and used a batch training algorithm with deep value networks to learn spatiotemporal state–value functions. Through a value-based policy search, planning and guidance were combined with the value network to generate the optimal real-time relocation policy.

Some scholars have applied deep reinforcement learning algorithms to the dynamic vehicle relocation problem of automatic car sharing systems. In order to realize the optimal charging, distribution, and scheduling task allocation decision-making process of shared autonomous driving electric vehicles, Wang & Guo [52] constructed the dynamic operation process for vehicles as a multi-agent and multi-task dynamic relocation problem based on a Markov decision process, and divided the decision-making process of each time step into three sub-processes. The task allocation process of charging and distribution was transformed into a maximum weight matching problem using a bipartite graph, the task allocation process of relocation was quantified as a maximum flow problem, and a Kuhn–Munkres algorithm and an Edmond–Karp algorithm were designed. Finally, an instant reward function was designed to balance order revenue and travel satisfaction, and a state–value function estimated by a backpropagation deep neural network was defined as the matching degree between each shared autonomous vehicle and each delivery task. Based on on-demand services, Kim et al. [53] applied a deep learning-based algorithm to solve the vehicle relocation problem in an autonomous electric vehicle sharing system.

3. Conclusions of Literature Review

From the above literature analysis, the following conclusions can be drawn:

(1) When researchers study the shared vehicle relocation problem in static scenarios with known demand and fixed parameters such as travel time, most of them construct a mixed integer programming model with split demand for pick-up and delivery, with profit maximization or cost minimization as the optimization objective, and most of them design heuristic algorithms matching the characteristics of the problem. A few scholars have designed exact algorithms.

(2) When scholars study the shared vehicle relocation problem under dynamic scenarios such as random demand and uncertain travel time, the dynamic demand or travel time becomes part of the mixed integer programming optimization model. Most scholars design complex heuristic algorithms or convert them into more tractable sub-problems, then construct simulation models of shared vehicle relocation to avoid the computational limitations of optimization models. However, the optimal or near-optimal solution is not produced, but the break-even point is generated through multiple simulation runs, or the solution is generated in a feasible way, and it is more dependent on the simulation framework, assumptions, and verification. Multi-stage methods can reduce the shortcomings of single methods and increase the effectiveness of the solution, but this is inefficient. Deep reinforcement learning algorithms combining the advantages of the above methods are currently mainly used to solve the dynamic vehicle relocation problem of autonomous car sharing systems.

(3) The existing research is limited to the maximization of operator profit, minimization of cost, and minimization of relocation completion time, and it does not consider the carbon emission index in the relocation process, which is not in line with the "double carbon" goal issued by the Chinese state.

(4) Scholars generally use complex and inefficient optimization simulation models or multi-stage methods when designing algorithms for the dynamic relocation problem of shared vehicles. A few scholars apply deep reinforcement learning algorithms to the vehicle relocation problem of autonomous driving based on dynamic demand.

4. Future Research Trends

The research trend of the shared electric vehicle relocation optimization problem mainly focuses on the following aspects:

(1) The nonlinear charging process of electric vehicles can be considered in the process of constructing relocation optimization models and designing algorithms for shared electric vehicles.

(2) In the process of relocation optimization model construction and algorithm design for shared electric vehicles, the conflict between the carbon emission target and the relocation completion time target can be considered to construct the optimal Pareto

set, and the coupling relationship between the targets can be fully considered in the algorithm design.

(3) In the design of shared electric vehicle relocation optimization algorithms, the advantages of the existing heuristic/simulation/multi-stage methods can be integrated, and a deep reinforcement learning algorithm combining the strong feature extraction ability of deep neural networks and the characteristics of reinforcement learning methods that are good at learning beneficial actions can be designed.

Author Contributions: Conceptualization, Q.Y., D.J. and Y.D.; Funding acquisition, Y.Z., Q.Y. and Y.D.; Methodology, Y.Z., Q.Y. and D.J.; Investigation, D.J.; Writing—original draft, Y.Z.; Writing—review & editing, Y.D. All authors have read and agreed to the published version of the manuscript.

Funding: This research was funded by the Humanities and Social Science Research of the Ministry of Education under Grant 22YJC630225, the Natural Science Foundation of Hunan Province under Grant 2022JJ40138, the Science Research Project of Education Department of Hunan Province under Grant 21B0585 and 22B0900, National Natural Science Foundation of China under Grant 52407146, and Natural Science Starting Project of SWPU under Grant 2023QHZ021.

Data Availability Statement: No new data were created or analyzed in this study. Data sharing is not applicable to this article.

Conflicts of Interest: The authors declare no conflicts of interest.

References

1. Qin, H.; Su, E.; Wang, Y.; Li, J. Branch-and-price-and-cut for the electric vehicle relocation problem in one-way carsharing systems. *Omega* **2022**, *109*, 102609. [CrossRef]
2. Zou, Y.; Yang, C.; Yang, J. P-D pair based EV-sharing vehicle relocation problem with route familiarity. *Comput. Ind. Eng.* **2021**, *159*, 107529. [CrossRef]
3. Barth, M.; Todd, M. Simulation model performance analysis of a multiple station shared vehicle system. *Transp. Res. Part C Emerg. Technol.* **1999**, *7*, 237–259. [CrossRef]
4. Shui, C.S.; James, C.; Chu, S.-C.L.; Shih, C.-H. Optimization of vehicle charging and dynamic relocation in free-floating electric carsharing systems with advanced reservations. *Comput. Ind. Eng.* **2024**, *190*, 110072. [CrossRef]
5. Tian, T.; Liu, Y.X.; Yan, B.C.; Li, J.L. Vehicle relocation and staff rebalance problem in one-way free-floating car-sharing system: A novel multi-iteration probabilistic greedy algorithm. *Expert Syst. Appl.* **2024**, *238*, 121528. [CrossRef]
6. Wang, N.; Tian, H.; Wu, G.; Tang, J.; Li, Y. Shared travel demand forecasting and multi-phase vehicle relocation optimization for electric carsharing systems. *Transp. Lett.* **2023**, *16*, 1002–1017. [CrossRef]
7. Li, H.; Hu, L.; Jiang, Y. Dynamic pricing, vehicle relocation and staff rebalancing for station-based one-way electric carsharing systems considering nonlinear charging profile. *Transp. Lett.* **2022**, *15*, 659–684. [CrossRef]
8. Brendel, A.B.; Brennecke, J.T.; Zapadka, P.; Kolbe, L.M. A decision support system for computation of carsharing pricing areas and its influence on vehicle distribution. In Proceedings of the Thirty Eighth International Conference on Information Systems, Seoul, Republic of Korea, 10–13 December 2017; pp. 1–21.
9. Jorge, D.; Molnar, G.; de Almeida Correia, G.H. Trip pricing of one-way station-based carsharing networks with zone and time of day price variations. *Transp. Res. Part B Methodol.* **2015**, *81*, 461–482. [CrossRef]
10. Willing, C.; Klemmer, K.; Brandt, T.; Neumann, D. Moving in time and space—Location intelligence for carsharing decision support. *Decis. Support Syst.* **2017**, *99*, 75–85. [CrossRef]
11. Cepolina, E.M.; Farina, A.; Holloway, C.; Tyler, N. Innovative strategies for urban car-sharing systems and a simulator to assess their performance. *Transp. Plan. Technol.* **2015**, *38*, 375–391. [CrossRef]
12. Brendel, A.B.; Lichtenberg, S.; Brauer, B.; Nastjuk, I.; Kolbe, L.M. Improving electric vehicle utilization in carsharing: A framework and simulation of an e-carsharing vehicle utilization management system. *Transp. Res. Part D Transp. Environ.* **2018**, *64*, 230–245. [CrossRef]
13. Angela, D.F.; Nicola, S.; Mahnam, S. One-way car-sharing profit maximization by means of user-based vehicle relocation. *IEEE Trans. Intell. Transp. Syst.* **2019**, *20*, 628–641.
14. Boyacı, B.; Zografos, K.G.; Geroliminis, N. An integrated optimization-simulation framework for vehicle and personnel relocations of electric carsharing systems with reservations. *Transp. Res. Part B Methodol.* **2017**, *95*, 214–237. [CrossRef]

15. Nourinejad, M.; Zhu, S.; Bahrami, S.; Roorda, M.J. Vehicle relocation and staff rebalancing in one-way carsharing systems. *Transp. Res. Part E Logist. Transp. Rev.* **2015**, *81*, 98–113. [CrossRef]
16. Bruglieri, M.; Pezzella, F.; Pisacane, O. Heuristic algorithms for the operator-based relocation problem in one-way electric carsharing systems. *Discret. Optim.* **2017**, *23*, 56–80. [CrossRef]
17. Bruglieri, M.; Pezzella, F.; Pisacane, O. An adaptive large neighborhood search for relocating vehicles in electric carsharing services. *Discret. Appl. Math.* **2019**, *253*, 185–200. [CrossRef]
18. Ait-Ouahmed, A.; Josselin, D.; Zhou, F. Relocation optimization of electric cars in one-way car-sharing systems: Modeling, exact solving and heuristics algorithms. *Int. J. Geogr. Inf. Sci.* **2018**, *32*, 367–398. [CrossRef]
19. Kek, A.G.H.; Cheu, R.L.; Meng, Q.; Fung, C.H. A decision support system for vehicle relocation operations in carsharing systems. *Transp. Res. Part E Logist. Transp. Rev.* **2009**, *45*, 149–158. [CrossRef]
20. Fan, W.; Machemehl, R.B.; Lownes, N.E. Carsharing dynamic decision-making problem for vehicle allocation. *Transp. Res. Rec.* **2008**, *2063*, 97–104. [CrossRef]
21. Alfian, G.; Rhee, J.; Yoon, B. A simulation tool for prioritizing product-service system (PSS) models in a carsharing service. *Comput. Ind. Eng.* **2014**, *70*, 59–73. [CrossRef]
22. Gambella, C.; Malaguti, E.; Masini, F.; Vigo, D. Optimizing relocation operations in electric car-sharing. *Omega* **2018**, *81*, 234–245. [CrossRef]
23. Nourinejad, M.; Roorda, M.J. A dynamic carsharing decision support system. *Transp. Res. Part E Logist. Transp. Rev.* **2014**, *66*, 36–50. [CrossRef]
24. Weikl, S.; Bogenberger, K. A practice-ready relocation model for free-floating carsharing systems with electric vehicles—Mesoscopic approach and field trial results. *Transp. Res. Part C Emerg. Technol.* **2015**, *57*, 206–223. [CrossRef]
25. Correia, G.H.D.A.; Antunes, A.P. Optimization approach to depot location and trip selection in one-way carsharing systems. *Transp. Res. Part E Logist. Transp. Rev.* **2012**, *48*, 233–247. [CrossRef]
26. Deng, Y.; Cardin, M. Integrating operational decisions into the planning of one-way vehicle-sharing systems under uncertainty. *Transp. Res. Part C Emerg. Technol.* **2018**, *86*, 407–424. [CrossRef]
27. Boyacı, B.; Zografos, K.G.; Geroliminis, N. An optimization framework for the development of efficient one-way car-sharing systems. *Eur. J. Oper. Res.* **2015**, *240*, 718–733. [CrossRef]
28. Li, X.; Ma, J.; Cui, J.; Ghiasi, A.; Zhou, F. Design framework of large-scale one-way electric vehicle sharing systems: A continuum approximation model. *Transp. Res. Part B Methodol.* **2016**, *88*, 21–45. [CrossRef]
29. Correia, G.H.D.A.; Jorge, D.R.; Antunes, D.M. The added value of accounting for users' flexibility and information on the potential of a station-based one-way car-sharing system: An application in Lisbon, Portugal. *J. Intell. Transp. Syst.* **2014**, *18*, 299–308. [CrossRef]
30. Weikl, S.; Bogenberger, K. Relocation strategies and algorithms for free-floating car sharing systems. *IEEE Intell. Transp. Syst. Mag.* **2013**, *5*, 100–111. [CrossRef]
31. Cai, L.; Wang, X.; Luo, Z.X.; Liang, Y.J. A hybrid adaptive large neighborhood search and tabu search algorithm for the electric vehicle relocation problem. *Comput. Ind. Eng.* **2022**, *167*, 108005. [CrossRef]
32. Wang, L.; Liu, Q.; Ma, W. Optimization of dynamic relocation operations for one-way electric carsharing systems. *Transp. Res. Part C Emerg. Technol.* **2019**, *101*, 55–69. [CrossRef]
33. Xu, M.; Meng, Q.; Liu, Z. Electric vehicle fleet size and trip pricing for one-way carsharing services considering vehicle relocation and personnel assignment. *Transp. Res. Part B Methodol.* **2018**, *111*, 60–82. [CrossRef]
34. Zou, Y.; Yang, C.; Yang, J. EV-sharing station location and vehicle relocation with parking rewards and punishments mechanism. *Int. J. Sustain. Transp.* **2021**, *15*, 306–322. [CrossRef]
35. Wang, D.; Liao, F.X. Incentivized user-based relocation strategies for moderating supply-demand dynamics in one-way car-sharing services. *Transp. Res. Part E-Logist. Transp. Rev.* **2023**, *171*, 103017. [CrossRef]
36. Stokkink, P.; Geroliminis, N. Predictive user-based relocation through incentives in one-way car-sharing systems. *Transp. Res. Part B-Methodol.* **2021**, *149*, 230–249. [CrossRef]
37. Schiffer, M.; Hiermann, G.; Rüdel, F.; Walther, G. A polynomial-time algorithm for user-based relocation in free-floating car sharing systems. *Transp. Res. Part B-Methodol.* **2021**, *143*, 65–85. [CrossRef]
38. Curtale, R.; Liao, F.X.; van der Waerden, P. Understanding travel preferences for user-based relocation strategies of one-way electric car-sharing services. *Transp. Res. Part C-Emerg. Technol.* **2021**, *127*, 103135.
39. Eilertsen, U.; Falck-Pedersen, O.M.; Henriksen, J.V.; Fagerholt, K.; Pantuso, G. Joint relocation and pricing in electric car-sharing systems. *Eur. J. Oper. Res.* **2024**, *315*, 553–566. [CrossRef]
40. Yoshida, Y.; Takano, Y. Linear control policies for online vehicle relocation in shared mobility systems. *Expert Syst. Appl.* **2022**, *210*, 118417. [CrossRef]

41. Chang, X.M.; Wu, J.J.; Correia, G.H.D.; Sun, H.J.; Feng, Z.Y. A cooperative strategy for optimizing vehicle relocations and staff movements in cities where several carsharing companies operate simultaneously. *Transp. Res. Part E-Logist. Transp. Rev.* **2022**, *161*, 102711. [CrossRef]
42. Yang, J.; Hu, L.; Jiang, Y.S. An overnight relocation problem for one-way carsharing systems considering employment planning, return restrictions, and ride sharing of temporary workers. *Transp. Res. Part E-Logist. Transp. Rev.* **2022**, *168*, 102950. [CrossRef]
43. Yang, S.; Wu, J.J.; Sun, H.J.; Qu, Y.C.; Li, T.F. Double-balanced relocation optimization of one-way car-sharing system with real-time requests. *Transp. Res. Part C-Emerg. Technol.* **2021**, *125*, 103071. [CrossRef]
44. Lin, D.Y.; Kuo, J.K. The vehicle deployment and relocation problem for electric vehicle sharing systems considering demand and parking space stochasticity. *Transp. Res. Part E-Logist. Transp. Rev.* **2021**, *156*, 102514. [CrossRef]
45. Zhao, M.; Li, X.; Yin, J.; Cui, J.; Yang, L.; An, S. An integrated framework for electric vehicle rebalancing and staff relocation in one-way carsharing systems: Model formulation and Lagrangian relaxation-based solution approach. *Transp. Res. Part B Methodol.* **2018**, *117*, 542–572. [CrossRef]
46. Balac, M.; Becker, H.; Ciari, F.; Axhausen, K.W. Modeling competing free-floating carsharing operators—A case study for zurich, switzerland. Transportation Research. *Part C Emerg. Technol.* **2019**, *98*, 101–117. [CrossRef]
47. Wang, H.; Cheu, R.; Lee, D. Dynamic relocating vehicle resources using a microscopic traffic simulation model for carsharing services. In Proceedings of the 2010 Third International Joint Conference on Computational Science and Optimization, Huangshan, China, 28–31 May 2010; Volume 1, pp. 108–111.
48. Jorge, D.; Correia, G.H.A.; Barnhart, C. Comparing optimal relocation operations with simulated relocation policies in one-way carsharing systems. *IEEE Trans. Intell. Transp. Syst.* **2014**, *15*, 1667–1675. [CrossRef]
49. Liu, Y.; Wu, F.Y.; Lyu, C.; Li, S.; Ye, J.P.; Qu, X.B. Deep dispatching: A deep reinforcement learning approach for vehicle dispatching on online ride-hailing platform. *Transp. Res. Part E-Logist. Transp. Rev.* **2022**, *161*, 102694. [CrossRef]
50. Deng, Y.T.; Chen, H.; Shao, S.P.; Tang, J.C.; Pi, J.Z.; Gupta, A. Multi-objective vehicle rebalancing for ridehailing system using a reinforcement learning approach. *J. Manag. Sci. Eng.* **2022**, *7*, 346–364. [CrossRef]
51. Jiao, Y.; Tang, X.C.; Qin, Z.W.; Li, S.J.; Zhang, F.; Zhu, H.T.; Ye, J.P. Real-world ride-hailing vehicle repositioning using deep reinforcement learning. *Transp. Res. Part C-Emerg. Technol.* **2021**, *130*, 103289. [CrossRef]
52. Wang, N.; Guo, J.H. Multi-task dispatch of shared autonomous electric vehicles for Mobility-on-Demand services—Combination of deep reinforcement learning and combinatorial optimization method. *Heliyon* **2022**, *8*, e11319. [CrossRef]
53. Kim, S.; Lee, U.K.; Lee, I.K.; Kang, N.M. Idle vehicle relocation strategy through deep learning for shared autonomous electric vehicle system optimization. *J. Clean. Prod.* **2022**, *333*, 130055. [CrossRef]

World Electric Vehicle Journal

MDPI

Review

Fair Energy Trading in Blockchain-Inspired Smart Grid: Technological Barriers and Future Trends in the Age of Electric Vehicles

Sameer Qazi [1,*], Bilal A. Khawaja [2,*], Abdullah Alamri [3] and Abdulrahman AlKassem [2]

1 Department of Computer Science, Faculty of Engineering, Sciences, and Technology, Iqra University, Karachi 75500, Pakistan
2 Department of Electrical Engineering, Faculty of Engineering, Islamic University of Madinah, Madinah 42351, Saudi Arabia
3 Saudi Energy Efficiency Center (SEEC), An Nakheel, Riyadh 12382, Saudi Arabia
* Correspondence: sameer.hashmat@iqra.edu.pk (S.Q.); kjbmohammed@iu.edu.sa (B.A.K.)

Abstract: The global electricity demand from electric vehicles (EVs) increased by 3631% over the last decade, from 2600 gigawatt hours (GWh) in 2013 to 97,000 GWh in 2023. The global electricity demand from EVs will rise to 710,000 GWh by 2030. These EVs will depend on smart grids (SGs) for their charging requirements. Like EVs, SGs are a booming market. In 2021, SG technologies were valued at USD 43.1 billion and are projected to reach USD 103.4 billion by 2026. As EVs become more prevalent, they introduce additional complexity to the SG landscape, with EVs not only consuming energy, but also potentially supplying it back to the grid through vehicle-to-grid (V2G) technologies. The entry of numerous independent sellers and buyers, including EV owners, into the market will lead to intense competition, resulting in rapid fluctuations in electricity prices and constant energy transactions to maximize profit for both buyers and sellers. Blockchain technology will play a crucial role in securing data publishing and transactions in this evolving scenario, ensuring transparent and efficient interactions between EVs and the grid. This survey paper explores key research challenges from an engineering design perspective of SG operation, such as the potential for voltage instability due to the integration of numerous EVs and distributed microgrids with fluctuating generation capacities and load demands. This paper also delves into the need for a synergistic balance to optimize the energy supply and demand equation. Additionally, it discusses policies and incentives that may be enforced by national electricity carriers to maintain grid reliability and manage the influx of EVs. Furthermore, this paper addresses emerging issues of SG technology providing primary charging infrastructure for EVs, such as incentivizing green energy, the technical difficulties in integrating diverse hetero-microgrids based on HVAC and HVDC technologies, challenges related to the speed of energy transaction processing during fluctuating prices, and vulnerabilities concerning cyber-attacks on blockchain-based SG architectures. Finally, future trends are discussed, including the impact of increased EV penetration on SGs, advancements in V2G technologies, load-shaping techniques, dynamic pricing mechanisms, and AI-based stability enhancement measures in the context of widespread SG adoption.

Keywords: electric vehicles (EVs); smart grid; blockchain; internet of things (IoT); internet of energy (IoE); artificial intelligence (AI); distributed energy resources (DERs); renewable energy sources (RESs)

1. Introduction

The smart grid (SG) is the next-generation electrical grid, which, through an integrated smart sensor framework, can coordinate the integration of electrical energy through several decentralized energy generation plants and manage its distribution to consumers as per demand. Smart sensors manage energy metering, i.e., units of electricity produced by the distributed generation plant that is being consumed by users on the grid. It is worth noting that, besides the traditional roles of electricity producers and electricity consumers on the grid, there is also a third hybrid entity called "prosumer"; prosumers can selectively supply the grid with electrical energy when producing surplus energy after meeting their own domestic needs or consume energy from the grid when energy demand exceeds its production capacity. The increasing global demand for energy and the urgent need for sustainable practices have spurred significant interest in SG technologies [1–4], especially with the rise in the adoption of electric vehicles (EVs) and renewable energy generation. The data show that global electricity demand from EVs increased by 3631% over the last decade, from 2600 gigawatt hours (GWh) in 2013 to 97,000 GWh in 2023. The global electricity demand from EVs will rise to 710,000 GWh by 2030 [5]. SGs promise greater efficiency, reliability, and integration of various renewable energy sources (RESs) [2]. However, significant challenges remain, particularly in managing the intermittent nature of renewables, ensuring grid stability with fluctuating supply and demand equation, especially with the rise in the EV consumer market that is still emerging, and securely handling energy transactions [3,4]. In 2021, SG technologies were valued at USD 43.1 billion and are projected to reach USD 103.4 billion by 2026 [3]. Recently, many consumers have also become producers, known as "prosumers", through the installation of small green-energy generators at their premises, such as solar or wind power. This allows them to meet their own energy demands and sell surplus electricity back to the grid. This concept extends to microgrids [2–6], which can be viewed as miniature models of the national grid with their own generation, transmission, and distribution networks. Microgrids may operate independently (in islanded mode) [2] when their generation capacity matches their load, but when their capacity exceeds the target load, they can connect to the national grid, thus meeting their own energy needs and selling surplus electricity. Microgrids may typically be small electricity-generation plants using renewable energy, e.g., solar, wind farms, or fossil fuels.

With the growing use of EVs and their potential to act as mobile energy storage units, new challenges and opportunities arise within SGs [7]. EVs can significantly contribute to grid stability by storing and supplying energy when needed. As EVs become more prevalent, managing their integration into the grid and optimizing their use for energy transactions become crucial. Other independent power projects (IPPs) have also begun electricity generation using conventional (thermal) or green technologies (solar and wind). These projects can supply electricity to the national grid, breaking the traditional model of a single, monopolized national grid and introducing a dynamic electricity market akin to a financial stock market, where competition drives rapid fluctuations in electricity prices. In this emerging scenario, blockchain technology [6,8–11] offers a robust solution for secure data publishing and executing transactions. Blockchain provides a reliable framework for financial transactions, preventing fraud and ensuring secure data distribution and publishing.

The blockchain's potential extends to the EV sector, where it can facilitate secure and transparent transactions for energy trading, including EV charging and discharging activities. By integrating blockchain, the industry can ensure that energy transactions involving EVs are secure and efficiently managed, enhancing the overall reliability and stability of the grid. Figure 1 presents an overview of the roles that various producers, consumers, and prosumers (a combination of the two) play within the SG scenario, highlighting how blockchain technology can facilitate secure transactions among them.

The new SG paradigm promises increased flexibility and reliability for the national grid by enabling distributed energy generation and providing attractive energy buying options for consumers through careful planning. However, several technical challenges still impede widespread SG deployment. A primary challenge is the instability associated with distributed energy generation from renewable sources like wind and solar, which contrasts with more stable energy sources, such as thermal and nuclear power. These fluctuations can lead to overall SG instability if not managed effectively. Additionally, the rise in EVs, while beneficial due to their low carbon footprint, introduces fluctuating and unpredictable loads in different parts of the network. This requires careful planning to ensure that the SG remains disruption-free and reliable.

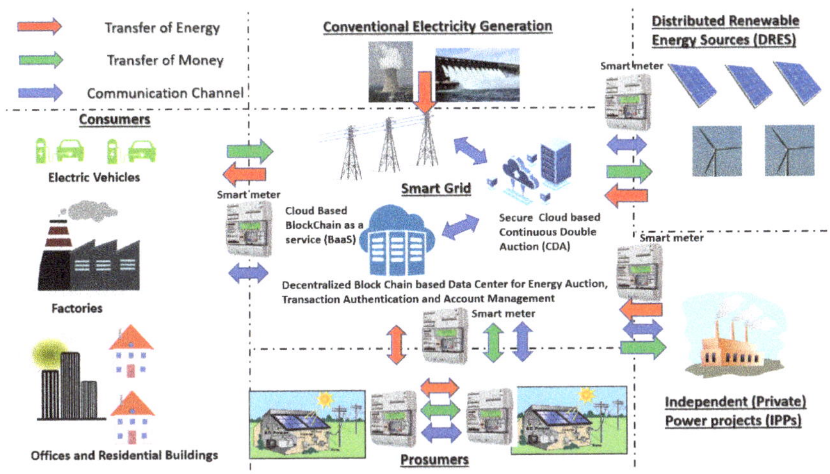

Figure 1. How different producers, consumers, and prosumers (combination) fit in the SG scenario and how blockchain can help them with secure transactions.

Another technical challenge is managing the synchronization of operating parameters, such as voltages and frequencies, with the national grid. This issue is exacerbated when DC sources, like solar power, are part of the energy distribution, necessitating complex algorithms to convert DC to AC and match the grid's frequency. The presence of multiple energy sellers and the integration of EVs also raise concerns about prioritizing energy sellers if their combined generation exceeds the total load demand.

Unwanted oscillations can occur when buyers rapidly shift their energy purchases to cheaper sellers, driving up prices and causing frequent changes in buyer behavior. Machine learning (ML) and artificial intelligence (AI) algorithms are expected to address these issues by managing and balancing distributed energy generation and consumption. Cyber-attacks pose a significant threat to the secure operation of SGs. With a decentralized architecture, there is an increased risk of malicious attacks disrupting SG operations.

This survey paper addresses key research challenges from an engineering design perspective of SG operation, including: (a) potential voltage instability in the SG caused by distributed renewable energy sources (DRES) and EVs, leading to rapid fluctuations in generation capacity and load demand; (b) the need for a synergistic balance to optimize the energy supply and demand equation, including incentives for green energy producers/consumers; (c) the challenge of national electricity carriers maintaining regulatory control over small market players while ensuring grid reliability; (d) integration issues with diverse microgrids using HVAC and HVDC technologies; (e) challenges related to the processing speed of energy transactions amid fluctuating prices and cyber-attacks on blockchain-based SG architectures; and (f) future trends such as increased EV penetration, which will drive advancements in vehicle-to-grid (V2G) technologies, innovative load-shaping solutions, and dynamic energy pricing models. Additionally, this paper explores how AI-based techniques can stabilize SGs, particularly with extensive blockchain technology integration.

The remainder of this paper is structured as follows: Section 2 discusses the related work and recent developments. Section 3 discusses the concept and modalities of using parallel sources for SGs. Section 4 discusses the method of blockchain and smart contract-based technology for a multi-buyer seller environment. Section 5 discusses the role of blockchain in the SG network. Section 6 discusses the technological barriers and current challenges in SG operations after the integration of the EV industry. Section 7 discusses the future trends in next-generation SGs after the integration of the EV market. Finally, Section 8 concludes this paper. The roadmap of this paper is also depicted in Figure 2. Table 1 shows the list of acronyms used in this paper.

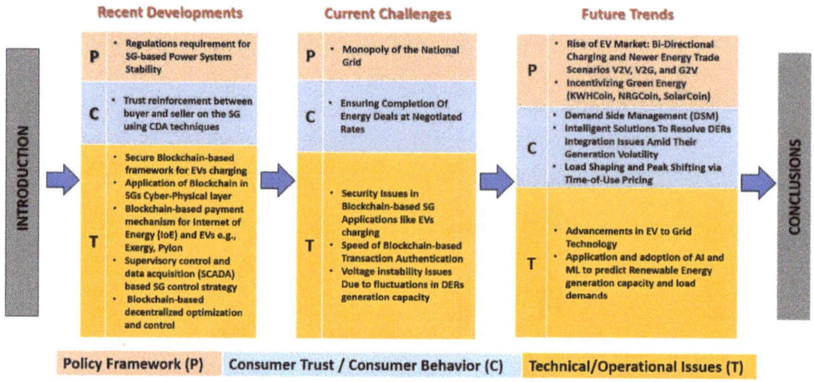

Figure 2. Roadmap of this paper.

Table 1. List of acronyms.

Acronyms	Definitions
EVs	Electric Vehicles
SG	Smart Grid
DERs	Distributed Energy Resources
EI	Energy Internet
AI	Artificial Intelligence
ML	Machine Learning
DL	Deep Learning
CESs	Crowdsourced Energy Systems
CPSs	Cyber–Physical Systems
ETTs	Energy Trading Transactions
FDIAs	False Data Injection Attacks
IPPs	Independent Power Projects
DRERs	Distributed Renewable Energy Resources
DRESs	Distributed Renewable Energy Sources
DESDs	Distributed Energy Storage Devices
RESs	Renewable Energy Sources
HVAC	High Voltage Alternating Current
HVDC	High Voltage Direct Current
DLT	Distributed Ledger Technology
PsR	Power-side Resilience
CsR	Cyber-side Resilience
SC	Smart Contract
PoW	Proof of Work
P2P	Peer-to-Peer
V2V	Vehicle-to-Vehicle
V2R	Vehicle-to-Renewable
R2V	Renewable-to-Vehicle
V2X	Vehicle-to-Everything
V2G	Vehicle-to-Grid
G2V	Grid-to-Vehicle
SCADA	Supervisory Control and Data Acquisition
FTs	Fungible Tokens
NFTs	Non-Fungible Tokens
DSM	Demand-Side Management

Table 1. *Cont.*

Acronyms	Definitions
BTC	BitCoin
CDA	Continuous Double Auction
PAER	Peak-to-Average Energy Ratio
TPS	Transactions Per Second
ToU	Time-of-Use
GDPR	General Data Protection Regulation

2. Related Work

The integration of blockchain technology into SG systems presents a promising solution for enabling fair and decentralized energy trading, particularly in the era of EVs. Hussain et al. [4], in their work, emphasized the significance of the energy internet (EI) revolving around SG, drawing its parallel with the legacy internet, where conventional routing and packets are replaced with energy routing and packets. The EI is an integration of conventional distributed energy resources (DERs), distributed renewable energy resources (DRERs), distributed energy-storage devices (DESDs), real-time energy monitoring, information sharing, real-time pricing, and energy transactions. EVs, with their inherent energy storage capabilities and potential for bidirectional energy flow in the V2G mode, represent both a significant disruption and an opportunity within this ecosystem. Musleh et al. [6] and Li et al. [8] make a case for blockchain applications in SG's cyber-physical layer. They presented a comprehensive framework with individual blockchains for sensors, microgrids, EVs, energy trading, and cyber security. In their seminal work, Mollah et al. [9] presented a thorough survey paper on the above topic covering various aspects of blockchain-based SGs. Similarly, Li et al. [10] addressed energy trading issues over the Industrial Internet of Things (IIoT). They concluded that security analysis on a real-world dataset confirmed that their proposed energy blockchain and credit-based payment scheme was meeting the security needs of IIoT. Gai et al. [11] considered the security issues of permissioned blockchain networks for SG applications and proposed an architecture based on a smart contract that met the security of the smart grid infrastructure and the privacy of stakeholders.

Mollah et al. [9] also presented several practically implemented prototype systems utilizing blockchain-based SG technology, such as KWHcoin [12], NRGcoin [13], and Solarcoin [14], etc., which reward customers for using green energy. The growing adoption of EVs further amplifies this push toward a more sustainable energy landscape. EVs, with their ability to not only consume, but also potentially supply energy back to the grid, represent a critical component of this evolving scenario. Several other blockchain-based SG platforms offer a favorable environment for purchasing energy from various stakeholders, including EV owners. These platforms facilitate energy transactions and ensure transparency, security, and efficiency in the process. Most notable of these include Exergy [15] and Pylon [16]. Exergy [15] is the first practical system implemented to link a microgrid through a blockchain-based energy trading platform. Similarly, Pylon [16] is a practically implemented blockchain-based platform for several energy providers, including small prosumers and potential EV owners utilizing V2G technology to make their information available to other energy stakeholders and make buying/selling decisions accordingly.

Hassan et al. [17] and Wang et al. [18] extensively discussed the scenario of energy trading in the blockchain-based SG network. They optimized the robustness issues of the conventional power systems focused on 'long-distance, one-way energy transmission' by catering to distributed bidirectional energy management, improving the system efficiency and stability, and allowing for more players (prosumers/IPPs) to enter the energy market.

Bielecki et al. [19] and Feng et al. [20] conducted comprehensive surveys on the intricacies of a blockchain-based payment system for managing energy trading on the EI. They introduced the concept of a cryptocurrency-based payment system with fungible tokens (FTs) and non-fungible tokens (NFTs). Blockchain technology and smart contracts allow for automatic and instant cash flows and the implementation of reward rules for specific desired activities on the SG.

Danzi et al. [21] highlighted the problem of DER operating power levels. The DERs can optimize their revenue by producing power at maximum capacity. However, when consumption on the grid is lower, this surplus power may lead to higher voltages on the grid. To solve this problem, the authors proposed a solution in which a subset of the DERs produced lower power levels (lower than available capacity), thus sacrificing their revenue in solving the high voltage problem. Blockchain mechanisms can be used to store a smart contract to create trust between the DERs so that this subset selection is performed each time using a special proportional-fairness algorithm. Another related study by Wang et al. [22] addressed a similar problem in which they developed an optimization model and blockchain-based architecture to manage the operation of crowdsourced energy systems (CESs) with peer-to-peer (P2P) energy trading transactions (ETTs). They present a two-phase operation that focuses on the day-ahead scheduling of generation and controllable DERs, and on hour-ahead or real-time operation of distribution networks. The developed approach supports seamless P2P energy trading between individual prosumers and/or the utility. A multi-agent deep reinforcement learning approach was proposed by Wang et al. [23] to optimize the P2P energy trading and demand-side management (DSM) strategies to improve the maximum cost savings for households and achieve savings of up to 14.03% compared with other state-of-the-art deep reinforcement learning methods. Munsing et al. [24] presented another blockchain-based decentralized optimization and control procedure in which loads are scheduled from a mix of batteries, shapable loads, and deferrable loads on an electricity distribution network. Fu et al. [25] and Sharma et al. [26] also discussed a collaborative optimization algorithm for SGs.

Mrabet et al. [27] exhaustively explored the issue of cyber security and a potential attack on a blockchain-based SG system. Naderi and Arash [28,29] investigated the potential cybersecurity risks associated with false data injection attacks for EVs connected to SGs. They concluded from their study that any interruption of a small part of the SG network could potentially result in higher instability, leading to the shutdown of the entire system at times. Karanfil et al. [30] designed and implemented a security monitoring platform for microgrids based on ML for anomaly detection in real-time, whereas Rahiminejad et al. [31] proposed an innovative recovery framework to support the power-side resilience (PsR) and cyber-side resilience (CsR). The developed metric captures load, available reserve capacity, available line capacity, and reliability, and determines the physical-side resilience and the possible damage to the system. Su et al. [32] presented a safe blockchain-based framework for the secure charging of EVs. Gunduz and Das [33] studied various attack scenarios and their potential solution for SG cyber–physical systems (CPS). Musleh et al. [34] applied a spatio-temporal signal processing approach to detect false data injection attacks (FDIA) in power distribution systems in SGs.

Li et al. [35], Liu et al. [36], and Gai et al. [11] assessed the security issues of blockchains that could potentially be used in an SG application. They investigated the security vulnerabilities of the blockchain system that have been practically exploited, as evidenced by attacks on the BitCoin (BTC) network from 2009 to 2017. Conti et al. [37] also investigated security vulnerabilities related to BTC. Moreover, Wan et al. [38] researched the impact of network latency on blockchain security. They concluded that networks with heterogeneous latency profiles must monitor fairness amongst nodes and, based on that, must preclude some nodes from being miners.

Gao et al. [39] discussed the trust issue between grid service providers and regular consumers, proposing a secure blockchain-based mechanism called grid monitoring. This mechanism offers verifiable provider-centric and consumer-centric information on regular energy consumption and applicable tariff rates to create trust between consumers and providers. The authors also presented comprehensive layer-based architecture with semantics for the systems, which guaranteed transparency, provenance, and immutability.

Kumar et al. [40] presented another study in which they considered the security issues of energy monitoring in SG networks. Similarly, Liang et al. [41] fused blockchain security with attribute-based access control (ABAC) to prevent unauthorized data access over the SG with favorable results.

Several other researchers have focused on the dynamic load management of EV charging and other related concerns, such as security and privacy. Lopes et al. [42] showed that uncoordinated EV charging, for even 10% of the total EVs, can give rise to unacceptable levels of instability in the grid.

Rassaei et al. [43] modeled the statistical patterns of EV owner charging and load optimization for the grid by utilizing V2G capability, where an EV adjusts its charging rate or returns stored energy to the system. The authors found that it was possible to accommodate a large number of EVs using their proposed algorithm without causing instability in the SG and that the algorithm performance further improved with V2G capability in the system.

Sayed et al. [44] presented an attacker's perspective and proposed a strategy for dynamic load-altering by manipulating EV charging to destabilize the grid using the feedback control theory. In a similar survey, Habib et al. [45] also described bidirectional charging as a means to tackle the issue of massive EV deployment. Knirsch et al. [46] addressed privacy concerns and dynamic charging tariffs through their blockchain-based smart contract publishing. Their study served as an honest and verifiable record of current electricity tariffs on individual charging stations and customers' bidding processes without revealing their location data for privacy.

Quite recently, Rastogi et al. [47], Chen et al. [48], and Rimal et al. [49] also considered the problems of EV charging from the electric grid as they become ubiquitous. The proposed algorithms optimized the EV charging behaviors, reduced the costs associated with charging, and effectively balanced the regional load profiles of the grids. On the other hand, Zeadally et al. [50] and Qazi et al. [51] considered the communications plane of autonomous, connected vehicles for vehicle-to-vehicle (V2V), vehicle-to-renewable (V2R), and renewable-to-vehicle (R2V) communication scenarios. The authors argued that EVs will use similar communication technologies to communicate with centralized and roadside equipment, e.g., 802.11bd (for DSRC) and 5G-NR V2X (for C-V2X), to transmit their current battery levels and the intended destination of travel, so that the system can calculate the best optimum charging schedule for their commute.

Huang et al. [52] discussed reliable wireless technologies for smart metering applications in blockchain-based SGs. Espinoza et al. [53], Mak and So [54], Gupta C.R. et al. [55], and Jakaria et al. [56] discussed the integration of supervisory control and data acquisition (SCADA) technology with SG. Moreover, Jakaria et al. [56] highlighted various factors for its successful integration, like real-time performance reporting by sensors for safeguarding against cyber-attacks, for which resilient network infrastructure is required. They proposed a hybrid system using legacy message forwarding and software-defined networking (SDN) switches. The detailed literature review presented in this section is also summarized in Table 2.

Table 2. Summary of the contributions of other researchers.

Referenced Work	Major Contributions
Hussain et al. [4]	Comparison of SG with legacy Internet
Musleh et al. [6], Agung and Handayani [57], Orazgaliyev [58], Raza et al. [59], Guo et al. [60]	Review of blockchain-based smart grid frameworks
Musleh et al. [6], Li et al. [8], Cao et al. [61]	Application of blockchain in SGs cyber–physical layer
Mollah et al. [9]	Review of several practically implemented blockchain SG networks, such as KWHCoin [12], NRGCoin [13], SolarCoin [14], Pylon [16], and Greeneum [62].

Table 2. *Cont.*

Referenced Work	Major Contributions
Gai et al. [11], Li et al. [35], Liu et al. [36], Hasan et al. [63]	Security vulnerabilities of SG-based applications
Mihaylov et al. [13], Hassan et al. [17], Wang et al. [18]	Energy trading scenarios in SG network
Mihaylov et al. [13], Greeneum [62], Wen et al. [64], Zhang et al. [65], Oad et al. [66]	Clean, green energy incentives on SG
Bielecki et al. [19], Feng et al. [20]	Blockchain-based payment system to manage energy trading on the energy internet (EI).
Danzi et al. [21], Wang et al. [22], Munsing et al. [24], Fu et al. [25], Sharma et al. [26], Cui et al. [67], Babayomi et al. [68]	Blockchain-based decentralized optimization and control procedure in which scheduling of loads is performed from a mix of batteries, shapable loads, and deferrable loads on an electricity distribution network
Mrabet et al. [27], Gunduz and Das [33], Paula et al. [69]	Various attack scenarios and robust architecture for SG network
Su et al. [32], Zielinska et al. [70], Kong et al. [71]	Secure blockchain-based framework for charging EVs
Musleh et al. [34]	Spatiotemporal techniques to detect false data injections in SG
Wan et al. [38]	Study on the effect of network latency on the security of blockchain in SG applications
Gao et al. [39], Kumar et al. [40], Zhou et al. [72], Ma et al. [73], Akkad et al. [74]	Trust reinforcement between buyer and seller on the SG using CDA techniques; security issues of energy monitoring
Lopes et al. [42], Rassaei et al. [43], Park and Moon [75]	Model for statistical patterns of EV vehicle owner charging patterns and load optimization; effects of uncoordinated EV charging on the grid; intelligent EV charging algorithms
Habib et al. [45], Rastogi et al. [47], Chen et al. [48], Rimal et al. [49], Ullah et al. [76], ElGhanam et al. [77], Guo et al. [78]	Problem of charging EVs from SG, bidirectional energy models for EVs, and other prosumers
Zeadally et al. [50], Qazi et al. [51], Huang et al. [52]	Communication plane technologies for connected EVs and blockchain-based SG
Espinoza et al. [53], Mak and So [54], Gupta C.R. et al. [55], Jakaria et al. [56]	Supervisory control and data acquisition (SCADA)-based SG control strategy
Chen et al. [79]	Regulations requirement for SG-based power system stability
Refaat et al. [80], Esenogho et al. [81], Pilz and Al-Fagih. [82], Zhang et al. [83], Panthi and Kanti Das [84], Sivarajan et al. [85], Pop et al. [86], Zhang et al. [87]	Application of AI-based algorithms such as neural networks, game theory, and other approaches to optimize demand side management and stabilize SG as multiple volatile energy sources such as EVs will inject power into it.
Karasu et al. [88]	Prediction of energy prices over SG using time series data using linear regression (LR) and support vector machine (SVM)
Hossain et al. [89], Diaba and Elmusrati [90], Mostafa et al. [91], Haq et al. [92]	Application of big data and deep learning (DL) of SG application

3. Smart Grid (SG) Architecture: How Multiple Power Sources like EVs Can Inject Energy into the Nationally Deployed Grid and Distribution Networks

The conventional idea in electrical energy is that a source, such as a battery or a generator, can supply current to drive electrical loads if a closed circuit is formed between the battery's terminals and the electrical load. Multiple batteries can be attached to the primary battery to increase the load capacity. The same principle applies to SG, but with some variations. The notable differences are that (a) the energy sources to be parallelized may not be physically sitting next to each other, and (b) the energy sources may be heterogeneous, like pure AC or converted AC (usually using inverter technology from real-time DC generation systems, such as solar farms), or when the energy from green technologies, such as wind and solar, is pre-stored in batteries. Unlike the traditional power grid, where energy flows one-way from centralized sources, the SG, empowered by EVs, enables a two-way exchange of electricity. So, during the day, an EV charges using clean energy from the workplace's solar array. Later, as evening demand peaks, the parked EV intelligently sends stored energy back to the grid, supporting its stability and reducing reliance on fossil fuels. This concept is typically referred to as V2G technology [42–44,47–49], which turns millions of EVs into mobile energy-storage units, supporting grid resilience and accelerating the transition to RES. However, this dynamic energy ecosystem presents new challenges. Coordinating the charging and discharging of millions of EVs, alongside the fluctuating output of solar and wind power, requires a sophisticated, decentralized control system. This is where the SG, with its intelligent communication and control infrastructure, comes into play, coordinating this complex energy exchange and unlocking the full potential of EVs as cornerstones of a sustainable energy future.

Running multiple DC voltage sources is easier than connecting them in parallel with the primary (original) voltage sources. However, when it comes to running two or more AC voltage sources in parallel, things get a little tricky. We have varying sinusoidal voltages with AC sources, and each AC generator could have as many as three different voltage sources, termed as three-phase AC supply. When a second generator is to be connected in parallel with a running system. The following steps must be followed:

1. The field current in the second generator should be modified so that its terminal voltage aligns with the line voltage of the initial generator;
2. The phase sequence of the second generator should be compared with that of the already running system;
3. Ensure that the phase angles of both AC sources are equal;
4. Subsequently, the frequency of the oncoming generator is fine-tuned to be marginally higher than that of the active system. This adjustment allows it to initiate as a generator that provides energy, rather than act as a motor that consumes it. Then, a synchroscope is used to match the systems until they are exactly in phase. In a large system, a computer is used for this purpose.

The interconnection of distributed AC generation units or DRES to a large power grid often results in a negligible impact on the grid, regardless of the actions taken by the generator or RES operator. For instance, the connection of a single generator to the expansive US power grid is unlikely to yield any observable changes in the overall grid frequency, given the grid's immense scale.

The concept of an infinite-bus can be idealized as a power system so vast that its voltage and frequency remain constant, regardless of the real and reactive power drawn or supplied. To analyze the behavior of a generator connected to such a large system, one can examine a setup with a generator and an infinite-bus operating in parallel to supply a load, as illustrated in Figure 3.

When multiple generators or a generator and a large power system, such as an electrical grid, are connected in parallel, the frequencies and terminal voltages of all the interconnected machines must be equivalent, since their output conductors are coupled. Consequently, their power–frequency graphs can be depicted adjacently with a shared vertical axis, as depicted in Figure 4a,b, respectively.

Under this scenario, the generator will effectively be operating in a floating mode, supplying a negligible amount of power. This situation is illustrated in Figure 4a. Conversely, if the generator had been connected to the grid at a slightly lower frequency than the system's operating frequency, it would begin drawing power from the grid, rather than delivering it, thereby behaving as a motor, rather than a generator. In this case, the generator frequency is slightly increased, as shown in Figure 4b. Once the generator has been connected, the no-load frequency of the generator is increased. The system's frequency remains constant (the frequency of an infinite-bus cannot be altered), resulting in an increase in the power supplied by the generator. In this case, the load supplied by the bus will decrease, and the load shared by the oncoming generator will increase. When paralleling is completed, the power supplied by the new generator becomes as per the desired value.

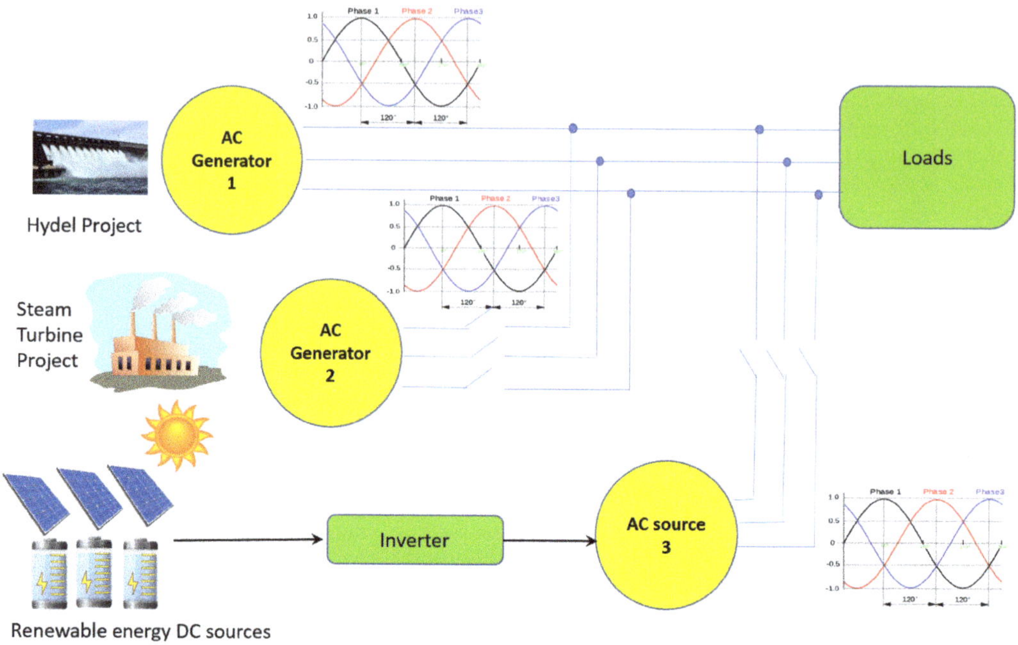

Figure 3. Transmission line of grid as an infinite bus on which multiple three-phase AC energy sources can be connected.

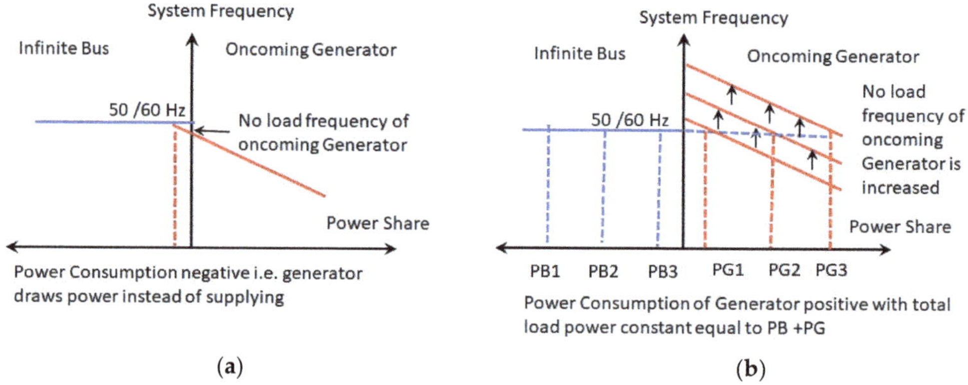

Figure 4. (**a**) Oncoming generator receiving power from the bus when its no-load frequency is less than the system frequency (50/60 Hz); (**b**) oncoming generator sharing the greater load as the no-load frequency is increased.

Chen et al. [79] suggested that the grid and distributed energy source should pre-negotiate the load and frequency requirements of the SG system in the blockchain-based trading framework.

Connecting DC systems based on DRES to the grid follows a similar process, with the addition of an extra inverter module to convert the DC power into AC power.

Different countries have introduced schemes to remunerate market participants for their available capacity through capacity markets, aiming to ensure resource adequacy and supply reliability. These capacity costs are determined based on the price of providing necessary energy when required, with long-term considerations in mind. For instance, the 15-year capacity market agreement for new generators in the United Kingdom incentivizes market participants to make new investments with an eye toward long-term supply security. Initially, capacity markets were proposed for conventional generation power plants, which produce a constant and predictable amount of energy. Supply flexibility is essential and could be encouraged through capacity measures in future power systems, which would be distinguished by a high percentage of variable renewable energy (VRE). Additionally, improving demand-side flexibility by shifting consumer consumption from peak to off-peak times in response to price signals could further enhance the system.

New market participants should be permitted to participate in a capacity adequacy process based on their business capability. Demand response and other flexible capacities, including battery storage technologies (when exporting energy to the grid), EVs, and interconnections, could improve the system's dependability during periods of supply shortfall.

4. Smart Grid (SG) Software Components: A Tutorial on Cloud-Based Blockchain, Distributed Ledger Technology, and Smart Contracts

Blockchain is the bloodline of modern decentralized financial transaction tracking, and it uses the cloud computing platform to generate irrevocable transaction certificates through distributed ledger technology (DLT). The first study to propose blockchain-based BTC [93] foresaw financial transactions solely using a P2P-based distributed framework, enabling electronic cash flow through online payments to be sent directly from one party to another without going through a financial institution. Digital signatures employing efficient cryptography techniques provide part of the solution, but the main benefits are lost if a trusted third party is still required to prevent double-spending. Double spending is defined as the customer initiating two quick transactions so that the account does not reflect an updated financial state while the second transaction is made. This problem is resolved using a P2P network, as shown in Figure 5. The P2P network timestamps transactions by hashing them into a contiguous chain of hash-based proof-of-work (PoW), called DLT, endorsed by majority node approval. On endorsement of the transaction establishing proof of authenticity, the prosumer or DRES will be awarded financial remuneration in the form of digital currency. This DLT forms an immutable record that cannot be changed without redoing the PoW and again obtaining majority vote approval. This is theoretically possible only if 51% of the participating blockchain nodes are compromised by an attack. The incentive for nodes is carrying out transaction block validation through PoW calculation. Smart contracts are automated computer programs running on the blockchain that conclude a transaction based on pre-agreed conditions and notify all participants through broadcast communication.

Since then, researchers have proposed blockchain-based architecture for several other applications besides the popular digital currency platform (BitCoin) [93]; these include the art trading platform [94], and SG [1–3,6,7], among others.

The Brooklyn Microgrid project is an example of an application of the microgrid system adopting blockchain. A network of residents is connected through a microgrid and adopts DERs through PV systems. A mobile application was developed for participants to access the local energy marketplace and choose to sell excess solar energy to the marketplace or continue to net metering [95,96].

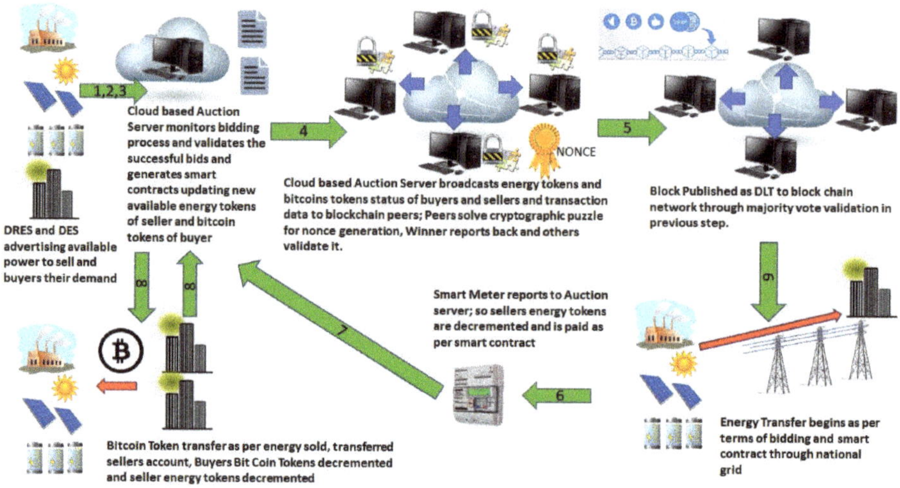

Figure 5. SG Operation in which multiple energy providers can negotiate energy deals and transfer power over the grid.

5. Working Operation of Blockchain in Smart Grid (SG)

The integration of blockchain into the SG allows the energy players (producers, consumers, and prosumers) to maintain transactions in the decentralized system through mutual consensus and trust, as depicted in Figure 5. The various stages of the energy transaction are outlined below:

Step-1: Different private sector energy sources (both green and conventional fossil-fuel-based) contact decentralized cloud-based energy auction services advertising their energy capacity with digital signature proof.

Step-2: The decentralized cloud-based energy continuous double auction (CDA) updates the current energy costings based on the continuous change in supply and demand equations. It also matches buyers and sellers with each other.

Step-3: After the buyer and seller commit to an energy deal at specified prices and other terms and conditions, the transaction agreements will be digitally verified by both the buyer and seller before an official transaction record is created to be authenticated and verified by the decentralized blockchain network.

Step-4: Transactions are performed through fungible energy and BTC tokens being assigned to the energy seller and buyer, respectively. The decentralized blockchain network will verify these transactions through a cryptographic puzzle-based nonce generation to authenticate the sub-transaction happening on the SG network. The nonce is difficult to compute, requiring significant computing resources to provide evidence of PoW, but it is very easy to verify by other nodes on the blockchain network.

Step-5: The blockchain and smart contracts will store the transaction history in the blockchain using DLT and securely distribute it to all nodes through specialized broadcast mechanisms. The blockchain ensures the immutability of smart contracts and transaction data by preventing records from being modified or deleted. Smart contracts will guarantee the completion of transactions between generators and consumers, providing assurance that producers will always deliver the electricity once a consumer has made the required payment.

Step-6: Energy transfer starts/continues between the distributed energy source and the transmission line. Smart meters continuously log the amount of energy injected into the system from the seller.

Step-7: The smart contract-based automated computer program in the cloud-based auction service will automatically keep the update of energy transfer from seller and buyer financial tokens and again publish these transactions after cryptographic verification to the blockchain network to avoid problems such as duplicate spending.

Step-8: The smart contract-based automated computer program will also maintain a record of FT (debit/credit) to update account management in the decentralized energy auction trading.

Steps 4–8 will be continually performed until energy transfers received by the cloud-based auction service have been successfully terminated and no new requests have been received. The above blockchain-based SG framework provides immutability and traceability, which is good for auditing or solving a transaction dispute. The authentication and immutability aspect of the blockchain, e.g., the Ethereum blockchain [97], runs on the PoW consensus mechanism. In this mechanism, blocks of transactions are verified by miners around the world in exchange for a reward (e.g., BTCs).

How blockchain can prevent a fraudulent transaction is depicted in Figure 6. Seller Sx initially has 100 energy units, and Buyer Bx has 500 BTC tokens. The seller, Sx, agrees to sell 10 energy units to Buyer Bx, which will cost 100 BTC tokens. The remaining energy units left with Seller Sx (=90) and the remaining BTC tokens left with Buyer Bx (=400) are updated by the smart contract application, and the transaction record and updated state of seller Sx energy units and buyer Bx BTC tokens are reflected too. This block (Block-1, as shown in Figure 6) is cryptographically verified by the miners in the BTC network and verified through the federated consensus mechanism as an authentic transaction.

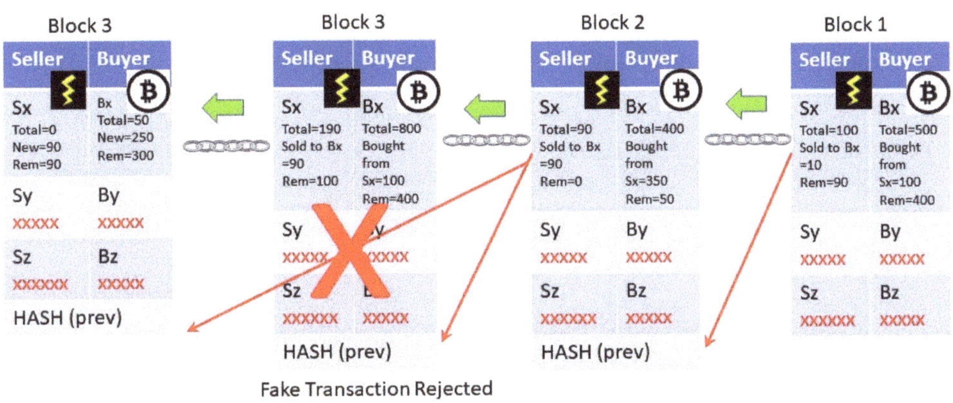

Figure 6. Blockchain records the SG transactions in a sequential manner. Wrong or fraudulent transactions are rejected.

The verified cryptographic signature is sent to the next block (Block-2) as proof of authentication of the previous transaction in Block-1. The transaction recorded in Block-2 includes another deal between the same buyer and seller, Seller Sx and Buyer By, which sees that the energy tokens of the seller are completely exhausted, and the remaining BTC tokens of the buyer are just 50. A fraudulent transaction between the same buyer and seller, as shown in Figure 6 in Block-3, will be rejected as solving its cryptographic puzzle with a verifiable answer (hash) is impossible to compute given its link with previous Block-2. When a certain mandatory number of peers fail to verify Block-3, it will be rejected by the network. The new transaction will first see Seller Sx announce with verifiable claims of more energy units, and Buyer Bx announces with a verifiable claim of more BTC tokens, after which any further transactions involving Buyer Bx and Seller Sx could be verified and endorsed by the BTC network, as in new Block-3.

Current research indicates that, in order to successfully propagate a fake (fraudulent) block on the blockchain network, the adversary will have computing power equivalent to at least 51% of the entire hashing (mining) power of all participating nodes on the blockchain network [98].

6. Technological Barriers and Current Challenges in Smart Grid (SG) Operation After Integration of EV Industry

Although SG architecture is based on previously available and implemented technologies such as blockchain, distributed energy generation, RES, IoT, and cloud computing, their synergy within SG applications, particularly with the integration of EVs, introduces new implementation challenges. These challenges require innovative solutions. In this context, we outline several research challenges that are specific to the integration of EVs into the SG ecosystem.

6.1. How National Grid May Still Hold Monopoly over Players in the EVs Energy Market

The national electric supply of any country has preferential dominance over the electrical energy sector. This is due to the reason that DES and DRES rely upon the physical infrastructure of the national grid to support the sale of their electrical energy to potential customers. DRES will likely emerge as a smaller player in the EV market in comparison; DRES would be the preferred energy generation option for the EV industry, given that its primary aim is to reduce the carbon footprint. The national SG will only seek their fair contribution when the national grid faces a shortfall, such as an increased gap between demand and supply during very hot or very cold weather. This could potentially hinder the financial benefits to DRES. However, on a national level, one advantage of this, as mentioned by Agung and Handyani [57], is that government-owned plants in the national grid can interfere with the price of electricity. When the price of electricity increases due to an increased gap between demand and supply, the government plant can sell its electricity at a lower price to lower the market price by obtaining increased contributions for DRES and other smaller IPPs. Colak et al. [99] considered the impact of SG on the national grid.

To promote free price formation in wholesale energy markets, including the occurrence of negative prices and price spikes, it is essential to eliminate price limitations. Negative prices are observed when power generation is highly inflexible and demand is low, while price spikes occur during periods of extremely high demand and relatively low generation. For example, Denmark has implemented negative pricing to facilitate the integration of wind power by incentivizing wind turbines to reduce output when there is an excess of wind power [100].

Zhao et al. [101] presented a novel distributed energy trading mechanism based on reputation. This system not only verifies transactions, but also monitors the reputation of potential buyers/sellers by considering their past history of defaults. Such a mechanism can foster trust between buyers/sellers in a decentralized trading platform.

6.2. Ensuring Completion of Energy Deals at Negotiated Rates for EV Consumers

Using blockchain to verify financial transactions when buying merchandise is a lot different than buying or selling energy on the internet. Energy transfer can be time-consuming; for instance, selling 1 MWh of power to an EV charging station(s) may take several hours to complete within a day. This situation presents various potential issues, including fluctuations in energy rates during the transfer due to rapid changes in digital currency values used by these platforms. Smart contracts would require additional temporal information regarding the commencement and conclusion of energy transfers based on the initial agreement between the energy seller and buyer. Agung and Handyani [57] have raised similar pertinent questions in their research paper. While this might not be a critical issue in smaller private blockchains, it could pose challenges in large national public blockchains where energy prices are solely determined by supply and demand and highly susceptible to speculation. For example, Ethereum-based digital cryptocurrency reached a price of USD 1432.88 on 13 January 2018, before dropping to USD 477.49 on 13 June that same year [57,97]. This condition is not desirable for an SG application, as energy price fluctuations in the midst of energy transfers may result in disputes. The only solution for such a problem is if blockchain-based payments are not linked with digital currency, but fiat currency. For example, the Tether cryptocurrency platform maintains a peg to the USD through the implementation of a proof-of-reserves system. This mechanism only generates Tether tokens in exchange for fiat currency.

6.3. Speed of Blockchain-Based Transaction Authentication for EV Consumers

Although the proposed blockchain-based system can manage transactions in the SG, there are some issues regarding its speed of execution. Current blockchain mechanisms, like the Ethereum blockchain, can currently only handle up to fifteen transactions per second. This is due to the heavy processing involved in solving cryptographic puzzles for 'nonce' calculation in a trustworthy decentralized authentication mechanism. This limitation will have to be revisited if implemented in a full-scale SG catering to the needs of EV consumers. Calvagna et al. [102] considered the pertinent problem of renewable energy instability in the blockchain-based SG framework. They tested whether the blockchain can scale to implement real-time power modulation commands through smart contracts on the scale of a few seconds to a few minutes, which will be necessitated in view of the highly volatile EV market.

6.4. Security and Its Underlying Energy Consumption Issues in Blockchain-Based SG Applications for the EV Market

Public blockchains pose a challenge in maintaining the privacy of energy market participants, including EV owners, as all transactions and account balances are publicly visible. For instance, a publicly accessible web resource listing Ethereum account information and their balance (Etherscan) [57] can compromise the privacy of these EV market players and end-users. To address this, a government-based organization could maintain a mapping between account addresses and owners' identities. However, this approach raises the risk of malicious actors targeting wealthy accounts and tracking their owners. An alternative solution is to implement a hybrid blockchain architecture, which separates transaction records from owner identities. In a hybrid model, participation remains open, but block validation and hosting are restricted to a set of trusted nodes, enhancing the network's security [103].

Another security concern is that the resource-intensive PoW consensus mechanism contrasts with the carbon footprint reduction potential of EVs. In a public blockchain, individuals with high-performance computing equipment could potentially dominate the system. A potential research direction is to explore consensus mechanisms that consider a node's overall system contribution when selecting eligible block validators, such as the proof-of-importance approach.

Two important terms in cyber-attacks on SGs are jamming attacks and false data attacks. Jamming attacks involve occupying the channel and interfering with the broadcast communication reception, while false data attacks impact the control center's computations. It was reported that a large-scale cyber-attack on the London energy network could cost up to GBP 111 million every day [104]. Cyber-attacks have been carried out regularly on energy or power network providers, with at least one cyber-attack affecting 75% of energy firms in the past year. AI models can be utilized to identify these cyber threats by examining erroneous data and enabling preemptive measures before they affect energy systems. Between accounts and fraudsters, passwords act as a barrier, while AI technology offers biometric authentication for secure and dependable energy operations. Various schemes have been proposed to develop AI applications to detect cyber-attack threats on the SG, including improving the cyber-physical security of IoTs, transient energy-based screening technology, wide-area monitoring protection, and protecting the industrial control system. Wide-area monitoring entails conveying specific local information to a remote site while also using system-wide data to prevent widespread disruption and lessen the likelihood of potential catastrophic outages [105]. We believe that future cyber security algorithms for EV applications in SG must consider privacy and computational costs as the foremost concerns while being robust to cyber threats. Alternate mining algorithms replacing proof of work (PoW) may be employed, such as proof of stake and proof of capacity.

6.5. Exacerbation of Voltage Instability Issues in SGs by Volatile Energy Demands of EV Market

Along with the advanced control and automation in the electrical grid provided by SG comes the potential for instability, which can cause serious problems. The intermittent nature of various RESs relying on renewable energy (solar and wind) necessitates the criticality of overcoming instability. Voltage instability problems are also compounded by a sudden increase or decrease in electrical load, such as one due to EV consumers. In SG, voltage instability can occur due to an incorrect configuration of the automated systems or faulty programming. The possible reasons for voltage instability in SG include:

- A sudden increase or decrease in the electrical load due to a spike in consumer demand or a disruption in the electric supply;
- Incorrect configuration of the automated systems and controls, such as the use of inverters and storage systems;
- Faulty programming of the automated systems;
- Poor placement of devices within the grid, such as transformers and capacitors.

Voltage instability can have a severe effect on the power supply in the SG. The most noticeable effect is a sudden drop in voltage, which can cause power outages and disruption. In addition, some devices can be permanently damaged due to sudden changes in voltage levels.

To reduce the risk of voltage instability in SGs, a number of solutions have been proposed, including ensuring the correct configuration of the automated systems and controls, reducing the load on the grid when demand is high, e.g., from EV consumers, through structured tariffs or when disruption is expected, implementing proper programming of the automated systems, and properly placing devices in the grid, to ensure an even distribution of voltage [106].

7. Future Trends in Next-Generation Smart Grids (SGs) After the Integration of EV Market

SG technology has already sparked new research questions that need to be addressed as its widespread deployment begins. These questions will increasingly involve the integration of EVs into the system alongside the idiosyncrasies of various participants. For instance, incentivizing green energy remains crucial, despite the volatile nature of renewable energy sources (RESs), which can disrupt SG operations. Similarly, the burgeoning EV market introduces load fluctuations that could vary unpredictably and have largely been explored through simulations without incorporating real-world data on user behavior patterns. Ensuring transparent energy transactions is another challenge that is essential for building trust among EV owners, energy buyers, and sellers in the EV industry. These participants aim to maximize their profits while optimizing grid stability and performance. Several futuristic problems are anticipated as blockchain technology becomes more widely adopted across different countries. Many of these issues will necessitate complex algorithms to address multi-player optimization problems, potentially using game theory-like approaches to effectively integrate EVs into the SG.

7.1. Intelligent Solutions to Resolve DER Integration Issues Amid Their Generation Volatility Keeping SG Stable for Expanding EV Consumers

SG supports increasing the integration of DERs and adopting a decentralized approach to complement centralized generation. DERs facilitate meeting the ever-increasing demand and power delivery to intermittent on-grid EV consumers and rural communities. However, stabilizing the SG is crucial to mitigate the instability introduced by DERs. This can be achieved through algorithms that analyze data from the grid and determine the best way to control the input from DERs, as well as dynamically varying EV loads. By managing the power flow from DERs, the grid can be made more stable, allowing it to operate without disruption. SG stabilization also enables the grid to absorb sudden fluctuations in energy production, thus avoiding voltage and frequency fluctuations.

SG stabilization technology further enhances the grid's responsiveness to sudden changes in energy demand, which is a common scenario among EV consumers. This technology can identify areas of high energy demand and automatically adjust the energy flow from DERs to meet EV charging demands. This helps to ensure the grid's stability and reduces the risk of blackouts and power surges. SG stabilization, with contributions from DERs, utilizes a combination of algorithms and technologies, such as linear programming, optimal power flow, and fuzzy logic. The technologies used include sensors, remote control systems, load forecasting, and demand response. Sensors monitor the power flow from DERs, while remote control systems manage their input. Additionally, load forecasting predicts future energy demand, and demand response systems adjust the DER output to meet EV charging in real-time based on changes in demand.

Another benefit of SG stabilization is its ability to proactively manage DERs. This technology can shift the input from DERs to match the grid's demand from randomly fluctuating EV loads. The transition from load-following to generation-following supports the integration of renewable energy sources (RES) into the grid. DERs offer many benefits, such as being closer to demand, which reduces the need for upgrading transmission and distribution networks. This proximity to demand also enhances efficiency by minimizing power loss in transmission.

As SGs transition to decentralized generation models through DERs, to meet fluctuating load demands due to EVs, they require a platform to enable energy trading. Blockchain technology could be a solution, providing a secure, trusted P2P platform for data and energy transactions, as well as a faster decentralized framework [107]. Despite the potential advantages of blockchain, challenges related to integrating a high penetration of DERs persist. These challenges include the curtailment of DERs due to the physical constraints of power systems and unit commitment problems [108–111]. For example, if a photovoltaic (PV) system's output suddenly decreases, other DERs need to compensate for the reduced generation. While blockchain could facilitate better management and trading of DERs, these operational challenges will remain. Nevertheless, blockchain offers benefits such as decentralization and improved DER management in real-time to meet EV demand response with matching generation capacity, market pricing, peak load shaving, and support for ancillary services [107].

Fan et al. [112] proposed a framework for P2P energy trading using blockchain technology. Their simulation study confirmed the system's capability to handle energy transactions effectively and efficiently within the Alberta electricity wholesale market.

7.2. Intelligent Solutions for Demand-Side Management (DSM) of the EV Market in SG

Demand-side management (DSM) is an actively researched field in the area of SG. The idea is for customers to vary their load profiles for the mutual benefit of the grid and themselves. Thus, its global objectives are electricity cost minimization, minimizing customer discomfort, maximizing the usage of local energy generation, minimizing peak demand, and minimizing the peak-to-average energy ratio (PAER). Many optimization algorithms have been proposed to tackle this problem [113]. For non-flexible loads, a fuzzy controller and integer linear programming technique have been applied [114]. Genetic algorithms for multi-objective optimization have been applied to situations to meet the different needs of residential and industrial customers. For optimum switching of flexible loads for highly fluctuating energy prices when integrating large-scale fluctuating RESs like solar and wind, a multistage stochastic mixed-integer linear programming (SMILP) is proposed [115]. The clonal selection algorithm (CSA) is also proposed for optimal day-ahead load scheduling for large-scale integration of PV plants by the projection of anticipated load and generated energy forecasting [116]. For the rapidly fluctuating load demands of the emerging EV market, several strategies, like covariance matrix adaption evolution strategy (CMAES)-based optimization algorithms, are proposed to maximize the EV charging capacity while minimizing the cost for owners and EV battery providers [117]. Currently, the problem could be tackled by ensuring consumers' acceptance and participation in DSM programs, as shown by a simulation study [118].

7.3. Rise of EV Market: Bi-Directional Charging and Newer Energy Trade Scenarios V2V, V2G, and G2V

The number of EVs on the roads is increasing; for instance, 16.5 million EVs in 2021 were estimated to be on the roads, which is triple the amount in 2018 [119]. These EVs are considered an added demand on existing power system components, and they might have an adverse impact on the grid. The study in [120] showed that increasing EVs penetration could lead to higher short-circuit currents, off-limits voltage levels, and a shortened lifespan of the equipment. Also, the study demonstrated how fast charging of EVs caused higher losses and voltage drops when compared with regular charging. Nevertheless, with the increasing number of EVs and their potential to be used for energy trading and bidirectional energy exchange, blockchain could be the framework to automate the energy trade.

The energy trade could be V2V, V2G, or G2V. These energy trade scenarios are depicted in Figure 7a,b, respectively. Figure 7a shows the V2G and G2V charging scenarios for EVs where bi-directional energy trade can happen between the SG and Smart Homes. Figure 7b highlights both the static and dynamic EV wireless charging solutions and shows that V2V wireless charging can also happen under static wireless charging. Nevertheless, it is important to note that creating such an environment using the blockchain introduces new challenges.

Namely, the scalability of blockchain data, interoperability between different blockchain platforms, privacy, data protection, and confidentiality are the key challenges in implementing P2P for EV energy trade [121,122]. Firstly, scalability is a challenge correlated with the number of EVs. As more EVs are connected, real-time processing and communication through the chains are required. Blockchain platforms are capable of a few transactions per second (TPS); for instance, BTC is capable of 4 TPS [123].

This is a very low number compared with credit card platform rates. Hence, a higher rate is needed to adopt more EV penetration in the SG using the blockchain. Secondly, interoperability is needed for EVs to share data across heterogeneous blockchain platforms. The work in [122,124] argues that existing platforms do follow a defined standard. The wide adoption of blockchain for EVs is highly dependent on assuring interoperability across all platforms. In addition, security is important as there will be a large number of trades between EVs and other EVs or the SG.

Even though blockchain favors security at the expense of scalability, security performance needs to be assessed. Reference [125] discusses different security issues encountered in utilizing IoT and blockchain, such as inherently vulnerable particularities, programming vulnerabilities, and attacks against smart contracts. Introducing a lightweight model to increase scalability could hinder security, and this is a challenge faced in implementing it for EVs. Finally, privacy, data protection, and confidentiality are crucial requirements for the successful implementation of blockchain for EV applications.

Compliance with regulations is indeed mandatory. According to reference [122], regulations like the general data protection regulation (GDPR) are intended to limit accessibility to authorized users only. Due to the immutability of blockchain technology, which means that it cannot be erased even by authorized users, there may be a risk of compromising privacy and confidentiality. This could make blockchain non-compliant with data privacy regulations, such as GDPR [125].

7.4. Advancements in Vehicle-to-Grid (V2G) Technologies for EVs

V2G capabilities are put forth to allow EVs to support the demand-side auxiliary services. EV batteries are made to be able to quickly charge for drivers' convenience when necessary and quickly discharge during rapid acceleration and other situations that call for a power surge. In order to support the grid through peak shaving and load balancing, they can engage in supply ramping and power regulation [126]. EVs can help to balance out swings in renewable energy. Many issues have to be considered besides cybersecurity when considering EVs for supporting energy market schemes, including degradation brought on by frequent charging and discharging cycles and the impact of EV owners' behavior on grid reliability, because their main concerns would be driving their cars, rather than meeting grid needs [127].

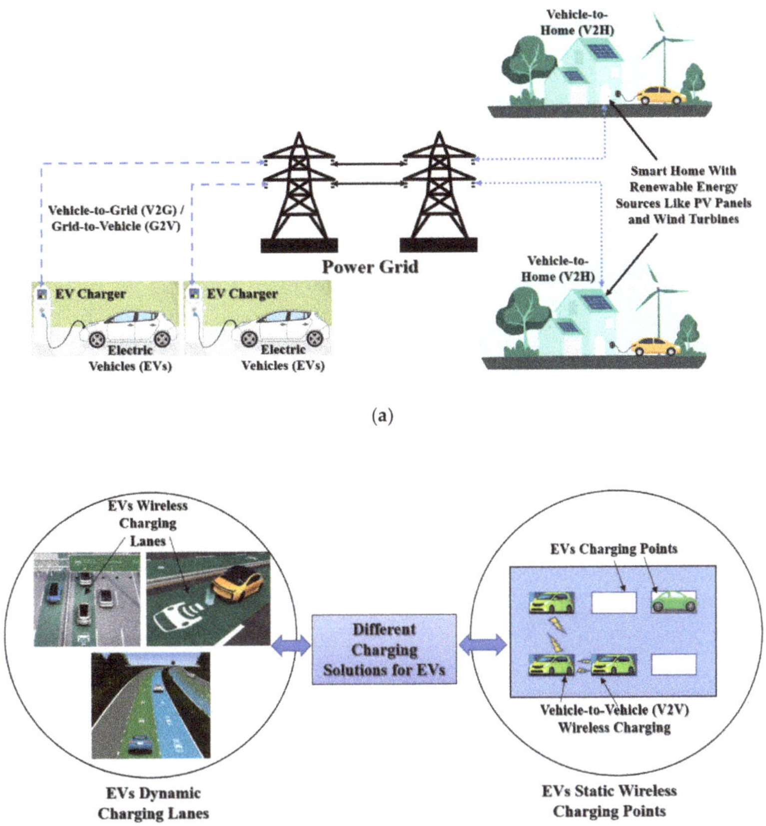

Figure 7. (**a**) V2G and G2V charging scenarios for EVs where bi-directional energy trade can happen between the SG and smart home, (**b**) static and dynamic wireless charging of EVs, and also V2V wireless charging between EVs.

7.5. Load Shaping and Peak Shifting via Time-of-Use (ToU) Pricing to Meet Fluctuating EV Loads

Time-of-use (ToU) pricing is adopted in various regions to incentivize load shaping and peak shifting. Seventeen European countries, in addition to the USA and India, have adopted different forms of ToU tariffs. These forms include:

- Static ToU pricing applies to all low-voltage consumers in Italy who do not select a supplier from the liberalized market, with different rates for day and night usage.
- Real-time pricing, which includes different forms such as dynamic real-time pricing (e.g., hourly pricing), has been adopted in countries like Spain, the UK, and Estonia. Other dynamic pricing forms based on monthly averages are utilized in countries like Denmark, Norway, and Sweden.

- Variable peak pricing, which is a hybrid of static and dynamic prices by means of pre-defined pricing, except at peak demand during the day, when real-time prices apply.
- Critical peak pricing, which is a hybrid of static and dynamic pricing by means of pre-defined pricing all year, except for a maximum of 22 days where the highest wholesale prices occur, and is adopted in countries like Lithuania, Romania, and France.

Figure 8 presents graphical representations of the different forms of ToU tariffs [128]. How pricing strategies could be integrated with blockchain-based SG architecture without causing disputes is a current challenge that will be resolved in the future. Such tariff rates would be necessitated for the stable operation of blockchain-based SG electricity trading operations in the presence of a highly dynamic EV consumer market.

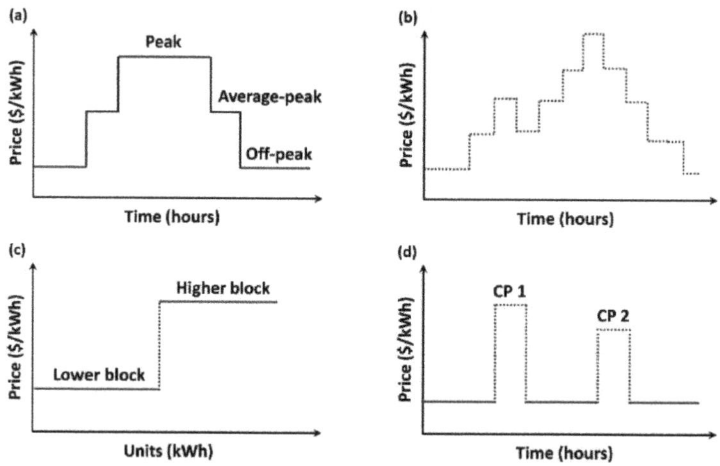

Figure 8. Types of ToU tariff source. (**a**) Normal ToU; (**b**) real-time pricing (RTP); (**c**) inclining block rate (IBR); and (**d**) critical peak pricing (CPP). Reprinted from Ref. [128].

7.6. Incentivising Green Energy for EVs

The blockchain records all transaction history, and if a device capable of detecting how the electricity was generated is included in the SG metering system, pricing can be adjusted based on this condition. This promotes green energy and limits electricity generated from dirtier sources. The EV market should be given incentives to use green energy for EV charging, given that their primary aim is the reduction of carbon footprints. Several pilot projects, e.g., Greeneum [62], provide a decentralized blockchain-based SG application to incentivize green energy production. It has AI algorithms to guide organizations to become carbon-neutral through the efficient generation and utilization of green energy. This will, however, require government-based subsidies to keep the price of green energy lower, especially if it is offered entirely by the private sector. For example, Tether, a blockchain-based SG platform, uses a proof-of-reserves, which then creates tokens only in exchange for fiat money. This system can provide buyer and seller security, but cannot forbid people from selling cheap electricity produced by dirty mechanisms, e.g., using fossil fuels instead of renewable resources. People, including the burgeoning EV market, will tend to buy cheaper electricity, regardless of its source, if proper incentives are not given. In comparison, Green energy is cheaper to produce than conventional fossil-fuel-based energy generation, but requires initial large investments and expensive maintenance and future upgradation costs, thus increasing its price until the project achieves break-even condition.

7.7. Application of Artificial Intelligence (AI) to SG Technology for EV Market

Hossain et al. [89] and Diaba and Elmusrati [90] explored the application of big data and deep learning (DL) in SG applications to predict solar and wind power generation, as well as load demand, facilitating more informed decisions for power generation optimization. A similar survey was carried out by Rigas et al. [129], focusing on EV charging issues. Given the intermittent nature of renewable energy sources, optimizing EV charging is crucial to balance the load distribution on the SG while ensuring that no EV owner is left without power. El Ghanam et al. [77] conducted a simulation-based study on EV charging scenarios for dynamic wireless charging (DWC) lanes in Dubai and Sharjah, employing various mixed-integer non-linear programming optimization algorithms to enhance load balancing and prevent overloads. Their approach resulted in a 44% reduction in PAER compared with using a shortest-distance algorithm for assigning EVs to charging stations. Esenogho et al. [81] investigated the application of AI algorithms for 5G-powered SG applications. The high data speeds of 5G networks are critical for supporting massive IoT (mIoT), which is necessary for transmitting the substantial data required by AI algorithms [130]. These algorithms enable effective load forecasting, real-time fault management, and grid stability assessment. Refaat et al. [80] addressed how AI could stabilize the SG amidst volatile power generation from renewable energy sources. Greeneum [62], a decentralized P2P blockchain system for green energy, uses AI algorithms to reward carbon-neutral businesses and blockchain for smart contract generation and sharing. AI algorithms are also utilized as third-party platforms for predicting energy prices; Karasu et al. [88] recently researched predicting Bitcoin values using time-series data and applying two popular ML algorithms: support vector machine (SVM) and linear regression (LR). Pilz and Al-Fagih [82] and Zhang et al. [83] employed game theory approaches for optimizing energy trading.

One controversial aspect of blockchain in SG applications is the computationally intensive consensus mechanism, such as PoW, which consumes significant power at miner nodes. Chibi et al. [131] proposed a novel green PoW (GPoW) mechanism that uses ML and Markov decision processes to reduce the number of miners competing based on their historical win/lose records, thereby decreasing the energy consumption of the consensus mechanism.

7.8. Oscillatory Load Behavior on SG Technology Due to Selfish User-Incentivized EV Transportation Schedules

The gradual rise in the EV market, including private EV users as well as EV-based public transport options (buses and taxis), will present important design and implementation challenges for SG management as consumers strategize their transportation schedules according to beneficial tariffs for EV charging, putting strain on existing DSM algorithms [132–137]. Also, their charging demands would be dependent on their driving style and available time for partial/complete charging [138–140]. ML algorithms [141] could particularly assist EV owners in selfishly optimizing EV distance range and battery management according to user preference for a particular charging station on their route. For example, EV owners could choose to charge from charging stations located at their offices during the day as compared with their homes according to their work schedules and tariff variation according to place-of-use and ToU. As a result, this would result in spikes in tariff rates in central business district (CBD) areas as compared with residential suburbs due to increased demand during specific times of day. An increase in the tariff rate due to increased demand in CBD areas would prompt EV users to opt for selling surplus energy at CBD locations instead of buying and choosing an alternative charging option at some sub-urban location or en route to their homes. As a result, the demand for CBD energy would again go down, lowering the tariff rates. This would often result in a complex selfish relationship between smart grid and EV users, giving rise to oscillatory behavior causing rapid cyclic fluctuations in demands and tariffs, which could be managed using a combination of centralized and decentralized scheduling algorithms at EV charging stations [142–144].

8. Conclusions

This paper provides a comprehensive review of the existing research literature and the latest developments in blockchain-based smart grid (SG) architecture, with a particular emphasis on the integration of electric vehicles (EVs). It begins by outlining the essential components of a working SG system model before shifting to the primary focus: identifying the technological barriers and current challenges, and exploring future trends in this evolving domain.

The article offers an in-depth analysis of the current implementation challenges associated with SG architecture, highlighting issues such as privacy concerns for participants, including EV owners, and the need to enhance blockchain transaction speeds to handle the rapid fluctuations in energy prices caused by the dynamic energy contributions from EVs. Balancing security and transaction speed is critical to the successful application of blockchain technology on a national scale within the energy market. Another significant challenge is managing the diverse objectives of EV owners alongside traditional energy sellers and buyers, while ensuring the stability of the national grid. This involves considering the extent of regulatory control national grid operators are prepared to cede to enable a more decentralized energy market inclusive of EVs.

Future research directions highlighted in this paper include addressing the challenges brought by new EV entrants into the decentralized energy market, such as unpredictable user behavior and fluctuating energy consumption patterns. Additionally, the need to incentivize green energy sources remains crucial, despite the volatility of renewable energy sources (RESs), which can impact SG operations. A critical area for future research is developing mechanisms to reinforce the trust of energy traders, including EV participants, in an environment where government policies might continue to exert monopolistic control over the grid, potentially hindering true decentralization.

Addressing these challenges will likely involve formulating complex multi-player optimization problems. Solutions will require the application of advanced technologies, such as big data and artificial intelligence (AI), essential for the efficient operation of decentralized energy markets and the integration of EVs. Future work should focus on developing robust AI-based algorithms for load forecasting, dynamic pricing, and real-time grid management, particularly in the context of EVs. Furthermore, enhancing cybersecurity measures to protect the integrity and reliability of blockchain-based transactions in SG systems will be paramount.

In conclusion, as discussed in this paper, while the path to fully realizing the potential of blockchain-enabled SGs in the context of EV integration is filled with challenges, the opportunities for innovation and improvement are vast. Continued interdisciplinary research and collaboration will be crucial in overcoming these obstacles and advancing the next generation of SG technologies and EV integration.

Author Contributions: Conceptualization, S.Q. and B.A.K.; methodology, A.A. (Abdullah Alamri) and A.A. (Abdulrahman AlKassem); formal analysis, S.Q., B.A.K., A.A. (Abdullah Alamri) and A.A. (Abdulrahman AlKassem); investigation, S.Q. and A.A. (Abdullah Alamri); resources, B.A.K. and A.A. (Abdulrahman AlKassem); writing—original draft preparation, S.Q. and B.A.K.; writing—review and editing, B.A.K., A.A. (Abdullah Alamri) and A.A. (Abdulrahman AlKassem); visualization, S.Q. and A.A. (Abdulrahman AlKassem); project administration, S.Q. and B.A.K. All authors have read and agreed to the published version of the manuscript.

Funding: This research received no external funding.

Data Availability Statement: No new data were created or analyzed in this study. Data sharing is not applicable to this article.

Conflicts of Interest: The authors declare no conflicts of interest.

References

1. Aoudia, M.; Alaraj, M.B.M.; Abu Waraga, O.; Mokhamed, T.; Abu Talib, M.; Bettayeb, M.; Nasir, Q.; Ghenai, C. Toward better blockchain-enabled energy trading between electric vehicles and smart grids in Internet of Things environments: A survey. *Front. Energy Res.* **2024**, *12*, 1393084. [CrossRef]
2. Adnan, M.; Ahmed, I.; Iqbal, S. Load flow balancing in super smart grids: A review of technical challenges, possible solutions and future trends from European perspective. *Comput. Electr. Eng.* **2024**, *117*, 109265. [CrossRef]
3. Smart Grid Market. Available online: https://www.marketsandmarkets.com/Market-Reports/smart-grid-market-208777577.html (accessed on 8 December 2023).
4. Hussain, S.M.S.; Nadeem, F.; Aftab, M.A.; Ali, I.; Ustun, T.S. The Emerging Energy Internet: Architecture, Benefits, Challenges, and Future Prospects. *Electronics* **2019**, *8*, 1037. [CrossRef]
5. Jacobs, F. EVs Will Push Up Global Electricity Demand by 630% by 2030. Available online: https://www.globalfleet.com/en/technology-and-innovation/global/features/evs-will-push-global-electricity-demand-630-2030?a=FJA05&t[0]=Electricity&t[1]=EVs&curl=1#:~:text=That%20data%20shows%20that%20global,to%2097,000%20GWh%20in%202023 (accessed on 10 October 2024).
6. Musleh, A.S.; Yao, G.; Muyeen, S.M. Blockchain Applications in Smart Grid–Review and Frameworks. *IEEE Access* **2019**, *7*, 86746–86757. [CrossRef]
7. İnci, M.; Çelik, Ö.; Lashab, A.; Bayındır, K.Ç.; Vasquez, J.C.; Guerrero, J.M. Power System Integration of Electric Vehicles: A Review on Impacts and Contributions to the Smart Grid. *Appl. Sci.* **2024**, *14*, 2246. [CrossRef]
8. Li, D.; Guo, Q.; Bai, D.; Zhang, W. Research and Implementation on the Operation and Transaction System Based on Blockchain Technology for Virtual Power Plant. In Proceedings of the 2022 International Conference on Blockchain Technology and Information Security (ICBCTIS), Huaihua, China, 15–17 July 2022; pp. 165–170. [CrossRef]
9. Mollah, M.B.; Zhao, J.; Niyato, D.; Lam, K.Y.; Zhang, X.; Ghias, A.M.; Koh, L.H.; Yang, L. Blockchain for Future Smart Grid: A Comprehensive Survey. *IEEE Internet Things J.* **2021**, *8*, 18–43. [CrossRef]
10. Li, Z.; Kang, J.; Yu, R.; Ye, D.; Deng, Q.; Zhang, Y. Consortium Blockchain for Secure Energy Trading in Industrial Internet of Things. *IEEE Trans. Ind. Inform.* **2018**, *14*, 3690–3700. [CrossRef]
11. Gai, K.; Wu, Y.; Zhu, L.; Xu, L.; Zhang, Y. Permissioned Blockchain and Edge Computing Empowered Privacy-Preserving Smart Grid Networks. *IEEE Internet Things J.* **2019**, *6*, 7992–8004. [CrossRef]
12. KWHCoin [White Paper]. Available online: https://ww1.prweb.com/prfiles/2018/01/19/15109995/KWHCoin-White-Paper-REVISED.pdf (accessed on 1 December 2023).
13. Mihaylov, M.; Jurado, S.; Avellana, N.; Van Moffaert, K.; de Abril, I.M.; Nowe, A. NRGcoin: Virtual currency for trading of renewable energy in smart grids. In Proceedings of the 11th International Conference on the European Energy Market (EEM14), Krakow, Poland, 28–30 May 2014; pp. 1–6. [CrossRef]
14. SolarCoin [White Paper]. Available online: https://www.allcryptowhitepapers.com/solarcoin-whitepaper/ (accessed on 1 December 2023).
15. Orsini, L.; Collins, B.; Webb, M.; Montgomery, C.; Conte, B.; Adamson, M.; Heitmann, P.; Kessler, S.; Brown, M. Exergy-Technical [White Paper]. Available online: https://www.truevaluemetrics.org/DBpdfs/Initiatives/Exergy/Exergy-2018-Business-Whitepaper-v10.pdf (accessed on 2 February 2024).
16. Pylon Network. The Energy Blockchain Platform [White Paper]. Available online: https://pylon-network.org/wp-content/uploads/2019/02/WhitePaper_PYLON_v2_ENGLISH-1.pdf (accessed on 12 February 2024).
17. Ul Hassan, N.; Yuen, C.; Niyato, D. Blockchain Technologies for Smart Energy Systems: Fundamentals, Challenges, and Solutions. *IEEE Ind. Electron. Mag.* **2019**, *13*, 106–118. [CrossRef]
18. Wang, N.; Zhou, X.; Lu, X.; Guan, Z.; Wu, L.; Du, X.; Guizani, M. When Energy Trading Meets Blockchain in Electrical Power System: The State of the Art. *Appl. Sci.* **2019**, *9*, 1561. [CrossRef]
19. Bielecki, S.; Skoczkowski, T.; Sobczak, L.; Wołowicz, M. Electricity Usage Settlement System Based on a Cryptocurrency Instrument. *Energies* **2022**, *15*, 7003. [CrossRef]
20. Feng, H.; Shaffer, J.L.; Yu, L.; Xu, R.; Chen, Y.; Ma, X. A Sustainable Energy-Tailored Transaction Mechanism for Smart Grids Based on Block Chains. In Proceedings of the 10th Renewable Power Generation Conference (RPG 2021), Online Conference, 14–15 October 2021; pp. 605–609. [CrossRef]
21. Danzi, P.; Angjelichinoski, M.; Stefanovic, C.; Popovski, P. Distributed proportional-fairness control in microgrids via blockchain smart contracts. In Proceedings of the 2017 IEEE International Conference on Smart Grid Communications (SmartGridComm), Dresden, German, 23–27 October 2017; pp. 45–51. [CrossRef]
22. Wang, S.; Taha, A.F.; Wang, J.; Kvaternik, K.; Hahn, A. Energy Crowdsourcing and Peer-to-Peer Energy Trading in Blockchain-Enabled Smart Grids. *IEEE Trans. Syst. Man Cybern, Syst.* **2019**, *49*, 1612–1623. [CrossRef]
23. Wang, J.; Li, L.; Zhang, J. Deep reinforcement learning for energy trading and load scheduling in residential peer-to-peer energy trading market. *Int. J. Electr. Power Energy Syst.* **2023**, *147*, 108885. [CrossRef]
24. Munsing, E.; Mather, J.; Moura, S. Blockchains for decentralized optimization of energy resources in microgrid networks. In Proceedings of the 2017 IEEE Conference on Control Technology and Applications (CCTA), Maui, HI, USA, 27–30 August 2017; pp. 2164–2171. [CrossRef]

25. Fu, X.; Wang, H.; Wang, Z. Research on Block-Chain-Based Intelligent Transaction and Collaborative Scheduling Strategies for Large Grid. *IEEE Access* **2020**, *8*, 151866–151877. [CrossRef]
26. Sharma, P.; Senapati, R.; Swetapadma, A. Mathematical Modeling for Matching the Prosumers in Decentralized Energy Trading System. In Proceedings of the 2022 11th International Conference on Renewable Energy Research and Application (ICRERA), Istanbul, Turkey, 18–21 September 2022; pp. 502–507. [CrossRef]
27. Mrabet, Z.E.; Kaabouch, N.; Ghazi, H.E.; Ghazi, H.E. Cyber-security in smart grid: Survey and challenges. *Comput. Electr. Eng.* **2018**, *67*, 469–482. [CrossRef]
28. Naderi, E.; Asrari, A.; Fajri, P. False Data Injection Cyberattacks Targeting Electric Vehicles in Smart Power Distribution Systems. In Proceedings of the 2024 IEEE Transportation Electrification Conference and Expo (ITEC), Chicago, IL, USA, 19 June 2024. [CrossRef]
29. Naderi, E.; Asrari, A. Integrated Power and Transportation Systems Targeted by False Data Injection Cyberattacks in a Smart Distribution Network. In *Electric Transportation Systems in Smart Power Grids;* CRC Press: Boca Raton, FL, USA, 2022. [CrossRef]
30. Karanfil, M.; Rebbah, D.E.; Debbabi, M.; Kassouf, M.; Ghafouri, M.; Youssef, E.N.S.; Hanna, A. Detection of Microgrid Cyberattacks using Network and System Management. *IEEE Trans. Smart Grid* **2022**, *14*, 2390–2405. [CrossRef]
31. Rahiminejad, A.; Plotnek, J.; Atallah, R.; Dubois, M.A.; Malatrait, D.; Ghafouri, M.; Mohammadi, A.; Debbabi, M. A resilience-based recovery scheme for smart grid restoration following cyberattacks to substations. *Int. J. Electr. Power Energy Syst.* **2023**, *145*, 108610. [CrossRef]
32. Su, Z.; Wang, Y.; Xu, Q.; Fei, M.; Tian, Y.-C.; Zhang, N. A Secure Charging Scheme for Electric Vehicles With Smart Communities in Energy Blockchain. *IEEE Internet Things J.* **2019**, *6*, 4601–4613. [CrossRef]
33. Gunduz, M.Z.; Das, R. Cyber-security on smart grid: Threats and potential solutions. *Comput. Netw.* **2020**, *169*, 107094. [CrossRef]
34. Musleh, A.S.; Chen, G.; Dong, Z.Y.; Wang, C.; Chen, S. Spatio-temporal data-driven detection of false data injection attacks in power distribution systems. *Int. J. Electr. Power Energy Syst.* **2023**, *145*, 108612. [CrossRef]
35. Li, X.; Jiang, P.; Chen, T.; Luo, X.; Wen, Q. A survey on the security of blockchain systems. *Future Gener. Comput. Syst.* **2020**, *107*, 841–853. [CrossRef]
36. Liu, D.; Li, D.; Liu, X.; Ma, L.; Yu, H.; Zhang, H. Research on a Cross-Domain Authentication Scheme Based on Consortium Blockchain in V2G Networks of Smart Grid. In Proceedings of the 2018 2nd IEEE Conference on Energy Internet and Energy System Integration (EI2), Beijing, China, 20–22 October 2018; pp. 1–5. [CrossRef]
37. Conti, M.; Kumar, E.S.; Lal, C.; Ruj, S. A Survey on Security and Privacy Issues of Bitcoin. *IEEE Commun. Surv. Tutor.* **2018**, *20*, 3416–3452. [CrossRef]
38. Wan, L.; Eyers, D.; Zhang, H. Evaluating the Impact of Network Latency on the Safety of Blockchain Transactions. In Proceedings of the 2019 IEEE International Conference on Blockchain (Blockchain), Atlanta, GA, USA, 14–17 July 2019; pp. 194–201. [CrossRef]
39. Gao, J.; Asamoah, K.O.; Sifah, E.B.; Smahi, A.; Xia, Q.; Xia, H.; Zhang, X.; Dong, G. GridMonitoring: Secured Sovereign Blockchain Based Monitoring on Smart Grid. *IEEE Access* **2018**, *6*, 9917–9925. [CrossRef]
40. Kumar, P.; Lin, Y.; Bai, G.; Paverd, A.; Dong, J.S.; Martin, A. Smart Grid Metering Networks: A Survey on Security, Privacy and Open Research Issues. *IEEE Commun. Surv. Tutor.* **2019**, *21*, 2886–2927. [CrossRef]
41. Liang, X.; An, N.; Li, D.; Zhang, Q.; Wang, R. A Blockchain and ABAC Based Data Access Control Scheme in Smart Grid. In Proceedings of the 2022 International Conference on Blockchain Technology and Information Security (ICBCTIS), Huaihua, China, 15–17 July 2022; pp. 52–55. [CrossRef]
42. Lopes, J.A.P.; Soares, F.J.; Almeida, P.M.R. Integration of Electric Vehicles in the Electric Power System. *Proc. IEEE* **2011**, *99*, 168–183. [CrossRef]
43. Rassaei, F.; Soh, W.-S.; Chua, K.-C. A Statistical modelling and analysis of residential electric vehicles' charging demand in smart grids. In Proceedings of the 2015 IEEE Power & Energy Society Innovative Smart Grid Technologies Conference (ISGT), Washington, DC, USA, 18–20 February 2015; pp. 1–5. [CrossRef]
44. Sayed, M.A.; Ghafouri, M.; Debbabi, M.; Assi, C. Dynamic Load Altering EV Attacks Against Power Grid Frequency Control. In Proceedings of the 2022 IEEE Power & Energy Society General Meeting (PESGM), Denver, CO, USA, 17–21 July 2022; pp. 1–5. [CrossRef]
45. Habib, S.; Khan, M.M.; Abbas, F.; Sang, L.; Shahid, M.U.; Tang, H. A Comprehensive Study of Implemented International Standards, Technical Challenges, Impacts and Prospects for Electric Vehicles. *IEEE Access* **2018**, *6*, 13866–13890. [CrossRef]
46. Knirsch, F.; Unterweger, A.; Engel, D. Privacy-preserving blockchain-based electric vehicle charging with dynamic tariff decisions. *Comput. Sci. Res. Dev.* **2018**, *33*, 71–79. [CrossRef]
47. Rastogi, S.K.; Sankar, A.; Manglik, K.; Mishra, S.K.; Mohanty, S.P. Toward the Vision of All-Electric Vehicles in a Decade [Energy and Security]. *IEEE Consum. Electron. Mag.* **2019**, *8*, 103–107. [CrossRef]
48. Chen, X.; Wang, H.; Wu, F.; Wu, Y.; Gonzalez, M.C.; Zhang, J. Multimicrogrid Load Balancing Through EV Charging Networks. *IEEE Internet Things J.* **2022**, *9*, 5019–5026. [CrossRef]
49. Rimal, B.P.; Kong, C.; Poudel, B.; Wang, Y.; Shahi, P. Smart Electric Vehicle Charging in the Era of Internet of Vehicles, Emerging Trends, and Open Issues. *Energies* **2022**, *15*, 1908. [CrossRef]
50. Zeadally, S.; Javed, M.A.; Hamida, E.B. Vehicular Communications for ITS: Standardization and Challenges. *IEEE Comm. Stand. Mag.* **2020**, *4*, 11–17. [CrossRef]

51. Qazi, S.; Sabir, F.; Khawaja, B.A.; Atif, S.M.; Mustaqim, M. Why is Internet of Autonomous Vehicles not as Plug and Play as We Think ? Lessons to Be Learnt From Present Internet and Future Directions. *IEEE Access* **2020**, *8*, 133015–133033. [CrossRef]
52. Huang, Z.; Suankaewmanee, K.; Kang, J.; Niyato, D.; Sin, N.P. Development of Reliable Wireless Communication System for Secure Blockchain-based Energy Trading. In Proceedings of the 2019 16th International Joint Conference on Computer Science and Software Engineering (JCSSE), Chonburi, Thailand, 10–12 July 2019; pp. 126–130. [CrossRef]
53. Espinoza, F.; Mar, M.; Ramirez, E.; Noel, J. Control of real power in a synchronous machine using a SCADA system in a smart grid. In Proceedings of the 2016 IEEE ANDESCON, Arequipa, Peru, 19–21 October 2016; pp. 1–4. [CrossRef]
54. Mak, S.T.; So, E. Integration of PMU, SCADA, AMI to accomplish expanded functional capabilities of Smart Grid. In Proceedings of the 29th Conference on Precision Electromagnetic Measurements (CPEM 2014), Rio de Janeiro, Brazil, 24–29 August 2014; pp. 68–69. [CrossRef]
55. Gupta, P.C.R.; Ramesh, A.; Satvik, D.; Nagasundari, S.; Honnavalli, P.B. Simulation of SCADA System for Advanced Metering Infrastructure in Smart Grid. In Proceedings of the 2020 International Conference on Smart Electronics and Communication (ICOSEC), Trichy, India, 10–12 September 2020; pp. 1071–1077. [CrossRef]
56. Jakaria, A.H.M.; Rahman, M.A.; Gokhale, A. Resiliency-Aware Deployment of SDN in Smart Grid SCADA: A Formal Synthesis Model. *IEEE Trans. Netw. Serv. Manag.* **2021**, *18*, 1430–1444. [CrossRef]
57. Agung, A.A.G.; Handayani, R. Blockchain for smart grid. *J. King Saud Univ.—Comput. Inf. Sci.* **2022**, *34*, 666–675. [CrossRef]
58. Orazgaliyev, D.; Lukpanov, Y.; Ukaegbu, I.A.; Sivanand Kumar Nunna, H.S.V. Towards the Application of Blockchain technology for Smart Grids in Kazakhstan. In Proceedings of the 2019 21st International Conference on Advanced Communication Technology (ICACT), PyeongChang, Republic of Korea, 17–20 February 2019; pp. 273–278. [CrossRef]
59. Raza, M.A.; Aman, M.M.; Abro, A.G.; Tunio, M.A.; Khatri, K.L.; Shahid, M. Challenges and potentials of implementing a smart grid for Pakistan's electric network. *Energy Strategy Rev.* **2022**, *43*, 100941. [CrossRef]
60. Guo, Y.; Wan, Z.; Cheng, X. When blockchain meets smart grids: A comprehensive survey. *High-Confid. Comput.* **2022**, *2*, 100059. [CrossRef]
61. Cao, Y.-N.; Wang, Y.; Ding, Y.; Guo, Z.; Wu, Q.; Liang, H. Blockchain-empowered security and privacy protection technologies for smart grid. *Comput. Stand. Interfaces* **2023**, *85*, 103708. [CrossRef]
62. Introducing Greeneum: A Blockchain Powered, Decentralized Platform Built to Incentivize Green Energy Production. Available online: https://aithority.com/technology/blockchain/introducing-greeneum-a-blockchain-powered-decentralized-platform-built-to-incentivize-green-energy-production/ (accessed on 23 April 2018).
63. Hasan, M.K.; Habib, A.A.; Shukur, Z.; Ibrahim, F.; Islam, S.; Razzaque, M.A. Review on cyber-physical and cyber-security system in smart grid: Standards, protocols, constraints, and recommendations. *J. Netw. Comput. Appl.* **2023**, *209*, 103540. [CrossRef]
64. Wen, Z.; Li, Y.; Zheng, Y.; Yang, Y. Virtual Power Plant Trading Strategy Based on Block-chain to Satisfy Clean Energy Partiality. In Proceedings of the 2020 IEEE 4th Conference on Energy Internet and Energy System Integration (EI2), Wuhan, China, 30 October–1 November 2020; pp. 3004–3010. [CrossRef]
65. Zhang, N.; Hu, Z.; Dai, D.; Dang, S.; Yao, M.; Zhou, Y. Unit Commitment Model in Smart Grid Environment Considering Carbon Emissions Trading. *IEEE Trans. Smart Grid* **2016**, *7*, 420–427. [CrossRef]
66. Oad, A.; Ahmad, H.G.; Talpur, M.S.H.; Zhao, C.; Pervez, A. Green smart grid predictive analysis to integrate sustainable energy of emerging V2G in smart city technologies. *Optik* **2023**, *272*, 170146. [CrossRef]
67. Cui, G.; Shi, K.; Qin, Y.; Liu, L.; Qi, B.; Li, B. Application of block chain in multi-level demand response reliable mechanism. In Proceedings of the 2017 3rd International Conference on Information Management (ICIM), Chengdu, China, 21–23 April 2017; pp. 337–341. [CrossRef]
68. Babayomi, O.; Zhang, Z.; Dragicevic, T.; Hu, J.; Rodriguez, J. Smart grid evolution: Predictive control of distributed energy resources—A review. *Int. J. Electr. Power Energy Syst.* **2023**, *147*, 108812. [CrossRef]
69. Paula, A.D.O.; Meneguette, R.I.; Giuntini, F.T.; Peixoto, M.L.; Gonçalves, V.P.; Rocha Filho, G.P. STRAYER: A Smart Grid adapted automation architecture against cyberattacks. *J. Inf. Secur. Appl.* **2022**, *67*, 103195. [CrossRef]
70. Zielinska, A.; Skowron, M.; Bien, A. The concept of the blockchain technology model use to settle the charging process of an electric vehicle. In Proceedings of the 2019 Applications of Electromagnetics in Modern Engineering and Medicine (PTZE), Janow Podlaski, Poland, 9–12 June 2019; pp. 271–274. [CrossRef]
71. Kong, P.Y. Autonomous Robot-Like Mobile Chargers for Electric Vehicles at Public Parking Facilities. *IEEE Trans. Smart Grid* **2019**, *10*, 5952–5963. [CrossRef]
72. Zhou, X.; Sun, Y.; Tu, F.; Su, H. Intelligent Substation Online Monitoring System Based on Block Chain Technology. In Proceedings of the 2022 IEEE International Conference on Mechatronics and Automation (ICMA), Guilin, China, 7–10 August 2022; pp. 18–23. [CrossRef]
73. Ma, Y.; Qiu, J.; Sun, X.; Tao, Y. A Multi-Stage Information Protection Scheme for CDA-Based Energy Trading Market in Smart Grids. *IEEE Trans. Smart Grid* **2022**, *13*, 2305–2317. [CrossRef]
74. Akkad, A.; Wills, G.; Rezazadeh, A. An information security model for an IoT-enabled Smart Grid in the Saudi energy sector. *Comput. Electr. Eng.* **2023**, *105*, 108491. [CrossRef]
75. Park, K.; Moon, I. Multi-agent deep reinforcement learning approach for EV charging scheduling in a smart grid. *Appl. Energy* **2022**, *328*, 120111. [CrossRef]

76. Ullah, Z.; Mokryani, G.; Khan, B.; Khan, I.; Mehmood, C.A.; Ali, S.M. Smart Grid Block-Chain (BC) Conceptual Framework: Bi-Directional Models for Renewable Energy District and Utility. In Proceedings of the 2019 15th International Conference on Emerging Technologies (ICET), Peshawar, Pakistan, 2–3 December 2019; pp. 1–5. [CrossRef]
77. ElGhanam, E.; Sharf, H.; Odeh, Y.; Hassan, M.S.; Osman, A.H. On the Coordination of Charging Demand of Electric Vehicles in a Network of Dynamic Wireless Charging Systems. *IEEE Access* **2022**, *10*, 62879–62892. [CrossRef]
78. Guo, S.; Wang, K.; Li, Y.; Yan, Z.; Xia, X. An Electric Vehicle Charging Transaction Model Based on Blockchain. In Proceedings of the 2022 International Conference on Blockchain Technology and Information Security (ICBCTIS), Huaihua City, China, 15–17 July 2022; pp. 147–150. [CrossRef]
79. Chen, R.; Sun, H.; Zhang, B.; Xue, Z. Load and Frequency Regulation Power Market Trading Model based on Block Chain. In Proceedings of the 2021 IEEE Sustainable Power and Energy Conference (iSPEC), Nanjing, China, 25–27 December 2021; pp. 4031–4036. [CrossRef]
80. Refaat, S.S.; Ellabban, O.; Bayhan, S.; Abu-Rub, H.; Blaabjerg, F.; Begovic, M.M. *Smart Grid and Enabling Technologies*, 1st ed.; Wiley: New York, NY, USA, 2021. [CrossRef]
81. Esenogho, E.; Djouani, K.; Kurien, A.M. Integrating Artificial Intelligence Internet of Things and 5G for Next-Generation Smartgrid: A Survey of Trends Challenges and Prospect. *IEEE Access* **2022**, *10*, 4794–4831. [CrossRef]
82. Pilz, M.; Al-Fagih, L. Recent Advances in Local Energy Trading in the Smart Grid Based on Game-Theoretic Approaches. *IEEE Trans. Smart Grid* **2019**, *10*, 1363–1371. [CrossRef]
83. Zhang, B.; Jiang, C.; Yu, J.L.; Han, Z. A Contract Game for Direct Energy Trading in Smart Grid. *IEEE Trans. Smart Grid* **2018**, *9*, 2873–2884. [CrossRef]
84. Panthi, M.; Kanti Das, T. Intelligent Intrusion Detection Scheme for Smart Power-Grid Using Optimized Ensemble Learning on Selected Features. *Int. J. Crit. Infrastruct. Prot.* **2022**, *39*, 100567. [CrossRef]
85. Sivarajan, S.; Sundarsingh Jebaseelan, S.D. Efficient adaptive deep neural network model for securing demand side management in IoT enabled smart grid. *Renew. Energy Focus* **2022**, *42*, 277–284. [CrossRef]
86. Pop, C.B.; Cioara, T.; Anghel, I.; Antal, M.; Chifu, V.R.; Antal, C.; Salomie, I. Review of bio-inspired optimization applications in renewable-powered smart grids: Emerging population-based metaheuristics. *Energy Rep.* **2022**, *8*, 11769–11798. [CrossRef]
87. Zhang, L.; Jiang, C.; Pang, A. Black-box attacks and defense for DNN-based power quality classification in smart grid. *Energy Rep.* **2022**, *8*, 12203–12214. [CrossRef]
88. Karasu, S.; Altan, A.; Sarac, Z.; Hacioglu, R. Prediction of Bitcoin prices with machine learning methods using time series data. In Proceedings of the 2018 26th Signal Processing and Communications Applications Conference (SIU), Izmir, Turkey, 2–5 May 2018; pp. 1–4. [CrossRef]
89. Hossain, E.; Khan, I.; Un-Noor, F.; Sikander, S.S.; Sunny, S.H. Application of Big Data and Machine Learning in Smart Grid, and Associated Security Concerns: A Review. *IEEE Access* **2019**, *7*, 13960–13988. [CrossRef]
90. Diaba, S.Y.; Elmusrati, M. Proposed algorithm for smart grid DDoS detection based on deep learning. *Neural Netw.* **2022**, *159*, 175–184. [CrossRef]
91. Mostafa, N.; Ramadan, H.S.M.; Elfarouk, O. Renewable energy management in smart grids by using big data analytics and machine learning. *Mach. Learn. Appl.* **2022**, *9*, 100363. [CrossRef]
92. Haq, E.U.; Pei, C.; Zhang, R.; Jianjun, H.; Ahmad, F. Electricity-theft detection for smart grid security using smart meter data: A deep-CNN based approach. *Energy Rep.* **2023**, *9*, 634–643. [CrossRef]
93. Nakamoto, S. Bitcoin [WhitePaper]. Available online: https://www.ussc.gov/sites/default/files/pdf/training/annual-national-training-seminar/2018/Emerging_Tech_Bitcoin_Crypto.pdf (accessed on 5 April 2024).
94. Wang, Z.; Yang, L.; Wang, Q.; Liu, D.; Xu, Z.; Liu, S. ArtChain: Blockchain-Enabled Platform for Art Marketplace. In Proceedings of the 2019 IEEE International Conference on Blockchain, Atlanta, GA, USA, 14–17 July 2019; pp. 447–454. [CrossRef]
95. Orsini, L.; Kessler, S.; Wei, J.; Field, H. How the Brooklyn Microgrid and TransActive Grid are paving the way to next-gen energy markets. In *The Energy Internet*; Elsevier: Amsterdam, The Netherlands, 2019; pp. 223–239. [CrossRef]
96. Zia, M.F.; Benbouzid, M.; Elbouchikhi, E.; Muyeen, S.M.; Techato, K.; Guerrero, J.M. Microgrid Transactive Energy: Review, Architectures, Distributed Ledger Technologies, and Market Analysis. *IEEE Access* **2020**, *8*, 19410–19432. [CrossRef]
97. Ethereum. Available online: https://ethereum.org/en/ (accessed on 5 April 2024).
98. Nahar, P. What Are 51% Attacks in Cryptocurrencies? *Economic Times*. Available online: https://economictimes.indiatimes.com/markets/cryptocurrency/what-are-51-attacks-in-cryptocurrencies/articleshow/85802504.cms?from=mdr (accessed on 31 August 2021).
99. Colak, I.; Bayindir, R.; Sagiroglu, S. The Effects of the Smart Grid System on the National Grids. In Proceedings of the 2020 8th International Conference on Smart Grid (icSmartGrid), Paris, France, 17–19 June 2020; pp. 122–126. [CrossRef]
100. Increasing Time Granularity in Electricity Markets Innovation Landscape Brief. 2019. Available online: https://www.irena.org/-/media/Files/IRENA/Agency/Publication/2019/Feb/IRENA_Increasing_time_granularity_2019.pdf (accessed on 5 April 2024).
101. Zhao, C.; Han, D.; Li, C.; Wang, H. A Blockchain Consensus Mechanism to Optimize Reputation-Based Distributed Energy Trading in Urban Energy System. *IEEE Access* **2024**, *12*, 53698–53712. [CrossRef]
102. Calvagna, A.; Marotta, G.; Pappalardo, G.; Tramontana, E. A Blockchain-Based Real-Time Power Balancing Service for Trustless Renewable Energy Grids. *Future Internet* **2024**, *16*, 149. [CrossRef]

103. Jo, B.; Khan, R.; Lee, Y.-S. Hybrid Blockchain and Internet-of-Things Network for Underground Structure Health Monitoring. *Sensors* **2018**, *18*, 4268. [CrossRef] [PubMed]
104. Oughton, E.J.; Ralph, D.; Pant, R.; Leverett, E.; Copic, J.; Thacker, S.; Dada, R.; Ruffle, S.; Tuveson, M.; Hall, J.W. Stochastic Counterfactual Risk Analysis for the Vulnerability Assessment of Cyber-Physical Attacks on Electricity Distribution Infrastructure Networks. *Risk Anal.* **2019**, *39*, 2012–2031. [CrossRef] [PubMed]
105. Ahmad, T.; Zhang, D.; Huang, C.; Zhang, H.; Dai, N.; Song, Y.; Chen, H. Artificial intelligence in sustainable energy industry: Status Quo, challenges and opportunities. *J. Clean. Prod.* **2021**, *289*, 125834. [CrossRef]
106. Khalili, M.; Ali Dashtaki, M.; Nasab, M.A.; Reza Hanif, H.; Padmanaban, S.; Khan, B. Optimal instantaneous prediction of voltage instability due to transient faults in power networks taking into account the dynamic effect of generators. *Cogent Eng.* **2022**, *9*, 2072568. [CrossRef]
107. Cantillo-Luna, S.; Moreno-Chuquen, R.; Chamorro, H.R.; Sood, V.K.; Badsha, S.; Konstantinou, C. Blockchain for Distributed Energy Resources Management and Integration. *IEEE Access* **2022**, *10*, 68598–68617. [CrossRef]
108. Alamri, A.; AlKassem, A.; Draou, A. Composite Demand-Based Energy Storage Sizing for an Isolated Microgrid System. *Sustainability* **2023**, *15*, 1517. [CrossRef]
109. Ostrowski, J.; Anjos, M.F.; Vannelli, A. Tight Mixed Integer Linear Programming Formulations for the Unit Commitment Problem. *IEEE Trans. Power Syst.* **2012**, *27*, 39–46. [CrossRef]
110. Atakan, S.; Lulli, G.; Sen, S. A State Transition MIP Formulation for the Unit Commitment Problem. *IEEE Trans. Power Syst.* **2018**, *33*, 736–748. [CrossRef]
111. Baker, K.; Hug, G.; Li, X. Energy Storage Sizing Taking Into Account Forecast Uncertainties and Receding Horizon Operation. *IEEE Trans. Sustain. Energy* **2017**, *8*, 331–340. [CrossRef]
112. Fan, C.; Khazaei, H.; Musilek, P. BPET: A Unified Blockchain-Based Framework for Peer-to-Peer Energy Trading. *Future Internet* **2024**, *16*, 162. [CrossRef]
113. Nasir, T.; Bukhari, S.S.H.; Raza, S.; Munir, H.M.; Abrar, M.; Muqeet, H.A.U.; Bhatti, K.L.; Ro, J.S.; Masroor, R. Recent Challenges and Methodologies in Smart Grid Demand Side Management: State-of-the-Art Literature Review. *Math. Probl. Eng.* **2021**, *2021*, 5821301. [CrossRef]
114. Abid, A.H.; Hasan, A. An approach for demand side management of non-flexible load in academic buildings. In Proceedings of the 2018 15th International Conference on Smart Cities: Improving Quality of Life Using ICT & IoT (HONET-ICT), Islamabad, Pakistan, 8–10 October 2018; pp. 26–28. [CrossRef]
115. Bohlayer, M.; Fleschutz, M.; Braun, M.; Zottl, G. Demand Side Management and the Participation in Consecutive Energy Markets—A Multistage Stochastic Optimization Approach. In Proceedings of the 2018 15th International Conference on the European Energy Market (EEM), Lodz, Poland, 27–29 June 2018; pp. 1–5. [CrossRef]
116. Elyas, S.H.; Sadeghian, H.; Alwan, H.O.; Wang, Z. Optimized household demand management with local solar PV generation. In Proceedings of the 2017 North American Power Symposium (NAPS), Morgantown, WV, USA, 17–19 September 2017; pp. 1–6. [CrossRef]
117. Zhang, L.; Zhou, S.; An, J.; Kang, Q. Demand-Side Management Optimization in Electric Vehicles Battery Swapping Service. *IEEE Access* **2019**, *7*, 95224–95232. [CrossRef]
118. Naja, R.; Soni, A.; Carletti, C. Electric Vehicles Energy Management for Vehicle-to-Grid 6G-Based Smart Grid Networks. *J. Sens. Actuator Netw.* **2023**, *12*, 79. [CrossRef]
119. Global EV Outlook 2022 Securing Supplies for an Electric Future. Available online: https://iea.blob.core.windows.net/assets/ad8fb04c-4f75-42fc-973a-6e54c8a4449a/GlobalElectricVehicleOutlook2022.pdf (accessed on 23 April 2018).
120. Dulău, L.I.; Bică, D. Effects of Electric Vehicles on Power Networks. *Procedia Manuf.* **2020**, *46*, 370–377. [CrossRef]
121. Appasani, B.; Mishra, S.K.; Jha, A.V.; Mishra, S.K.; Enescu, F.M.; Sorlei, I.S.; Bîrleanu, F.G.; Takorabet, N.; Thounthong, P.; Bizon, N. Blockchain-Enabled Smart Grid Applications: Architecture, Challenges, and Solutions. *Sustainability* **2022**, *14*, 8801. [CrossRef]
122. Al-Saif, N.; Ahmad, R.W.; Salah, K.; Yaqoob, I.; Jayaraman, R.; Omar, M. Blockchain for Electric Vehicles Energy Trading: Requirements, Opportunities, and Challenges. *IEEE Access* **2021**, *9*, 156947–156961. [CrossRef]
123. Habib, G.; Sharma, S.; Ibrahim, S.; Ahmad, I.; Qureshi, S.; Ishfaq, M. Blockchain Technology: Benefits, Challenges, Applications, and Integration of Blockchain Technology with Cloud Computing. *Future Internet* **2022**, *14*, 341. [CrossRef]
124. Hardjono, T.; Lipton, A.; Pentland, A. Toward an Interoperability Architecture for Blockchain Autonomous Systems. *IEEE Trans. Eng. Manag.* **2020**, *67*, 1298–1309. [CrossRef]
125. Peng, K.; Li, M.; Huang, H.; Wang, C.; Wan, S.; Choo, K.-K.R. "Security Challenges and Opportunities for Smart Contracts in Internet of Things: A Survey. *IEEE Internet Things J.* **2021**, *8*, 12004–12020. [CrossRef]
126. Abdul Kader, M.O.; Akindeji, K.T.; Sharma, G. Application of PHEVs Influence on Frequency Regulation of a Two Area Power System. In Proceedings of the 2022 10th International Conference on Smart Grid (icSmartGrid), Istanbul, Turkey, 27–29 June 2022; pp. 23–28. [CrossRef]
127. Kennel, F.; Gorges, D.; Liu, S. Energy Management for Smart Grids With Electric Vehicles Based on Hierarchical MPC. *IEEE Trans. Ind. Inf.* **2013**, *9*, 1528–1537. [CrossRef]
128. Silva, B.N.; Khan, M.; Han, K. Futuristic Sustainable Energy Management in Smart Environments: A Review of Peak Load Shaving and Demand Response Strategies, Challenges, and Opportunities. *Sustainability* **2020**, *12*, 5561. [CrossRef]

129. Rigas, E.S.; Ramchurn, S.D.; Bassiliades, N. Managing Electric Vehicles in the Smart Grid Using Artificial Intelligence: A Survey. *IEEE Trans. Intell. Transport. Syst.* **2015**, *16*, 1619–1635. [CrossRef]
130. Shafique, K.; Khawaja, B.A.; Sabir, F.; Qazi, S.; Mustaqim, M. Internet of Things (IoT) for Next-Generation Smart Systems: A Review of Current Challenges, Future Trends and Prospects for Emerging 5G-IoT Scenarios. *IEEE Access* **2020**, *8*, 23022–23040. [CrossRef]
131. Chibi, N.T.; Oualhaj, O.A.; Fihri, W.F.; Ghazi, H.E. A Novel Approach Based on Machine Learning, Blockchain, and Decision Process for Securing Smart Grid. *IEEE Access* **2024**, *12*, 33190–33199. [CrossRef]
132. Wang, L.; Ding, Y.; Chen, Z.; Su, Z.; Zhuang, Y. Heuristic Algorithms for Heterogeneous and Multi-Trip Electric Vehicle Routing Problem with Pickup and Delivery. *World Electr. Veh. J.* **2024**, *15*, 69. [CrossRef]
133. Wang, Z.; Liu, L.; Ran, L. Research on Strategy of Large-Scale Electric Vehicle Charging Scheduling System in Smart Community. In Proceedings of the 2023 7th International Conference on Electronic Information Technology and Computer Engineering. ACM, Xiamen, China, 20–22 October 2023. [CrossRef]
134. Han, Y.L. Research on Charging Strategy of Electric Vehicles Considering Bilateral Benefits. In Proceedings of the 2023 6th International Conference on Big Data Technologies (ICBDT '23), Qingdao, China, 22–24 September 2023; pp. 206–210. [CrossRef]
135. Suanpang, P.; Jamjuntr, P. Optimizing Electric Vehicle Charging Recommendation in Smart Cities: A Multi-Agent Reinforcement Learning Approach. *World Electr. Veh. J.* **2024**, *15*, 67. [CrossRef]
136. Kurukuru, V.S.B.; Khan, M.A.; Singh, R. Electric Vehicle Charging/Discharging Models for Estimation of Load Profile in Grid Environments. *Electr. Power Compon. Syst.* **2022**, *51*, 279–295. [CrossRef]
137. Mohammed, A.; Saif, O.; Abo-Adma, M.; Fahmy, A.A. Strategies and sustainability in fast charging station deployment for electric vehicles. *Sci. Rep.* **2024**, *14*, 283. [CrossRef]
138. Hesami, S.; Vafaeipour, M.; De Cauwer, C.; Rombaut, E.; Vanhaverbeke, L.; Coosemans, T. Proactive Eco-Driving Control of an Autonomous Electric Vehicle in Presence of Signalized Intersections and Preceding Vehicles. In Proceedings of the 2023 IEEE Vehicle Power and Propulsion Conference (VPPC), Milan, Italy, 24–27 October 2023; pp. 1–6. [CrossRef]
139. Husam Shaheen, I.; Ghamgeen Rashed, I.; Yang, B.; Yang, J. Optimal electric vehicle charging and discharging scheduling using metaheuristic algorithms: V2G approach for cost reduction and grid support. *J. Energy Storage* **2024**, *90 Pt A*, 111816. [CrossRef]
140. Sultan, V.; Aryal, A.; Chang, H.; Kral, J. Integration of EVs into the Smart Grid: A Systematic Literature Review. *Energy Inform.* **2022**, *5*, 65. [CrossRef]
141. Boudmen, K.; Eddaoudi, Z.; Aarab, Z.; Rahmani, M.D. Electric vehicles, the future of transportation powered by machine learning: A brief review. *Energy Inf.* **2024**, *7*, 80. [CrossRef]
142. Bhatti, A.R.; Tamoor, M.; Liaqat, R.; Rasool, A.; Salam, Z.; Ali, A.; Sherefa, A. Electric vehicle charging stations and the employed energy management schemes: A classification based comparative survey. *Discov. Appl. Sci.* **2024**, *6*, 503. [CrossRef]
143. Amir, M.; Zaheeruddin, H.A.; Bakhsh, F.I.; Kurukuru, V.B.; Sedighizadeh, M. Intelligent energy management scheme-based coordinated control for reducing peak load in grid-connected photovoltaic-powered electric vehicle charging stations. *IET Gener. Trans. Distrib.* **2024**, *18*, 1205–1222. [CrossRef]
144. Rehman, S.; Mohammed, A.B.; Alhems, L.M.; Alsulaiman, F.A. Smart car parks with EV charging for academic campus. *Clean. Energy Syst.* **2024**, *8*, 100121. [CrossRef]

World Electric Vehicle Journal

MDPI

Systematic Review

Optimization and Trends in EV Charging Infrastructure: A PCA-Based Systematic Review

Javier Alexander Guerrero-Silva [1,*], Jorge Ivan Romero-Gelvez [1,*], Andrés Julián Aristizábal [1] and Sebastian Zapata [2]

1 Faculty of Natural Sciences and Engineering, Universidad de Bogotá Jorge Tadeo Lozano, Bogotá 110311, Colombia; andresj.aristizabalc@utadeo.edu.co
2 Escuela de Ingenieria y Ciencias Basicas, Universidad EIA, Envigado 055422, Colombia; sebastian.zapata23@eia.edu.co
* Correspondence: javiera.guerreros@utadeo.edu.co (J.A.G.-S.); jorgei.romerog@utadeo.edu.co (J.I.R.-G.)

Abstract: The development of a robust and efficient electric vehicle (EV) charging infrastructure is essential for accelerating the transition to sustainable transportation. This systematic review analyzes recent research on EV charging network planning, with a particular focus on optimization techniques, machine learning applications, and sustainability integration. Using bibliometric methods and Principal Component Analysis (PCA), we identify key thematic clusters, including smart grid integration, strategic station placement, renewable energy integration, and public policy impacts. This study reveals a growing trend toward hybrid models that combine artificial intelligence and optimization methods to address challenges such as grid constraints, range anxiety, and economic feasibility. We provide a taxonomy of computational approaches—ranging from classical optimization to deep reinforcement learning—and synthesize practical insights for researchers, policymakers, and urban planners. The findings highlight the critical role of coordinated strategies and data-driven tools in designing scalable and resilient EV charging infrastructures, and point to future research directions involving intelligent, adaptive, and sustainable charging solutions.

Keywords: electric vehicle charging stations; charging infrastructure optimization; sustainable mobility; smart grid integration; machine learning; strategic placement algorithms; vehicle-to-Grid (V2G); urban mobility planning; bibliometric analysis; principal component analysis (PCA)

1. Introduction

The successful deployment of electric vehicle (EV) charging infrastructure is essential for promoting widespread EV adoption and reducing transportation emissions [1,2]. Several factors influence the optimal placement of charging stations, including geographic attributes, land use, and construction costs [3,4]. Researchers have applied various optimization algorithms to address placement and scheduling challenges [5], while market development requires a balance between public and private sector roles [6]. Charging infrastructure planning must consider key factors such as range anxiety, charging duration, and user convenience [7]. Different categories of charging stations, defined by location and technology, further complicate decision-making [8]. The integration of renewable energy sources and vehicle-to-grid (V2G) technology represents a growing trend in the sector [1].

However, increasing EV density raises concerns about grid reliability, necessitating effective countermeasures [2]. Despite advancements, several challenges remain in deploying EV charging infrastructure, including strategic location selection, grid integration, and economic feasibility [3]. Various modeling approaches focus on user behavior, route optimization, and destination-centric strategies [4]. Studies indicate that stakeholders tend to prioritize transportation-related features over land use considerations when determining charging station placement [7]. The development of fast-charging infrastructure along highways is critical, yet its deployment across EU countries remains inconsistent [6]. Optimization strategies emphasize maximizing station utilization while minimizing costs [8]. Government policies, standardization efforts, and the integration of renewable energy sources play pivotal roles in infrastructure expansion [5].

The global shift towards electric vehicles (EVs) has gained momentum due to environmental concerns and the commitment to reducing greenhouse gas emissions. EVs offer significant benefits such as a lower carbon footprint and enhanced energy efficiency [9,10]. However, the rapid increase in EV adoption presents challenges related to the development and optimization of charging infrastructures, including the strategic placement of charging stations and effective charging management [11,12]. Efficiently locating charging stations is crucial for improving accessibility, alleviating range anxiety, and minimizing electrical grid overloads [1,13]. Furthermore, optimizing the charging process is essential for maximizing energy efficiency and reducing costs for users and operators [14,15]. The electric vehicle market has experienced exponential growth, with EV sales reaching 10 million units in 2022, representing 14% of all new car sales. This growth is predominantly seen in China, Europe, and the United States, with China leading the market with a 60% share [16]. The anticipated increase in EV sales to 14 million units by the end of 2023 underscores the urgent need for a robust charging infrastructure to support this growing fleet [17,18]. As cities strive to decarbonize their transportation systems, addressing these infrastructure and environmental challenges through innovative solutions becomes imperative [19,20]. Emerging technologies, such as artificial intelligence (AI) and machine learning (ML), offer promising solutions to optimize the planning and operation of charging networks. These technologies enable predictive modeling of charging demands, dynamic energy management, and integration of renewable energy sources. Additionally, recent advancements in bibliometric analysis provide insights into research trends and highlight knowledge gaps, supporting the development of innovative strategies for sustainable transportation systems. Current projections estimate that EV sales could reach 14 million units globally by 2023, fueled by strong policy support in countries like China, various European nations, and the United States [16]. Despite these encouraging numbers, significant gaps remain in the literature and practical implementation regarding how to optimally design, locate, and manage charging networks [17,18]. Emerging approaches in artificial intelligence (AI) and machine learning (ML) show promise for forecasting demand and dynamically optimizing resource allocation [19,20], yet real-world adoption lags behind theoretical advancements. In this context, the objectives of the present work are threefold. First, it aims to identify and synthesize the primary methodologies used in EV charging infrastructure planning, with an emphasis on classification and bibliometric techniques for mapping research trends. Second, it seeks to highlight key challenges, such as the integration of renewables, cost minimization, and location optimization, while discussing how advanced algorithms (e.g., machine learning and metaheuristics) address these issues. Finally, it aims to formulate recommendations for policymakers, researchers, and industry stakeholders to enhance the sustainability, reliability, and scalability of EV charging networks. The findings also underscore the critical role of collaboration among academic, industrial, and governmental entities in achieving a seamless transition to sustainable mobility.

As shown in Figure 1, the global electric vehicle (EV) market has experienced significant growth between 2020 and 2023, with total sales increasing from approximately 4 million to over 14 million units. By 2024, sales effectively rose to 17 million units, marking a continued upward trajectory. This expansion reflects the combined impact of technological advancements, supportive policies, and shifting consumer preferences toward sustainable mobility. Regionally, China exhibits the highest market share growth, surpassing 40% in 2023, driven by strong government incentives, extensive charging infrastructure, and a well-established EV manufacturing ecosystem. Europe follows with a steady increase, reaching over 25% market penetration, largely influenced by stringent emission regulations and ambitious electrification policies. In contrast, the United States shows a more gradual adoption, with market share exceeding 10% by 2023, likely due to infrastructural and policy variations across states. These trends highlight the uneven global transition to electric mobility, underscoring the need for continued policy support, infrastructure expansion, and technological innovations to ensure a widespread and sustainable adoption of EVs. The development of an adequate and efficient charging station network is crucial for sustaining the growth of electric vehicles and ensuring their effective integration into the transportation network [21]. The strategic location of these stations is essential for improving accessibility, reducing range anxiety, and minimizing overloads on the electrical grid. Additionally, optimizing the charging process is vital for maximizing energy efficiency and minimizing costs for both users and operators.

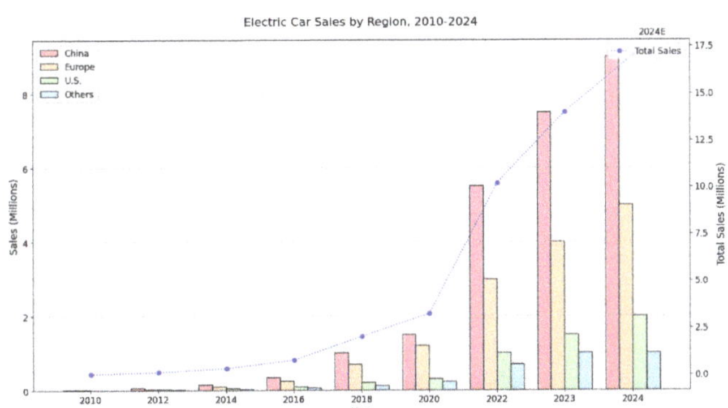

Figure 1. Global electric vehicle sales and market share in key regions (2020–2023). Source: [16].

1.1. Charging Networks and Smart Charging Strategies

The expansion and optimization of EV charging networks are crucial to support the growing adoption of electric vehicles. These networks must provide reliable and efficient charging options to alleviate range anxiety and promote widespread adoption [22]. Charging infrastructure is categorized into Level 1, Level 2, and DC Fast Charging (Level 3), each addressing different user needs and charging times [23]. While China, Europe, and the United States lead in deployment, expanding these networks presents challenges such as space limitations, installation costs, and grid capacity [24]. Smart charging strategies are vital for optimizing energy use and minimizing grid impact. Techniques such as load balancing, peak shaving, and vehicle-to-grid (V2G) technology help manage charging demand and integrate renewable energy sources [25]. Alongside environmental benefits, adopting renewables can also exacerbate thermal management challenges. Rapid charge–discharge cycles in battery storage or increased power throughput from solar and wind can raise operating temperatures. Consequently, sustainable station designs require not only solutions for intermittent generation but also robust cooling or heat-dissipation measures to maintain the health of both storage units and EV batteries. However, as charging power ramps up, thermal management becomes a crucial consideration for both station hardware and vehicle batteries. High-power sessions can generate significant heat, potentially degrading battery performance and reducing charging station reliability over time. Effective ventilation or active cooling systems, along with robust thermal monitoring, help mitigate these risks and maintain safe operating conditions. Despite the benefits, implementing smart charging infrastructure requires advanced grid management systems and user acceptance. Future innovations, such as AI and machine learning, have the potential to enhance these strategies [26].

As shown in Figure 2, the global distribution of charging stations by type in 2023 reveals significant regional disparities. China leads with the highest number of charging stations, particularly in DC fast charging, which facilitates rapid recharging and aligns with the country's aggressive EV adoption policies.

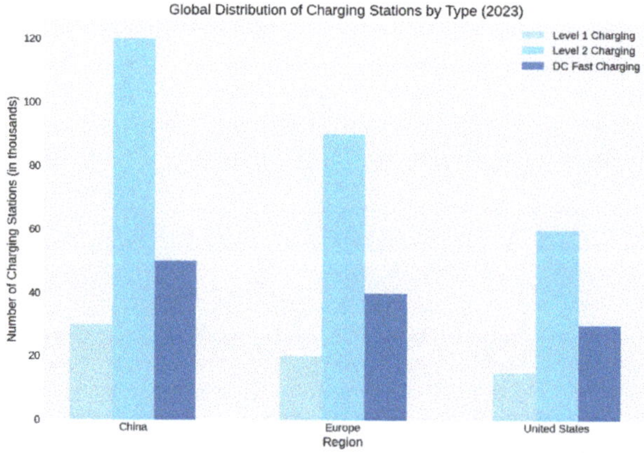

Figure 2. Global distribution of charging stations by type (2023). Source: [16].

The expansion of charging infrastructure and the adoption of smart charging strategies, as illustrated in Figure 3, are critical to supporting the growing transition to electric vehicles (EVs). Smart charging helps manage grid load effectively, ensuring stability and efficiency. Europe follows with a more balanced mix of Level 1, Level 2, and DC fast chargers, reflecting structured regulatory efforts and investment in public infrastructure. Meanwhile, the United States lags behind in overall charging station availability, with a stronger presence of Level 2 chargers, commonly used in residential and workplace settings. These differences highlight the varying levels of policy support, investment, and market maturity across regions. Beyond infrastructure, how EVs interact with the power grid is equally important. The right panel (Figure 3) demonstrates the impact of smart charging on grid demand throughout the day. Without smart charging (red curve), grid load experiences a sharp midday peak, likely driven by uncontrolled charging behavior overlapping with general electricity consumption. In contrast, when smart charging strategies are implemented (green curve), the demand curve flattens, reducing peak loads and distributing charging activity more evenly. This suggests that coordinated charging policies, such as demand response mechanisms and time-of-use pricing, can significantly enhance grid stability while ensuring efficient energy use.

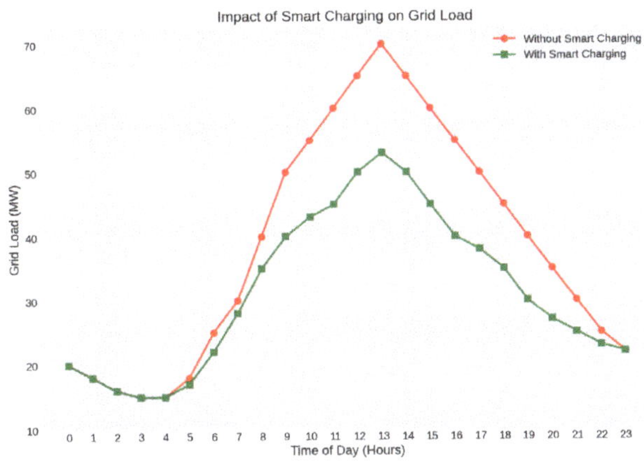

Figure 3. Impact of smart charging on grid load. Source: [16].

1.2. EV Charging Optimization and Location

Optimizing EV charging station placement involves using simulation techniques and mathematical models to address various challenges. Research highlights the use of Monte Carlo simulations for predicting peak hour energy demand and optimizing recharging points [27]. Optimization frameworks include a two-level approach that addresses both station location and driver behavior [28]. Techniques such as bat optimization and differential evolution algorithms are employed to enhance the efficiency and cost-effectiveness of charging stations [29,30]. Nature-inspired optimization algorithms and integration with renewable energy sources are explored to improve EV charging station planning and operation [31,32]. Studies show that integrating solar energy with optimized designs can significantly enhance the sustainability of EV charging infrastructure [33]. An additional factor in site selection is thermal load management, especially if charging stations support high-power or rapid-charging capabilities. Locations with high ambient temperatures or inadequate ventilation can amplify heat stress on station components and EV batteries alike. Incorporating thermal sensors and designing for efficient heat dissipation can extend system longevity and ensure user safety. The strategic placement of EV charging stations (EVCSs) is essential for user convenience and traffic network efficiency. Various optimization methods are used to address range anxiety and charging costs, including the OCEAN algorithm and queue theory models [34,35]. Simulation and optimization models, such as the simulated annealing immune Particle Swarm Optimization (SA-IPSO) algorithm, offer solutions for planning and improving EVCS deployment [36]. Research also focuses on EV travel patterns and the impact of station locations on vehicle-miles traveled (VMT) [37].

Artificial intelligence (AI) and machine learning (ML) have emerged as key technologies in optimizing electric vehicle (EV) integration within smart grids and dynamic energy networks. Recent studies emphasize their role in real-time energy management, predictive maintenance, load balancing, and infrastructure planning [38–47]. Machine learning algorithms, including deep learning, reinforcement learning, and decision trees, enhance demand forecasting and optimize EV charging schedules. Studies have demonstrated that clustering methods and predictive analytics can significantly improve the placement of charging stations, reducing congestion and optimizing energy use [43,45,47]. Reinforcement learning techniques, such as deep Q-networks, play a crucial role in coordinating charging processes by adjusting pricing models and grid demand responses. This dynamic approach to load shifting reduces grid strain and promotes sustainable energy usage [40,41].

Furthermore, AI-powered vehicle-to-grid (V2G) systems enable bidirectional energy flow, allowing EVs to return excess energy to the grid, stabilizing demand fluctuations [38,44]. Challenges remain in the large-scale adoption of AI in smart grids, including regulatory constraints, interoperability between charging infrastructures, and security concerns regarding data privacy and cyber threats [39,46]. Future research should focus on the integration of AI with decentralized blockchain networks to enhance transparency and security in energy transactions, as well as the development of predictive control systems for better renewable energy integration [42,44,45]. Addressing these challenges will facilitate the seamless implementation of AI-enhanced smart grid strategies, leading to more efficient, resilient, and sustainable EV charging infrastructures [38–47].

2. Materials and Methods

In this study, we employed a multi-step bibliometric analysis to identify and classify thematic areas of research. The process involves data collection, cleaning, and normalization, followed by keyword extraction and co-occurrence analysis. A Principal Component Analysis (PCA) was then applied to identify clusters of closely related topics. Finally, we conducted an influence analysis to determine the most impactful authors and references.

As shown in Figure 4, the workflow is divided into three main parts. The first part (blue box) encompasses the core bibliometric steps: data gathering from scientific databases, relevance filtering, data cleaning, and subsequent keyword extraction. The second part (green box) focuses on generating a keyword network map and reducing dimensionality through PCA, which facilitates the identification of thematic clusters. The third element (red box) addresses the analysis of key authors and references, allowing us to determine the most influential contributors within each research domain. This structured methodology ensures a comprehensive and reproducible bibliometric analysis, serving as a foundation for deeper insights into the evolving landscape of the field.

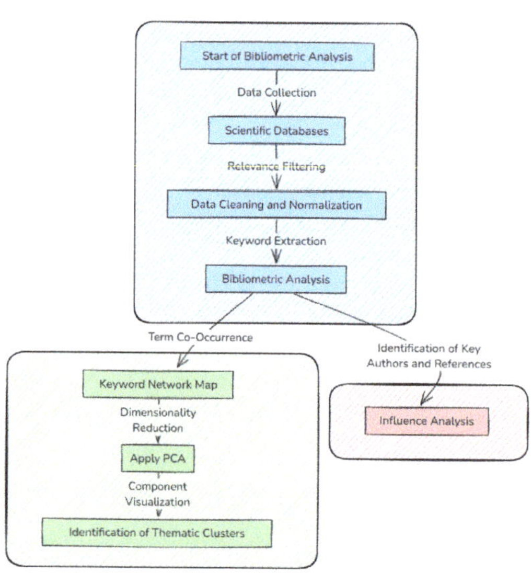

Figure 4. Overview of the bibliometric methodology. Colors are used exclusively to differentiate each phase of the process.

2.1. Data Preparation and Co-Word Analysis

The analysis began with the collection and preparation of data, focusing on the titles of research articles related to electric vehicle (EV) charging infrastructure. The titles were sourced from a comprehensive database, ensuring that the dataset was representative of the current research landscape.

- Article Sources and Selection Criteria: Research articles were gathered from major academic databases, including Web of Science, Scopus, and IEEE Xplore, covering the period from 2010 to 2023. Only full-length articles and peer-reviewed conference proceedings written in English were considered. To ensure relevance, initial keywords such as `(“electric vehicle” OR “EV”) AND (“charging station” OR “charging infrastructure”) AND (optimization OR planning OR location)` were used. The search results were then filtered by reviewing abstracts and, when necessary, full texts; studies explicitly focused on EV charging strategies, locations, or operational planning were retained. Any duplicates found across databases were removed, resulting in a final corpus that captured both historical and emerging research trends in EV charging infrastructure.
- Tokenization and Cleaning: Each title was first tokenized, breaking down the text into individual words (tokens). To enhance the relevance of the analysis, common stopwords (such as "and", "the", "of", etc.) were removed, along with any non-informative words that did not contribute to the thematic content. This step was crucial to isolate the core terms that reflect the primary focus of each research article. Where applicable, lemmatization or stemming was employed to consolidate variations of the same word (e.g., "optimization", "optimizing").
- Vectorization: Following the cleaning process, the tokenized titles were transformed into a term–document matrix using the *CountVectorizer* technique. This matrix quantifies the frequency of each term across all titles, providing a structured representation of the data that is suitable for further analysis. The resulting matrix captures the presence and significance of specific terms within the entire dataset, enabling a detailed exploration of thematic connections.
- Building the Co-Occurrence Matrix: To uncover the relationships between terms, a co-occurrence matrix was constructed. This matrix captures the frequency with which specific terms appear together within the same titles, revealing patterns of association that are indicative of underlying themes. By calculating cosine similarity between term vectors, the co-occurrence matrix serves as the foundation for identifying thematic clusters within the research corpus.

  The significance of co-occurrence analysis lies in its ability to highlight how different concepts are interconnected within the literature. Terms that frequently co-occur are likely to represent related concepts or themes, which can be grouped together to form a coherent cluster. This approach is particularly effective in uncovering the structure of research fields where multiple interrelated topics are explored.
- Visualization of Co-Word Analysis: The co-word analysis was visualized using a network graph, where each node represents a term, and edges between nodes indicate co-occurrences within the same titles. The size of the nodes is proportional to the frequency of the term, while the thickness of the edges corresponds to the strength of the co-occurrence relationship. The graph was generated using a spring layout algorithm, which positions the nodes in a way that emphasizes the clustering of related terms. The visualization in Figure 5 clearly delineates the thematic clusters within the research corpus. These clusters represent groups of terms that frequently appear together, indicating common topics or themes that dominate the literature.

The graph provides a visual representation of the key areas of focus within the field of EV charging infrastructure, facilitating an understanding of how different research themes are interconnected.

- Transitioning to PCA for Thematic Clustering: After constructing the co-occurrence matrix, it was further analyzed using Principal Component Analysis (PCA) in a subsequent step. By reducing the dimensionality of the co-occurrence data, PCA helped isolate the dominant axes of variation and revealed distinct thematic groupings. This process enabled the creation of simplified 2D or 3D plots (e.g., the 3D galaxy graph) that clustered articles according to prevalent research topics. Thus, the outputs of co-word analysis seamlessly informed the PCA-based classification, ensuring that the discovered themes were both data-driven and interpretable.

The resulting network structure, depicted in Figure 5, illustrates the co-word relationships derived from research article titles. Each node represents a distinct term, while the edges indicate co-occurrence links between words, revealing underlying thematic associations in the dataset. The dense connectivity in the graph suggests a highly interrelated research landscape, where frequently co-occurring terms form clusters corresponding to major research themes.

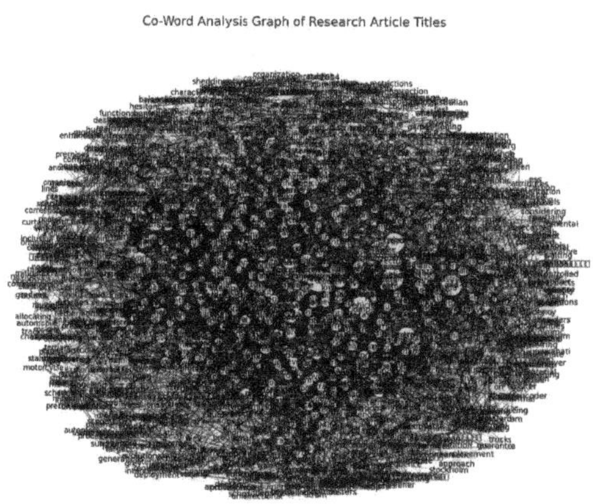

Figure 5. Co-word analysis graph of research article titles, highlighting the density of recurring terms across the dataset.

The distribution of nodes and edges highlights the presence of central terms that serve as thematic anchors, connecting multiple subfields. Highly connected nodes indicate key concepts that appear across various studies, potentially signaling fundamental research topics or emerging trends in the field. Peripheral nodes, on the other hand, may correspond to more specialized or niche areas of investigation.

By applying clustering techniques to this co-word network, it is possible to extract meaningful topic structures and identify dominant research directions. This visualization not only facilitates an intuitive understanding of the relationships between key concepts but also provides a foundation for deeper quantitative analyses, such as community detection and centrality measures, to further refine the thematic mapping of the research domain.

2.2. Principal Component Analysis

Classification methods are fundamental in the thematic analysis of scientific articles, enabling the identification and organization of underlying topics within large datasets. According to Goodfellow et al. [48], deep learning techniques, particularly neural networks, have revolutionized classification tasks by learning hierarchical representations of data. Hastie et al. [49] highlight the significance of statistical learning approaches, such as support vector machines and decision trees, in improving classification accuracy and interpretability. By integrating these advanced methodologies, we can enhance the classification of scientific articles, leading to more efficient information retrieval and knowledge discovery in extensive research corpora. Using PCA in this case is advantageous because it reduces the dimensionality of the scientific text data, simplifying the dataset into a manageable number of principal components without significant loss of information. This facilitates thematic analysis by eliminating redundancies and noise, improving the efficiency and accuracy of subsequent classification methods applied afterward. In this study, we employed a systematic approach to identify and visualize thematic clusters within the body of research on electric vehicle (EV) charging infrastructure. The methodology involved several key steps, including data preparation, co-word analysis, Principal Component Analysis (PCA), and clustering, culminating in a 3D galaxy graph visualization. This process allowed us to identify distinct thematic groups, each representing a different aspect of the research landscape.

Principal Component Analysis (PCA) is particularly advantageous in this context over other techniques. PCA reduces the dimensionality of large datasets by transforming the original variables into a smaller set of uncorrelated variables called principal components. Mathematically, given the dataset $\mathbf{X} \in \mathbb{R}^{n \times p}$, where n is the number of observations and p is the number of variables, PCA seeks to find a set of orthogonal vectors $\{\mathbf{w}_1, \mathbf{w}_2, \ldots, \mathbf{w}_k\}$ that maximize the variance captured in the data. This is achieved by solving the eigenvalue decomposition of the covariance matrix $\mathbf{C} = \frac{1}{n-1}\mathbf{X}^\top \mathbf{X}$:

$$\mathbf{C}\mathbf{w}_i = \lambda_i \mathbf{w}_i, \quad (1)$$

where λ_i are the eigenvalues representing the variance explained by each principal component, and $\mathbf{w}_i$ are the corresponding eigenvectors. By projecting the original data onto these eigenvectors, we obtain the following:

$$\mathbf{Z} = \mathbf{X}\mathbf{W}, \quad (2)$$

where $\mathbf{W} = [\mathbf{w}_1, \mathbf{w}_2, \ldots, \mathbf{w}_k]$. We obtain the transformed data in the reduced dimensional space. PCA does not require labeled data and effectively handles high-dimensional textual data from scientific articles, simplifying the dataset without significant loss of information. Unlike supervised classification methods, PCA's unsupervised nature makes it suitable for exploratory thematic analysis. Additionally, PCA can enhance the performance of subsequent classification algorithms by eliminating noise and redundancy, leading to more accurate and interpretable results.

As illustrated in Figure 6, the thematic clustering and visualization process begins with a structured data preparation phase, followed by co-word analysis techniques to uncover underlying patterns in textual data. The initial step involves *data preparation*, where raw text is collected, preprocessed, and formatted to ensure consistency. This includes removing stopwords, correcting misspellings, and standardizing terms to enhance analytical accuracy. Once the data is cleaned, the next step is *tokenization and cleaning*, where text is segmented into individual tokens—words or phrases—allowing for granular analysis. Additional preprocessing steps such as stemming and lemmatization can be applied to reduce words to their base forms, ensuring that semantically similar terms are grouped effectively. Following this, the *vectorization* process converts textual data into numerical representations, typically using frequency-based methods (e.g., TF-IDF) or distributed representations such as word embeddings. This transformation enables further computational analysis by making textual patterns more accessible to statistical and machine learning algorithms.

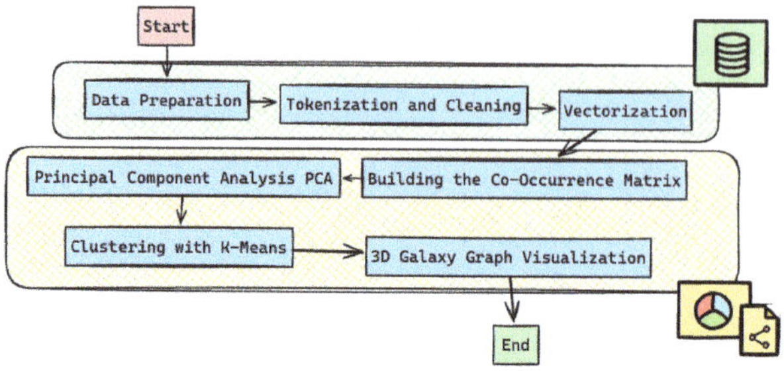

Figure 6. Thematic clustering and visualization.

To construct meaningful relationships between terms, a *co-occurrence matrix* is generated, capturing the frequency with which words appear together in textual contexts. This matrix serves as the foundation for subsequent dimensionality reduction and clustering techniques, helping to identify latent structures within the dataset.

3. Findings from Bibliometric and Cluster Analysis

In this section, the bibliometric analysis is presented in a segmented manner. Initially, it demonstrates an exponential growth in publications in this research area since 2010, along with an evaluation of the 20 most frequent authors and the most relevant keywords associated with the field. The objective is to illustrate publication trends, identify prominent authors and key terms, and synthesize the most influential research articles. Subsequently, a thematic clustering is performed to visualize research trends, in which Principal Component Analysis (PCA) is applied to delineate thematic clusterings and elucidate trends related to the topic.

3.1. Bibliometric and Content Analysis

This section provides an overview of the bibliometric trends and content analysis within the research on EV charging infrastructure.

As shown in Figure 7, the number of publications on EV charging infrastructure has increased significantly over the past two decades, with a sharp rise after 2010. This trend reflects growing global interest in sustainable transportation solutions, policy support for electric mobility, and advances in charging technology. The acceleration in research output aligns with the broader push for decarbonization and the integration of electric vehicles into power grids.

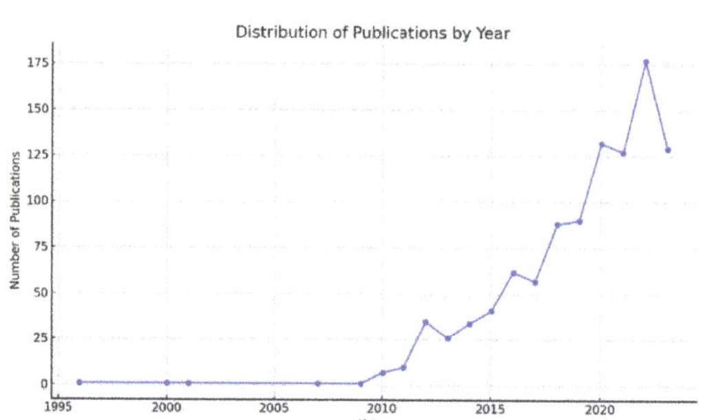

Figure 7. Distribution of publications by year.

A closer look at the leading contributors to this field, presented in Figure 8, reveals a concentration of publications among a small group of researchers. These authors, many of whom are affiliated with major research institutions and industry collaborations, play a central role in advancing charging strategies, optimization techniques (see Appendix A), and grid management solutions. The presence of multiple authors with overlapping research areas suggests the formation of specialized research clusters focused on addressing the technical and infrastructural challenges of large-scale EV adoption.

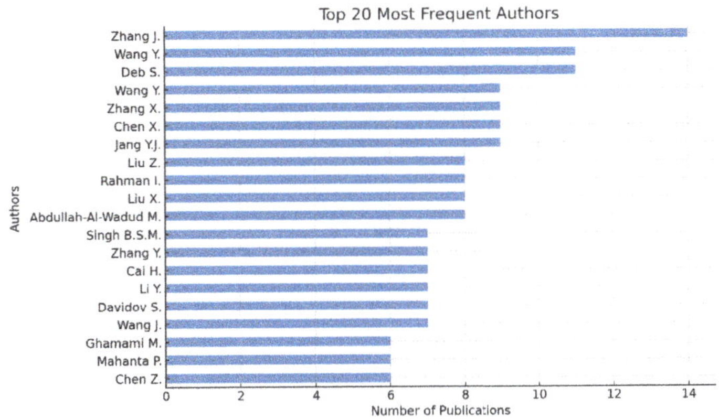

Figure 8. Top 20 most frequent authors.

The thematic focus of this body of research is further highlighted in Figure 9, which displays the most frequently occurring terms in EV charging literature. Keywords such as *charging (batteries)*, *electric vehicles*, and *charging station* emphasize the technical aspects of charging infrastructure, while others, including *optimization, vehicle-to-grid, and electric power distribution*, indicate a strong emphasis on improving efficiency and grid integration. The recurrent presence of terms related to *costs* and *investments* suggests a growing awareness of the economic implications of widespread EV deployment, reinforcing the need for scalable and cost-effective charging solutions.

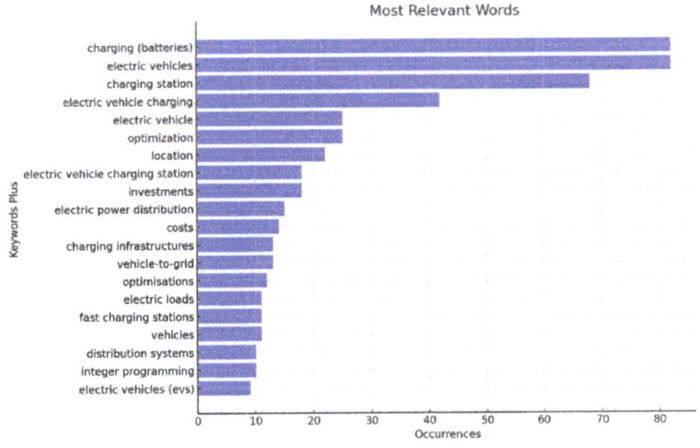

Figure 9. Most relevant words.

- Publication Trends: Analysis of annual publication growth reflects increasing research interest and advancements in EV infrastructure.

- Prominent Authors: Identifies key contributors and their impact on the field through publication counts and citation metrics.
- Keyword Analysis: Highlights core themes and research areas based on frequently used keywords.
- Synthesis of Influential Research Articles: Compiles significant contributions, including studies on fast-charging station planning, demand analysis, and optimization models for EVCS location and capacity [50–54].

While our bibliometric overview captures publication trends and thematic foci, several recent studies exemplify how advanced techniques can uncover critical research gaps. For instance, Zhang et al. [50] leveraged a combination of co-word analysis and citation network mapping to pinpoint insufficient attention to the economic trade-offs of EV station deployment. Similarly, Jia et al. [52] employed topic modeling on a large corpus of EV-related articles, revealing that the integration of renewable sources into charging networks remains underrepresented in high-impact journals. These examples illustrate how systematic bibliometric methodologies not only map established themes but also highlight emerging areas—such as addressing thermal loads in high-power charging or scaling vehicle-to-grid services—that warrant deeper investigation. By integrating similar approaches, our current review can provide policymakers and researchers with a clearer sense of where additional work is needed, thus guiding evidence-based improvements to EV charging infrastructure.

3.2. Thematic Clustering

- Principal Component Analysis (PCA): Given the high dimensionality of the co-occurrence matrix, we applied Principal Component Analysis (PCA) to reduce the data to a lower-dimensional space. PCA is a statistical technique that identifies the principal components accounting for the most variance within the data. By focusing on the first three principal components, we retained the most significant dimensions for visualization, facilitating a clearer understanding of the thematic structure within the dataset.
- Clustering with K-Means: To further refine our analysis, we employed K-Means clustering, a widely used unsupervised machine learning algorithm. The titles were grouped into five clusters, with each cluster representing a distinct thematic area. The clustering was based on the PCA components, which ensured that titles with similar thematic content were grouped together. The resulting clusters provided a meaningful categorization of the research topics, allowing for a nuanced exploration of the thematic diversity within the field.
- 3D Galaxy Graph Visualization: The final step involved visualizing the thematic clusters using a 3D galaxy graph. Each point in the graph represents a research paper, and the position of each point is determined by its PCA components. The colors of the points correspond to different clusters, highlighting the distinct thematic groups identified by the analysis.

3.3. 3D Galaxy Graph Visualization

The thematic clustering results are visualized through both 2D and 3D projections, providing complementary perspectives on the structure of research in EV charging infrastructure. Figure 10a presents a 2D PCA plot, where each point represents a research paper, and its position is determined by Principal Component Analysis. The color-coded clusters highlight distinct thematic groups, showing how certain topics form well-defined areas, while others exhibit a more dispersed distribution, suggesting interdisciplinary overlaps.

Expanding on this, Figure 10b represents the same thematic clusters in a 3D space, often referred to as a "galaxy graph". This visualization enhances the perception of relationships between clusters, making it possible to observe how thematic groups interact and how closely related certain research areas are. The spatial organization of the points reveals the degree of separation between clusters, with denser regions indicating strongly defined research themes and more scattered points reflecting broader, less specialized topics. By comparing both representations (Figure 10), it becomes evident that while PCA effectively reduces dimensionality for better interpretability, the 3D projection provides additional depth to explore inter-cluster relationships. These visualizations reinforce the structured yet evolving nature of EV charging research, capturing both established domains and emerging intersections that may shape future developments in the field.

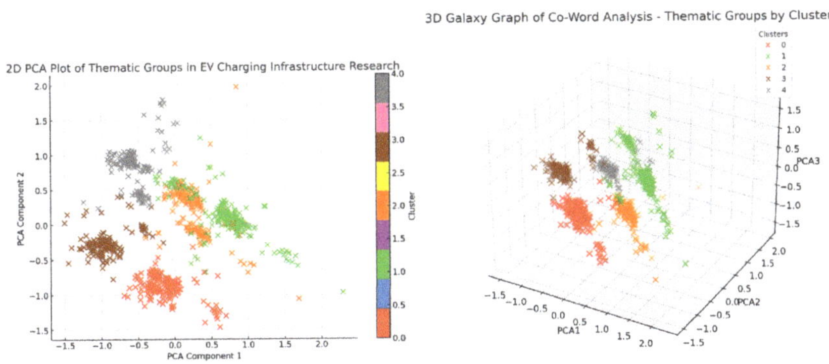

(**a**) (**b**)

Figure 10. Visualization of thematic groups using clustering techniques: (**a**) Two-dimensional PCA plot of thematic groups; (**b**) Three-dimensional galaxy graph of thematic groups. The colors indicate different thematic clusters identified through co-word analysis.

3.4. Thematic Groups Identified

The analysis revealed five key thematic clusters, each representing a different aspect of research in EV charging infrastructure:

- **Cluster 1—"Smart Grid and Communication Technologies":** This cluster includes research focused on the integration of EV infrastructure with smart grid technologies and communication systems. It encompasses topics related to energy management, communication protocols, and the optimization of grid resources. Research in this group provides methods—such as Genetic Algorithms or deep learning—to solve complex resource-allocation and scheduling problems. For planners, these techniques can significantly reduce costs and accelerate decision-making, whether it is selecting optimal station sites or anticipating high-demand hours. In practice, deploying algorithms from this cluster can help coordinate charging sessions to avoid grid overload, allocate power efficiently, and refine operational policies over time based on usage data.
- **Cluster 2—"Optimization Algorithms and Machine Learning":** This group is centered around the use of optimization techniques and machine learning for enhancing EV charging infrastructure. It covers a wide range of algorithms, including Genetic Algorithms, Particle Swarm Optimization, and computational methods for grid management. Research in this group provides methods—such as Genetic Algorithms or deep learning—to solve complex resource-allocation and scheduling problems.

For planners, these techniques can significantly reduce costs and accelerate decision-making, whether it is selecting optimal station sites or anticipating high-demand hours. In practice, deploying algorithms from this cluster can help coordinate charging sessions to avoid grid overload, allocate power efficiently, and refine operational policies over time based on usage data.

- **Cluster 3—"Sustainability and Renewable Energy Integration":** Papers in this cluster discuss the integration of renewable energy sources, such as solar and wind, with EV charging infrastructure. This group also addresses sustainability issues, including the reduction in carbon emissions and the efficient use of renewable energy in transportation. Studies within this cluster delve into the synergy between EVs and renewable energy sources, highlighting pathways to reduce carbon footprints. In infrastructure planning, this means accounting for local renewable resource availability (e.g., solar, wind) and ensuring that stations are designed to capture or store green energy. Planners can leverage this cluster's findings to achieve sustainability targets, comply with regulations (e.g., emissions limits), and potentially secure funding or incentives for environmentally friendly projects.
- **Cluster 4—"Urban Planning and Charging Station Placement":** This cluster emphasizes the strategic placement of charging stations within urban environments. It includes studies on traffic flow, urban planning, and the economic and environmental considerations of deploying charging infrastructure in cities. Papers in this cluster assess how traffic flow, population density, and land-use constraints affect charging station deployment. They address balancing user convenience (e.g., minimal detour) with cost-effectiveness and equity of service across different neighborhoods. Urban and regional planners can directly apply these insights to refine zoning laws, optimize station spacing, and ensure accessibility for diverse communities, thereby boosting EV adoption and alleviating range anxiety.
- **Cluster 5—"Policy, Standards, and Market Trends":** The final cluster focuses on policy-making, standardization, and market analysis related to EV charging infrastructure. It covers topics such as government policies, industry standards, and evolving market trends in electric mobility. Finally, this cluster reveals the crucial role played by policies, technical standards, and evolving market dynamics. In practical terms, planners must remain aware of regulations on safety, interoperability, and energy market structures. Findings from this cluster can guide legislative or financial mechanisms—such as subsidies, carbon taxes, or infrastructure rebates—that spur private-sector investment and consumer uptake of EV charging.

Table 1 summarizes some of the most prominent authors associated with each thematic cluster, outlining their main research contributions.

The clusters highlight distinct areas of expertise, ranging from smart grid technologies and optimization algorithms to sustainability, urban planning, and policy development. By identifying these key researchers and their focal points, the table offers a structured perspective on the intellectual landscape shaping the evolution of EV charging infrastructure.

Table 1. Prominent authors in each thematic cluster.

Cluster	Author(s)	Relevant Information and Contributions
Cluster 1	Zhang, Y.	Focuses on smart grid integration, communication protocols, and energy management strategies for EV infrastructure. Zhang's research includes the development of algorithms for optimizing grid resources in conjunction with EV charging demands.
"Smart Grid and Communication Technologies"	Wang, C.	Explores communication technologies for EV networks, including protocols for energy exchange and grid stabilization. Wang's work is instrumental in advancing smart grid technologies that support large-scale EV adoption.
Cluster 2	Li, J.	Specializes in the application of optimization algorithms, such as Genetic Algorithms and Particle Swarm Optimization, to enhance the efficiency of EV charging infrastructure. Li's research also covers machine learning techniques for predictive grid management.
"Optimization Algorithms and Machine Learning"	Chen, H.	Investigates the use of machine learning models to predict charging demands and optimize the deployment of EV chargers. Chen's contributions include developing algorithms that improve the performance and reliability of charging networks.
Cluster 3	Liu, X.	Focuses on the integration of renewable energy sources with EV infrastructure, emphasizing the sustainability aspects of EV adoption. Liu's research addresses the challenges of aligning renewable energy generation with EV charging needs.
"Sustainability and Renewable Energy Integration"	Huang, Y.	Investigates the environmental impacts of EV charging infrastructure, particularly in the context of reducing carbon emissions through the use of solar and wind energy. Huang's work contributes to the understanding of how renewable energy can be effectively integrated into transportation systems.
Cluster 4	Zhang, Z.	Examines the strategic placement of charging stations in urban environments, with a focus on optimizing traffic flow and minimizing environmental impact. Zhang's research provides insights into the economic and logistical challenges of deploying charging infrastructure in cities.
"Urban Planning and Charging Station Placement"	Li, Y.	Analyzes the interplay between urban planning and EV infrastructure development, offering models for efficient station placement that consider both economic and environmental factors. Li's work is critical for cities planning to expand their EV networks.
Cluster 5	Smith, J.	Focuses on policy-making and industry standards related to EV charging infrastructure. Smith's research includes the analysis of government policies that promote EV adoption and the development of international standards for charging technologies.
"Policy, Standards, and Market Trends"	Brown, T.	Specializes in market trends and economic analysis of the EV industry, with a particular emphasis on how policies influence market dynamics. Brown's work is valuable for understanding the broader economic implications of the transition to electric mobility.

4. Classical Optimization and Heuristics

The strategic placement of electric vehicle (EV) charging stations is essential to ensure the adoption of electric vehicles and maximize the efficiency of the charging network. Optimal placement not only improves accessibility for EV users but also helps in minimizing range anxiety and ensuring the efficient use of resources [13]. Several models and approaches have been developed to optimize the placement of charging stations. Flow-based models focus on traffic patterns and vehicle flow to determine the best locations for charging stations [24]. Node-based models, on the other hand, prioritize locations based on proximity to key points, such as highways or urban centers. Path-based and tour-based models consider the routes commonly taken by EV users and place charging stations along these paths to minimize detours [15]. Network equilibrium models aim to balance the distribution of charging stations across a network, ensuring that no single area is overburdened or under-served [55]. In practice, these models have been successfully implemented in various regions. For example, in Beijing, China, a node-based model was used to optimize the placement of fast-charging stations, resulting in a significant reduction in charging wait times and an increase in station utilization [37]. Similarly, in Europe, a flow-based model was used to place charging stations along major highways, ensuring that EV users could travel long distances without experiencing range anxiety [1]. Despite these successes, there are still challenges to be addressed in optimizing the placement of charging stations. Data availability and accuracy are critical for the success of these models, yet obtaining reliable data on traffic patterns, grid capacity, and user behavior can be difficult [56]. Additionally, urban planning constraints, such as space limitations and zoning regulations, can hinder the placement of charging stations in optimal locations. Future advancements in data analytics, machine learning, and artificial intelligence hold promise for improving the accuracy and efficiency of placement models, enabling more precise and dynamic optimization [26].

Determining the optimal placement of EV charging stations requires a balance between transportation demand and electrical grid capacity. Figure 11 illustrates a structured approach to this process, integrating key factors such as traffic flow data, grid capacity, and population density to guide infrastructure deployment.

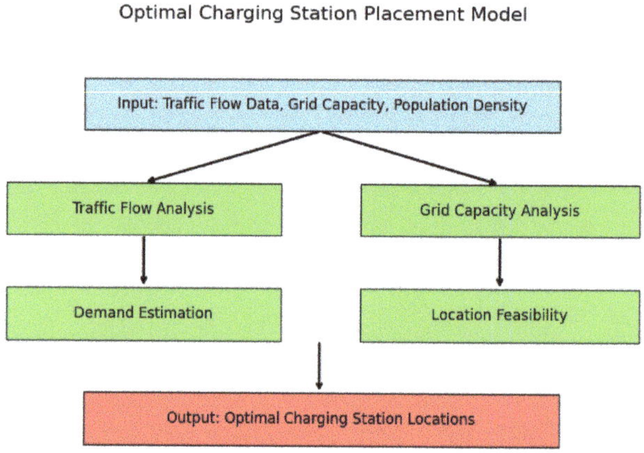

Figure 11. Optimal charging station placement model. Source: [16].

The model follows two parallel analyses. On one side, helps estimate demand by identifying high-mobility areas where charging stations are most needed. On the other hand, grid capacity analysis assesses whether the electrical infrastructure in these locations can support additional charging loads. Within these pathways, demand estimation refines the understanding of user needs, while location feasibility ensures that proposed sites align with grid limitations.

By integrating these components, the model generates optimal charging station locations, ensuring that new infrastructure is placed where it can be both highly utilized and sustainably supported by the grid. This approach enables more efficient network expansion, reducing congestion at existing stations and improving accessibility for EV users.

4.1. Objective Functions and Constraints of EV Charging Infrastructure

In the optimization of electric vehicle (EV) charging infrastructure, various optimization techniques are employed to achieve desired objectives, such as minimizing costs, maximizing accessibility, and optimizing energy efficiency. These techniques can be broadly classified into linear programming, nonlinear programming, dynamic programming, mixed-integer linear programming, and metaheuristic algorithms. The optimization of electric vehicle (EV) charging infrastructure is a critical aspect of ensuring the efficient and effective deployment of charging stations. The primary objectives of optimization include minimizing costs, maximizing accessibility, enhancing energy efficiency, and integrating renewable energy sources [14,57]. Various optimization techniques have been employed to achieve these goals, ranging from linear programming to advanced metaheuristic algorithms.

4.2. Recommendations from Relevant Authors

The techniques discussed range from exact approaches (LP, NLP, MILP, DP) to nature-inspired or approximate methods (PSO, GA, ACO). The selection of a suitable method depends on the following:

- Nature of the problem: Whether the relationships are linear or non-linear.
- Scale and complexity: The presence of binary variables, large-scale instances, or real-time requirements.
- Optimality and runtime needs: Whether an exact solution is necessary or an approximate solution in less time is sufficient.

The application of various optimization techniques in the planning and operation of electric vehicle (EV) charging infrastructure has been extensively studied. Several authors have proposed the use of specific methods depending on the context and objectives.

Linear Programming (LP): Linear programming is widely recommended for cost optimization problems where the relationships between variables can be modeled linearly. Liu et al. [31] suggest the use of LP for optimizing the placement of EV charging stations to minimize installation costs, especially in urban settings where resources are limited. Similarly, Cruz-Zambrano et al. [19] apply LP to optimize the allocation of charging stations while considering grid constraints.

Nonlinear Programming (NLP): Nonlinear programming is preferred when dealing with complex interactions within the electrical grid. Yang et al. [9] recommend NLP for scenarios where voltage stability and power flow are critical, such as in dense urban networks. The authors emphasize the importance of accurate modeling in capturing the nonlinear relationships inherent in these problems.

Dynamic Programming (DP): Dynamic programming is particularly effective for problems involving sequential decision-making, such as real-time charging management. Aghaei et al. [20] suggest the use of DP in scenarios where decisions need to be made over time, for instance, in managing charging loads during peak hours. The authors highlight the method's ability to optimize resource allocation dynamically.

Mixed-Integer Linear Programming (MILP): MILP is recommended for problems that involve both discrete and continuous decision variables. Funke et al. [15] propose using MILP for optimizing the placement and sizing of charging stations in large urban areas, balancing economic and environmental factors. Bilal et al. [57] also support the use of MILP for combining location and capacity decisions in a comprehensive framework.

Particle Swarm Optimization (PSO): PSO is advocated for scenarios with large solution spaces and multi-objective optimization requirements. Gao et al. [58] suggest PSO for optimizing EV charging schedules, noting its effectiveness in exploring complex, high-dimensional spaces. The authors recommend PSO for cases where the problem structure is difficult to define precisely, but approximate solutions are acceptable.

Genetic Algorithm (GA): Genetic Algorithms are highly recommended for problems requiring robust optimization across diverse scenarios. Ullah et al. [11] propose GA for optimizing the placement of charging stations in heterogeneous environments where multiple conflicting objectives must be balanced. The authors emphasize the method's adaptability to various problem types.

Ant Colony Optimization (ACO): ACO is often applied to routing problems and scenarios where the optimization process benefits from mimicking natural processes. Gao et al. [58] recommend ACO for optimizing the paths of EVs to minimize travel distances and energy consumption, particularly in dynamic environments where conditions change over time.

Selecting an appropriate optimization approach for EV charging infrastructure planning requires balancing computational efficiency, accuracy, and scalability. Table 2 provides a comparative overview of various optimization techniques, highlighting their key characteristics, advantages, and limitations.

Linear programming (LP) is well-suited for large-scale problems with linear relationships, offering high speed and accuracy, though it may struggle with complex interactions. Nonlinear programming (NLP) and dynamic programming (DP) extend these capabilities by handling more intricate constraints, but at the cost of increased computational complexity and sensitivity to initialization conditions. Mixed-integer linear programming (MILP) provides a balance by incorporating both discrete and continuous decision variables, making it applicable to real-world scenarios, though at the expense of computational efficiency.

Metaheuristic approaches, such as Particle Swarm Optimization (PSO), Genetic Algorithms (GAs), and Ant Colony Optimization (ACO), offer flexibility in solving multi-objective problems while requiring careful parameter tuning. These techniques are particularly useful in dynamic environments where exact solutions are computationally prohibitive. However, their convergence rates and robustness vary depending on the problem structure.

By comparing these approaches, it becomes evident that no single method universally outperforms the others. Instead, the choice of optimization technique should be guided by the specific constraints, problem scale, and computational resources available in a given EV infrastructure planning scenario. These recommendations provide a valuable framework for selecting the appropriate optimization technique based on the specific needs and context of the EV charging infrastructure project. By understanding the strengths and limitations of each method, planners and engineers can make informed decisions to enhance the efficiency and effectiveness of EV charging networks.

The various optimization techniques described above have been applied in numerous case studies related to the planning of the EV charging infrastructure. For example, a combination of mixed-integer linear programming (MILP) and Genetic Algorithms (GAs) has been used to optimize the placement of charging stations in urban areas, considering both economic and environmental factors [37]. Another case study implemented Particle Swarm Optimization (PSO) to manage real-time charging demands, significantly reducing peak load and improving grid stability [56].

Table 2. Comparison of optimization approaches for EV charging infrastructure.

Technique	Key Variables	Objectives	Pros	Cons	Speed	Accuracy	Other Measures
Linear Programming (LP)	Decision variables (e.g., number of stations, power allocation)	Minimize costs, optimize resource allocation	Efficient for large-scale problems, simple to implement	Limited to linear relationships, may not capture complex interactions	High	High for linear problems	Scalability: Good for large problems
Nonlinear Programming (NLP)	Decision variables, nonlinear constraints (e.g., voltage stability)	Optimize grid stability, minimize power losses	Handles complex interactions, applicable to realistic scenarios	Computationally intensive, may converge to local optima	Moderate to Low	High	Robustness: Sensitive to initial conditions
Dynamic Programming (DP)	Sequential decisions, state variables (e.g., charging times, load)	Optimize resource allocation over time, manage real-time demand	Suitable for time-dependent problems, handles sequential decisions	High computational cost, sensitive to dimensionality	Low	High	Scalability: Poor for large problems
Mixed-Integer Linear Programming (MILP)	Binary and continuous decision variables (e.g., station placement, power allocation)	Minimize costs, optimize station placement and sizing	Combines discrete and continuous optimization, applicable to real-world scenarios	Complex implementation, high computational demand	Moderate	High	Scalability: Moderate
Particle Swarm Optimization (PSO)	Particles' positions and velocities, global and local best positions	Find global optima, optimize multi-objective problems	Effective for large solution spaces, easy to implement	May converge prematurely, sensitive to parameter settings	Moderate to High	Moderate	Robustness: Sensitive to parameters
Genetic Algorithm (GA)	Chromosomes, fitness values, crossover and mutation rates	Evolve optimal solutions, optimize multi-objective problems	Robust against local minima, handles diverse problems	Requires careful tuning of parameters, can be slow to converge	Moderate to Low	Moderate	Robustness: High
Ant Colony Optimization (ACO)	Pheromone levels, heuristic information, ant paths	Find optimal paths, minimize travel distances	Effective for routing problems, adaptable to dynamic environments	Slow convergence, requires fine-tuning of pheromone evaporation rate	Low	Moderate to High	Scalability: Limited by convergence rate

5. Recent Advances in Machine Learning for EV Charging Optimization

The rapid proliferation of electric vehicles (EVs) has sparked substantial research on optimal placement and operation of charging infrastructure. Recent studies have leveraged advanced machine learning (ML) and deep learning (DL) techniques to address the spatial, temporal, and economic complexities involved. Alhussan et al. [59] and Shukla et al. [60] have proposed hybrid optimization models combining classification algorithms with nature-inspired metaheuristics for EV charging station siting. These models aim to minimize congestion and maximize accessibility in urban environments. Kalakanti and Rao [61], as well as Alsarhan et al. [62], analyzed charging patterns and consumption behaviors using supervised learning, contributing to more effective demand response. Similarly, Alanazi et al. [63] formulated scheduling algorithms for smart cities. Deb [64] provided a comprehensive review of ML applications in this domain, while Brahmachary and Ahmed [65] implemented ML-based forecasting for distribution network planning.

Spatiotemporal modeling is another growing area. Xi et al. [66] introduced graph neural networks to predict urban charging demand, and Heo and Chang [67] applied deep reinforcement learning for fast-charger location optimization. Elomiya et al. [68] integrated GIS, ML, and multi-criteria decision analysis to optimize service quality. Other novel contributions include Zhang's [69] use of clustering and ray casting techniques, Deb et al.'s [64] biologically inspired algorithms, and Roy and Law's [70] assessment of spatial disparities in infrastructure deployment. Open-source models have been proposed by Elattar et al. [71] to evaluate expansion needs, while Richard et al. [72] presented automated workflows for pattern discovery. Forecasting and demand prediction remain key areas, with Mohammad et al. [73] deploying ConvLSTM architectures and Rashid et al. [74] reviewing prediction techniques. Rao and Sodhi [75] introduced hybrid quantum-classical approaches, while Pourvaziri et al. [76] emphasized fairness and user preferences in planning. Sommer and Hossain [77] provided a literature review on V2G and en-route selection strategies using AI-based optimization. On the forecasting side, authors like Xi et al. [66] have introduced spatiotemporal graph models to predict urban EV charging demand, while Pourvaziri et al. [76] emphasize fairness and user preferences in multi-objective optimization. Contributions by Mousaei and Naderi [78] and Sabzi and Vajta [79] focus on site-specific predictions and user satisfaction analytics, enhancing adaptive planning frameworks. Collectively, these studies demonstrate the evolution of data-driven methods and hybrid frameworks—including GIS, ML, and MCDM tools—positioning them as foundational for next-generation EV infrastructure systems.

The surge in electric vehicle (EV) adoption has intensified the need for advanced methods to forecast charging demand, optimize station placement, and manage grid impacts dynamically. Machine learning (ML) has emerged as an essential tool, enabling data-driven approaches that effectively capture complex, nonlinear behaviors inherent in EV user patterns and energy consumption. Techniques including regression models, decision trees, clustering, and neural networks have been widely applied to predict temporal–spatial EV charging demand with high accuracy [43,45,46,54,59,62]. For instance, Alhussan et al. [59] implemented dynamic Grey-Lag Goose optimization allied with ML models to enhance charging station placement, yielding increased prediction accuracy while integrating user preferences for improved accessibility.

Beyond demand forecasting, the integration of deep reinforcement learning (DRL) techniques has substantially advanced real-time management and operational optimization for EV charging infrastructures. DRL algorithms such as Deep Q Networks (DQN), Soft Actor-Critic (SAC), and Advantage Actor-Critic (A2C) have been applied to enable adaptive, autonomous decision-making in complex environments characterized by demand uncertainty and network constraints [80–82]. These models facilitate dynamic allocation of charging stations, optimize load distribution, and improve overall energy efficiency. For example, Salehpour et al. [38] demonstrated that leveraging DRL for charging station placement yielded significant improvements in system throughput and power balancing compared to static heuristics. Moreover, multi-agent DRL frameworks coordinate between multiple charging points and grid operators, enabling scalable and decentralized solutions suitable for growing urban EV markets [41]. Such coordination ensures peak load management and enhances the operational resilience of charging infrastructures. However, despite these advances, challenges remain in ensuring model interpretability, addressing data privacy concerns, and integrating heterogeneous data sources for more robust performance [46,60]. Ongoing research is focusing on hybridizing ML models with classical optimization and control techniques to blend predictive capabilities with theoretical guarantees, thereby achieving more reliable and scalable solutions for future EV charging networks [59,83].

Machine learning (ML) and deep reinforcement learning (DRL) have become prominent tools in optimizing the allocation of electric vehicle (EV) charging infrastructure in data-rich, dynamic urban contexts. These techniques can forecast spatial demand, manage uncertainty, and adapt to real-time environmental fluctuations. Recent implementations of deep Q-networks have demonstrated reductions of up to 97% in waiting time and 497% improvements in benefits, indicating their efficacy in managing urban congestion and improving service quality [80].

Long short-term memory (LSTM) networks with attention mechanisms, paired with policy gradient methods, have also been used to minimize investment costs while outperforming classical models in both accuracy and computational efficiency [82]. Reinforcement learning frameworks adjust station deployment dynamically by responding to traffic and grid constraints, enhancing network responsiveness [81]. Clustering and predictive analytics have been applied for spatial optimization, and decision trees are used for estimating charging behavior under varying policy incentives [8,54]. Despite their advantages, these techniques face challenges in interpretability, data dependency, and regulatory constraints [6,60]. Machine learning and deep reinforcement learning have transformed EV charging infrastructure planning and management, from static forecasting to dynamic, online optimization, supporting more sustainable and user-centric mobility systems [38,43,59].

6. Hybrid Approaches

Hybrid computational strategies integrate the forecasting capabilities of ML with the decision-making robustness of optimization models, offering a comprehensive framework for EV infrastructure planning. For example, combining gradient boosting for demand estimation with deep Q-network reinforcement learning for charger placement yielded up to 30% better placement rewards and an R^2 score of 0.87 in demand prediction [84].

A more complex hybrid model—using mixed-integer nonlinear programming (MINLP), queueing theory, deep learning, and the Non-dominated Sorting Genetic Algorithm II (NSGA-II)—reduced deviation error from 13% to 1.6% and improved Pareto efficiency by 12.77% [83]. Other studies incorporated Gaussian Process Regression with capacitated maximal covering models, balancing spatial coverage with grid limitations [81]. Nature-inspired algorithms like Particle Swarm Optimization and Ant Colony Optimization are often embedded in hybrid frameworks to explore complex solution spaces efficiently [32,58].

These approaches have shown remarkable scalability and adaptability in heterogeneous urban environments, especially when addressing competing objectives such as cost, coverage, and energy efficiency [3,33]. However, integration complexity and data modeling requirements remain significant challenges that limit widespread deployment in policy-constrained environments [2,56].

7. Taxonomy and Structure of Computational Approaches for Electric Vehicle Charging Station Allocation

The proposed taxonomy Figure 12 for electric vehicle charging station allocation methodologies comprises three primary categories, each characterized by distinct computational paradigms and operational contexts.

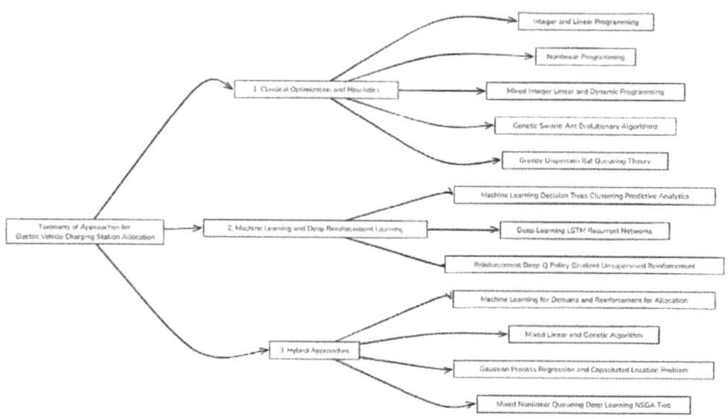

Figure 12. Taxonomy of computational approaches for EV charging station allocation.

The first category encompasses classical optimization and heuristic approaches. This includes methodologies such as integer and linear programming, nonlinear programming, mixed-integer linear programming, and dynamic programming, as well as a broad set of heuristic and metaheuristic techniques, including Genetic Algorithms, Particle Swarm Optimization, Ant Colony Optimization, and bat optimization. These methods are generally well-suited for problems with well-defined constraints and static configurations. Their primary strengths lie in their transparency and ease of interpretation, making them appropriate for small to medium-scale applications. However, they may face limitations in terms of scalability and adaptability when applied to dynamic or uncertain urban environments.

The second category involves machine learning (ML) and deep reinforcement learning (DRL) techniques, which are particularly effective in high-dimensional, data-intensive, and dynamically evolving settings. Approaches such as deep learning with long short-term memory (LSTM) networks and attention mechanisms, deep Q-networks, policy gradient methods, and unsupervised reinforcement learning are employed to support predictive modeling and adaptive decision-making. Complementary ML techniques, including decision trees, clustering, and predictive analytics, are frequently used to forecast demand patterns and optimize scheduling strategies. Despite their effectiveness, these methods often require substantial computational resources and large volumes of training data, and they may be less interpretable in operational contexts.

The third category comprises hybrid approaches that integrate machine learning models with optimization techniques. These methods are particularly advantageous in heterogeneous urban contexts that require both predictive accuracy and operational efficiency. Representative examples include the use of gradient boosting for demand forecasting in conjunction with reinforcement learning for station placement, as well as combinations of Gaussian process regression with capacitated maximal covering location models. More advanced configurations integrate mixed-integer nonlinear programming with queueing theory, deep learning architectures, and multi-objective optimization algorithms such as NSGA-II. These hybrid frameworks offer enhanced scalability and flexibility, though they often entail increased model complexity and integration challenges.

Collectively, these three categories provide a structured framework for addressing the multifaceted problem of EV charging station allocation. The selection of an appropriate computational strategy is contingent upon specific factors such as the complexity of the urban infrastructure, the availability and granularity of data, and the desired balance between interpretability, scalability, and adaptability. This taxonomy facilitates the alignment of methodological choices with practical and policy-driven objectives, contributing to the development of efficient, resilient, and sustainable transportation systems.

8. Discussion of Results and Relevant Measures in Cluster Analysis

The cluster analysis revealed five main themes in the research on electric vehicle (EV) charging infrastructure: Smart Grid and Communication Technologies, Optimization Algorithms and Machine Learning, Sustainability and Renewable Energy Integration, Urban Planning and Charging Station Placement, and Policy, Standards, and Market Trends. These clusters showcase the diversity of approaches and areas of interest in the field.

Cluster 1-Smart Grid and Communication Technologies:

This group emphasizes the integration of smart grid technologies and communication systems. The findings underscore the need for robust communication protocols to efficiently manage energy resources in charging networks. Innovations such as vehicle-to-grid (V2G) technology not only stabilize the grid but also promote the integration of renewable energy sources.

Cluster 2-Optimization Algorithms and Machine Learning: This cluster reflects the growing adoption of advanced algorithms and machine learning techniques. The studies analyzed demonstrate that these tools are crucial for solving complex optimization problems, such as strategic location of stations and prediction of demand patterns. This is particularly relevant in dense urban networks, where space and energy constraints limit traditional solutions.

Cluster 3-Sustainability and Renewable Energy Integration: This group highlights the role of sustainability in the design of charging infrastructure. The results indicate that the integration of renewable sources such as solar and wind energy significantly reduces carbon footprints. However, operational challenges, such as the intermittency of these sources, remain a key area for future research.

Practical Considerations for Renewable Integration

Although research on Sustainability and Renewable Energy Integration underscores the importance of leveraging solar, wind, and other clean energy sources for electric vehicle (EV) charging, real-world implementation faces multiple hurdles. Chief among these challenges is the inherent intermittency of renewables: unpredictable fluctuations in power generation can complicate load balancing for charging stations, particularly during peak demand periods or when weather conditions sharply reduce energy output. Mitigating strategies, such as on-site storage systems (e.g., lithium-ion batteries or hydrogen-based solutions), are often capital-intensive and demand sophisticated energy management algorithms to effectively smooth out supply–demand mismatches.

Additionally, policy barriers can limit large-scale integration of renewables into EV infrastructure. In many regions, energy regulations and market structures have not fully adapted to account for distributed generation or the two-way flow of electricity (e.g., with vehicle-to-grid technology). Fragmented permitting processes and the lack of unified incentives for renewable charging installations further complicate adoption. To address these issues, stronger policy frameworks that incorporate streamlined permitting, feed-in tariffs, tax credits, or carbon pricing may be required. In essence, achieving a robust synergy between clean energy and EV charging infrastructure demands not only technical innovations but also supportive legislation, long-term planning, and collaborative initiatives among energy providers, municipalities, and private stakeholders.

Cluster 4-Urban Planning and Charging Station Placement: This cluster examines strategies for positioning charging stations in urban environments. The analysis suggests that combining traffic models and mobility patterns with environmental and economic considerations is essential for reducing range anxiety and optimizing the use of existing infrastructure.

Cluster 5-Policy, Standards, and Market Trends: The results highlight the influence of public policies and standards on the development of charging networks. Studies suggest that well-designed governmental policies can catalyze infrastructure investments, while uniform standards facilitate interoperability and enhance user acceptance.

Several key metrics were employed to quantify and evaluate the results of the cluster analysis:

- **Co-occurrence Strength of Keywords:** Cosine similarity was used to measure the frequency of terms co-occurring in article titles and abstracts. Key terms such as *optimization*, *sustainability*, and *renewable energy* were central to forming thematic clusters.
- **Variance Explained by Principal Components:** Principal Component Analysis (PCA) revealed that the first three components explained over 85% of the variance in the data, validating the model's ability to capture significant thematic relationships.
- **Silhouette Coefficient in Clustering:** The average silhouette coefficient for the clusters was 0.72, indicating a clear separation between thematic groups and strong cohesion within each cluster.
- **Publication Trends:** A temporal analysis showed consistent growth in cluster-related themes, with significant increases in research on optimization algorithms and sustainability since 2019.
- **Bibliometric Impact of Key Authors:** Fifteen leading authors with high citation indices were identified within each cluster, reflecting their influence on research trends.

The analysis provides a structured view of research priorities in EV charging infrastructure, identifying mature and emerging areas. The employed metrics not only allow for quantitative evaluation but also highlight thematic connections that inform future research directions. These findings are critical for researchers, policymakers, and industry stakeholders aiming to optimize and expand EV charging networks sustainably.

This review has examined various optimization techniques for the planning and operation of electric vehicle (EV) charging infrastructure. Linear programming (LP), nonlinear programming (NLP), dynamic programming (DP), mixed-integer linear programming (MILP), and metaheuristic algorithms such as Particle Swarm Optimization (PSO), Genetic Algorithms (GA), and Ant Colony Optimization (ACO) have been explored. Each of these methods offers distinct advantages depending on the specific objectives and constraints involved in the problem [37,56,85].

The application of these optimization techniques in real-world scenarios can lead to significant improvements in the efficiency and effectiveness of EV charging infrastructure. By optimizing the placement and operation of charging stations, cities and regions can better accommodate the growing demand for electric vehicles, reduce costs, and enhance user satisfaction. For example, the use of MILP in urban planning can help identify the most cost-effective locations for new charging stations, while PSO can be employed to manage real-time charging loads, thereby reducing strain on the electrical grid [24,58].

Despite the potential benefits, several challenges and limitations must be addressed when applying these optimization techniques. One significant challenge is the availability and quality of data, which is crucial for accurate modeling and optimization. Additionally, the computational complexity of certain methods, particularly those involving large-scale nonlinear or dynamic systems, can be a barrier to their practical implementation. Furthermore, integrating different optimization techniques, such as combining LP with metaheuristics, requires careful consideration to ensure compatibility and effectiveness [55].

Future research should focus on addressing these challenges and exploring new avenues for optimization in EV infrastructure planning. Emerging technologies, such as artificial intelligence (AI) and machine learning (ML), offer promising opportunities to enhance the accuracy and efficiency of optimization models. For instance, AI-based algorithms could be used to predict future demand for charging stations, allowing for more proactive and dynamic planning. Additionally, the integration of renewable energy sources with smart charging infrastructure is an area that warrants further investigation, particularly in terms of optimizing energy consumption and reducing the environmental impact of EVs [26,58].

9. Conclusions and Recommendations

9.1. Conclusions

This study provided a systematic review of recent research on electric vehicle (EV) charging infrastructure, focusing on optimization techniques, machine learning applications, and thematic trends identified through bibliometric analysis and Principal Component Analysis (PCA). The main conclusions are as follows:

1. Thematic Clustering: Five dominant research clusters were identified: (i) smart grid and communication technologies, (ii) optimization algorithms and machine learning, (iii) sustainability and renewable energy integration, (iv) urban planning and charging station placement, and (v) policy, standards, and market trends. These clusters reveal a multidisciplinary approach to EV charging infrastructure planning and highlight both mature and emerging research directions.
2. Methodological Evolution: The field has evolved from classical optimization methods (e.g., linear, nonlinear, and mixed-integer programming) to advanced metaheuristics and hybrid models that incorporate machine learning and deep reinforcement learning. Hybrid approaches—integrating demand prediction with adaptive optimization—demonstrate superior performance in dynamic, data-rich urban environments.
3. Sustainability Emphasis: There is a growing emphasis on integrating renewable energy sources into charging networks, with recent studies addressing the technical, economic, and policy-related challenges. However, practical deployment still faces barriers due to intermittency, policy gaps, and infrastructure limitations.
4. Policy and Standard Impact: Policy support and standardization are critical for scaling up charging networks. Studies highlight the role of well-designed incentives, regulatory frameworks, and interoperability standards in accelerating infrastructure development and user adoption.
5. Limitations and Research Gaps: Although significant progress has been made, challenges remain in data availability, model interpretability, and the incorporation of real-world factors such as thermal management and regional diversity. Few studies address the operational impacts of ultra-fast charging or the unique needs of emerging markets.

9.2. Recommendations

Based on these findings, the following recommendations are proposed:

- For policymakers: Strengthen regulatory frameworks and incentive structures to support the expansion of charging networks and promote the integration of renewables. Encourage the adoption of interoperability standards to ensure system scalability and user convenience.
- For researchers: Prioritize hybrid and data-driven approaches that combine predictive modeling with robust optimization. Address gaps related to real-world operational factors, such as thermal management, and broaden the geographical scope of studies to include emerging markets.
- For urban planners and industry stakeholders: Use integrated decision-support tools that combine traffic, grid, and sustainability criteria for siting and managing charging stations. Foster cross-sector collaboration to align technological solutions with local needs and policy objectives.

These actions are essential to achieve a resilient, accessible, and sustainable electric vehicle charging infrastructure aligned with future mobility demands.

9.3. Limitations and Data Constraints

Although the use of co-word analysis, PCA, and clustering techniques provides valuable insights into thematic trends, several limitations must be acknowledged:

- PCA and Dimensional Reduction Biases: PCA projects high-dimensional data onto a smaller number of components, thereby simplifying the dataset. While this helps to reveal dominant themes, it may also result in the loss of nuanced or subtle patterns not captured by the principal components. This bias is especially notable if only the first two or three principal components are used for clustering. Researchers should interpret the clusters with the understanding that certain low-variance but thematically important dimensions could be suppressed.
- Clustering Algorithm Choices: Each clustering algorithm (e.g., K-Means, hierarchical clustering, DBSCAN) relies on assumptions regarding distance metrics, density thresholds, or cluster shapes. Using a single algorithm can, therefore, limit the scope of the analysis. Different methods might yield slightly different groupings, indicating that a multimethod approach could offer a more robust view of thematic structure.
- Dataset Size and Composition: The final corpus was drawn from specific databases and limited to English-language, peer-reviewed articles, potentially excluding regional studies or grey literature that might offer unique perspectives. Consequently, certain geographies or niche research areas may be underrepresented, which can bias conclusions about global trends. Moreover, smaller datasets or unbalanced distributions of topics can skew cluster formation, making some themes appear less significant than they truly are.
- Regional Focus and Generalizability: Most studies included in this dataset center on regions with established EV markets—such as North America, Europe, and parts of Asia. This emphasis may overlook challenges faced by emerging markets or rural areas, where infrastructural and socio-economic factors differ substantially. Consequently, policy recommendations or siting strategies inferred from these clusters might not generalize well to other contexts without further adaptation.
- Temporal Evolution of Research Themes: Although the dataset ranges from 2010 to 2023, EV technology is rapidly evolving. The distribution of topics may shift over time as infrastructure matures or new technologies emerge. Researchers should periodically

update their analyses to capture changes in focus—such as increasing attention to ultra-fast charging, wireless systems, or integrated renewable solutions.

Interestingly, few articles in the dataset address thermal considerations, despite the growing trend toward ultra-fast charging. Future research could explore more specific modeling of thermal loads, both in station hardware and within battery systems, as part of standard planning and optimization frameworks. Incorporating these factors would enhance the real-world applicability of solutions derived from the identified clusters.

Despite these constraints, the combined use of co-word analysis and PCA remains a valuable approach for synthesizing large bodies of research. Future work might involve integrating supplementary clustering approaches (e.g., hierarchical methods) or employing time-series bibliometrics to track the evolution of themes. In parallel, broadening the scope of databases or considering non-English sources could lead to more inclusive and globally representative insights.

Lastly, there is a notable gap in high-resolution thermal data. Many large-scale bibliometric or quantitative analyses do not capture crucial variables such as ambient temperature ranges, station cooling capacity, or battery heat-tolerance thresholds. Without such data, optimization studies may overestimate the performance and reliability of stations operating at higher power levels, especially in warmer climates.

Author Contributions: All authors contributed equally to the conceptualization, methodology, investigation, and writing of this manuscript. All authors have read and agreed to the published version of the manuscript.

Funding: This research was funded by Universidad de Bogotá Jorge Tadeo Lozano. The APC was funded by Universidad de Bogotá Jorge Tadeo Lozano.

Data Availability Statement: The data presented in this study are available on request from the corresponding author.

Conflicts of Interest: The authors declare no conflict of interest.

Appendix A. Optimization Techniques

Appendix A.1. Linear Programming (LP)

Linear programming (LP) is commonly employed to minimize costs and allocate resources when planning electric vehicle charging infrastructure, leveraging the linear nature of both the objective function and constraints [19,31]. In its standard formulation, LP seeks to minimize an objective function, typically expressed as $Z = \sum_{i=1}^{n} c_i x_i$, subject to a set of linear inequalities $\sum_{j=1}^{m} a_{ij} x_j \leq b_i$ and non-negativity constraints $x_i \geq 0$. One of the main strengths of this method lies in its solid theoretical foundation and the availability of efficient solvers—such as the simplex method—that can handle large-scale instances. Nevertheless, because LP is restricted to linear relationships among variables, it cannot directly capture nonlinear behaviors, potentially limiting its scope in more complex models.

Appendix A.2. Nonlinear Programming (NLP)

Nonlinear programming (NLP) is employed when the objective function or constraints include nonlinear relationships, as commonly encountered in voltage stability or power flow studies within electrical grids [9,11]. Its general formulation seeks to minimize a function $Z = f(x)$ subject to inequality and equality constraints $g_i(x) \leq 0,\ h_j(x) = 0$. By capturing more complex real-world interactions than linear methods, NLP can represent phenomena such as network losses or nonlinear cost structures. Among its main advantages are the ability to model realistic behaviors with flexible constraint types, albeit at the cost of higher computational complexity and potential convergence to local optima rather than a single global minimum.

Appendix A.3. Dynamic Programming (DP)

Dynamic programming (DP) is especially suitable for sequential decision-making, such as real-time charging management across multiple time periods [17,20]. Its standard approach defines a value function

$$V_t(s) = \max_{a_t}\left[R(s, a_t) \; + \; \gamma \sum_{s'} P(s' \mid s, a_t)\, V_{t+1}(s')\right] \tag{A1}$$

where each action a_t depends on the current state s and anticipated future outcomes via transition probabilities $P(s' \mid s, a_t)$. DP's principal strength lies in guaranteeing a globally optimal solution for discrete or discretized problems, yet the computational burden scales exponentially with the size of the state space (often termed the "curse of dimensionality"). Consequently, while DP can systematically handle multi-stage decisions, it may become infeasible for very large or continuous problems.

Appendix A.4. Mixed-Integer Linear Programming (MILP)

Mixed-integer linear programming (MILP) combines both continuous and binary decision variables in a single model, making it well-suited for decisions that involve facility siting (binary, e.g., whether to build a charging station) alongside resource allocation (continuous, e.g., power levels) [15,57]. Typically, the objective to be minimized is

$$Z = \sum_{i=1}^{n} c_i x_i \; + \; \sum_{j=1}^{m} d_j y_j \tag{A2}$$

subject to constraints of the form

$$\sum_{k=1}^{p} a_{ik} x_k \; + \; \sum_{l=1}^{q} b_{jl} y_l \; \leq \; r_i, \quad x_i \in \mathbb{R}^+, \quad y_j \in \{0, 1\}. \tag{A3}$$

The inclusion of integer variables increases the complexity compared to purely linear problems, demanding more sophisticated branch-and-bound or cutting-plane techniques. Despite this, powerful commercial and open-source solvers are available, enabling MILP to handle moderately large-scale problems where discrete location choices and continuous operational decisions must be optimized together.

Appendix A.5. Metaheuristics

Metaheuristic algorithms, such as Particle Swarm Optimization (PSO), Genetic Algorithms (GA), and Ant Colony Optimization (ACO), are particularly useful when the search space is large or when multiple conflicting objectives must be balanced [11,58]. Unlike exact methods, they iteratively explore potential solutions without requiring a strictly defined analytical formulation, offering flexibility and broad applicability even in situations with limited prior knowledge of the problem. Another key benefit is their ability to escape local optima in many cases, enhancing the likelihood of finding competitive solutions. However, they do not guarantee convergence to the global optimum and are highly sensitive to parameter settings such as population size or mutation rates, which can significantly influence their performance.

Appendix A.5.1. Particle Swarm Optimization (PSO)

Inspired by the social behavior of flocks of birds and schools of fish. A set of particles adjusts their positions and velocities according to individual and collective performance.

Algorithm A1: PSO Framework

Data: A population of particles with initial positions and velocities
for *each particle i* **do**
    Evaluate its fitness $f(x_i)$;
    Update personal best position p_i and global best g;
while *stopping criterion not met* **do**
    for *each particle i* **do**
        Update velocity v_i based on p_i and g;
        Move particle: $x_i \leftarrow x_i + v_i$;
        Evaluate the new fitness;
        Update p_i and g if improved;
return global best position g;

Appendix A.5.2. Genetic Algorithm (GA)

Based on natural selection and genetic principles. Solutions are encoded as chromosomes and selection, crossover, and mutation are applied to evolve the population.

Algorithm A2: GA Framework

Data: Initial population of chromosomes
while *stopping criterion not met* **do**
    Evaluate fitness of each chromosome;
    Select parents based on fitness;
    Apply crossover to generate offspring;
    Apply mutation with a certain probability;
    Replace current population with the new offspring;
return the best chromosome as the optimal solution;

Appendix A.5.3. Ant Colony Optimization (ACO)

Inspired by the pheromone trails ants use to find paths between their colony and food sources. Several ants construct solutions, and the best paths receive higher pheromone levels, guiding subsequent ants to promising solutions.

Algorithm A3: ACO Framework

Data: Graph whose nodes represent potential solutions
Initialize pheromone levels on all edges;
while *stopping criterion not met* **do**
    for *each ant* **do**
        Construct a path guided by pheromone and heuristic information;
        Evaluate the solution quality;
        Update pheromones along the path;
    Evaporate pheromones globally;
return best path found;

References

1. Pagany, R.; Ramirez Camargo, L.; Dorner, W. A review of spatial localization methodologies for the electric vehicle charging infrastructure. *Int. J. Sustain. Transp.* **2018**, *13*, 433–449. [CrossRef]
2. Metais, M.; Jouini, O.; Perez, Y.; Berrada, J.; Suomalainen, E. Too much or not enough? Planning electric vehicle charging infrastructure: A review of modeling options. *Renew. Sustain. Energy Rev.* **2022**, *153*, 111719. [CrossRef]
3. Majhi, R.C.; Ranjitkar, P.; Sheng, M.; Covic, G.A.; Wilson, D.J. A systematic review of charging infrastructure location problem for electric vehicles. *Transp. Rev.* **2020**, *41*, 432–455. [CrossRef]
4. Karolemeas, C.; Tsigdinos, S.; Tzouras, P.G.; Nikitas, A.; Bakogiannis, E. Determining Electric Vehicle Charging Station Location Suitability: A Qualitative Study of Greek Stakeholders Employing Thematic Analysis and Analytical Hierarchy Process. *Sustainability* **2021**, *13*, 2298. [CrossRef]
5. Yousuf, A.K.M.; Wang, Z.; Paranjape, R.; Tang, Y. An In-Depth Exploration of Electric Vehicle Charging Station Infrastructure: A Comprehensive Review of Challenges, Mitigation Approaches, and Optimization Strategies. *IEEE Access* **2024**, *12*, 51570–51589. [CrossRef]
6. Mastoi, M.S.; Zhuang, S.; Munir, H.M.; Haris, M.; Hassan, M.; Usman, M.; Bukhari, S.S.H.; Ro, J.S. An in-depth analysis of electric vehicle charging station infrastructure, policy implications, and future trends. *Energy Rep.* **2022**, *8*, 11504–11529. [CrossRef]
7. Szumska, E.M. Electric Vehicle Charging Infrastructure along Highways in the EU. *Energies* **2023**, *16*, 895. [CrossRef]
8. Chen, T.; Zhang, X.P.; Wang, J.; Li, J.; Wu, C.; Hu, M.; Bian, H. A Review on Electric Vehicle Charging Infrastructure Development in the UK. *J. Mod. Power Syst. Clean Energy* **2020**, *8*, 193–205. [CrossRef]
9. Yang, Z.; Guo, X.; Zhang, L. Optimal placement of electric vehicle charging stations: A review. *Renew. Sustain. Energy Rev.* **2017**, *80*, 509–519.
10. Sanguesa, J.; Torres-Sanz, V.; Garrido, P.; Martínez, F.; Marqués-Mateu, J. A review on electric vehicles: Technologies and challenges. *Smart Cities* **2021**, *311*, 127847. [CrossRef]
11. Ullah, I.; Zheng, J.; Jamal, A.; Zahid, M.; Almoshageh, M.; Safdar, M. Electric vehicles charging infrastructure planning: A review. *Int. J. Green Energy* **2024**, *21*, 1710–1728. [CrossRef]
12. Deb, S.; Tammi, K.; Kalita, K.; Mahanta, P. Impact of electric vehicle charging station load on distribution network. *Energies* **2018**, *11*, 178. [CrossRef]
13. Giménez-Gaydou, D.; García-Villalobos, J.; Zamora, I.; San Martín, J. Optimal location of battery electric vehicle charging stations in urban areas: A new approach. *Renew. Energy* **2016**, *88*, 616–629. [CrossRef]
14. Hosseini, S.; MirHassani, S. Optimal allocation of electric vehicle charging stations considering network constraints. *J. Power Sources* **2015**, *295*, 170–182.
15. Funke, S.; Nussler, S.; Storandt, S. Placement of fast charging stations for electric vehicles in urban areas. *Transp. Res. Part C Emerg. Technol.* **2016**, *65*, 97–110.
16. IEA. *Global Electric Vehicle Outlook 2023*; IEA: Paris, France, 2023.
17. Wang, Y.; Wang, Y. A review of optimal location and sizing for charging stations of electric vehicles. *Renew. Sustain. Energy Rev.* **2017**, *68*, 52–63.
18. Shamami, N.; Torabi, H.; Mazloum, J. Smart placement of charging stations for plug-in electric vehicles considering load demand uncertainties. *Energy* **2020**, *198*, 117308.
19. Cruz-Zambrano, M.; Corchero, C.; Heredia, F.; Pagés, E. Optimal location of fast charging stations in urban areas. *Energy* **2013**, *55*, 101–109.
20. Aghaei, J.; Shafie-Khah, M.; Catalão, J. Robust multi-objective model for simultaneous distributed generation and electric vehicle charging station planning. *Appl. Energy* **2016**, *165*, 517–530.
21. Tungom, C.; Niu, B.; Wang, H. Hierarchical framework for demand prediction and iterative optimization of EV charging network infrastructure under uncertainty with cost and quality-of-service consideration. *Expert Syst. Appl.* **2024**, *237*, 121761. [CrossRef]
22. Yilmaz, M.; Krein, P. Review of the impact of vehicle-to-grid technologies on distribution systems and utility interfaces. *IEEE Trans. Power Electron.* **2012**, *28*, 5673–5689. [CrossRef]
23. Yan, X.; Li, H.; Cui, J. Planning of Electric Vehicle charging station based on hierarchic genetic algorithm. In Proceedings of the 2014 IEEE Conference and Expo Transportation Electrification Asia-Pacific (ITEC Asia-Pacific), Beijing, China, 31 August–3 September 2014; pp. 1–5.
24. Kchaou-Boujelben, S.; Jouini, O. Charging station location problem: A comprehensive review on models and solution approaches. *Transp. Res. Part C Emerg. Technol.* **2021**, *125*, 102954. [CrossRef]
25. Pirouzi, K.; Shaker, H. Exploring potential storage-based flexibility gains of electric vehicles in smart distribution grids. *J. Energy Storage* **2022**, *40*, 104066. [CrossRef]

26. Wang, T.; Zhang, Y.; Li, M. Machine Learning-Based Electric Vehicle Charging Optimization: A Survey. In Proceedings of the 2020 IEEE International Conference on Communications, Control, and Computing Technologies for Smart Grids (SmartGridComm), Tempe, AZ, USA, 11–13 November 2020; pp. 1–7.
27. Elkasrawy, M.; Makeen, P.; Abdellatif, S.; Ghali, H. Optimizing electric vehicles station performance using AI-based decision maker algorithm. In Proceedings of the Emerging Topics in Artificial Intelligence 2020, SPIE Nanoscience + Engineering; Online, 24 August–4 September 2020; Volume 11469, p. 114691W. [CrossRef]
28. Tran, C.; Ngoduy, D.; Keyvan-Ekbatani, M.; Watling, D. Bi-level optimization for locating fast-charging stations in large-scale urban networks. In Proceedings of the 2020 Forum on Integrated and Sustainable Transportation Systems, FISTS, Delft, The Netherlands, 3–5 November 2020; pp. 205–210. [CrossRef]
29. Singh, K.; Mistry, K.; Patel, H. Optimal Placement of Electric Vehicle Charging Station and DG in a Distribution System for Loss Minimization. In Proceedings of the 2023 IEEE 3rd International Conference on Sustainable Energy and Future Electric Transportation, SeFet, Bhubaneswar, India, 9–12 August 2023. [CrossRef]
30. Zheng, Y.; Dong, Z.; Xu, Y.; Meng, K.; Zhao, J.; Qiu, J. Electric vehicle battery charging/swap stations in distribution systems: Comparison study and optimal planning. *IEEE Trans. Power Syst.* **2014**, *29*, 221–229. [CrossRef]
31. Liu, Z.; Wen, F.; Ledwich, G. Optimal planning of electric-vehicle charging stations in distribution systems. *IEEE Trans. Power Deliv.* **2013**, *28*, 102–110. [CrossRef]
32. Ahmad, F.; Al-Fagih, L.; Qadir, S.; Khalid, M. EV Charging Station Placement using Nature-Inspired Optimisation Algorithms. In Proceedings of the 2023 International Conference on Power, Instrumentation, Energy and Control, PIECON, Aligarh, India, 10–12 February 2023; p. 166. [CrossRef]
33. Clemente, S.; Escandon, W.; Iparraguirre-Villanueva, O.; Zapata-Paulini, J.; Cabanillas-Carbonell, M. Optimization of a Photovoltaic Station for Charging Electric Vehicles. *Lect. Notes Netw. Syst.* **2024**, *695*, 963–973. [CrossRef]
34. Xiong, Y.; Gan, J.; An, B.; Miao, C.; Bazzan, A. Optimal electric vehicle fast charging station placement based on game theoretical framework. *IEEE Trans. Intell. Transp. Syst.* **2018**, *19*, 2493–2504. [CrossRef]
35. Xiang, Y.; Liu, J.; Tang, S.; Zhou, H.; Dai, S.; Li, T.; Wang, X. A traffic flow based planning strategy for optimal siting and sizing of charging stations. In Proceedings of the Asia-Pacific Power and Energy Engineering Conference, APPEEC, Xi'an, China, 25–28 October 2016. [CrossRef]
36. Sun, J.; Che, Y.; Yang, T.; Zhang, J.; Cai, Y. Location and Capacity Determination Method of Electric Vehicle Charging Station Based on Simulated Annealing Immune Particle Swarm Optimization. *Energy Eng. J. Assoc. Energy Eng.* **2023**, *120*, 367–384. [CrossRef]
37. Shahraki, N.; Cai, H.; Turkay, M.; Xu, M. Optimal locations of electric public charging stations using real world vehicle travel patterns. *Transp. Res. Part D Transp. Environ.* **2015**, *41*, 165–176. [CrossRef]
38. Salehpour, M.J.; Hossain, M. Leveraging machine learning for efficient EV integration as mobile battery energy storage systems: Exploring strategic frameworks and incentives. *J. Energy Storage* **2024**, *92*, 112151. [CrossRef]
39. Khan, M.R.; Haider, Z.M.; Malik, F.H.; Almasoudi, F.M.; Alatawi, K.S.S.; Bhutta, M.S. A Comprehensive Review of Microgrid Energy Management Strategies Considering Electric Vehicles, Energy Storage Systems, and AI Techniques. *Processes* **2024**, *12*, 270. [CrossRef]
40. Mousaei, A.; Naderi, Y.; Bayram, I.S. Advancing State of Charge Management in Electric Vehicles With Machine Learning: A Technological Review. *IEEE Access* **2024**, *12*, 43255–43283. [CrossRef]
41. Maguluri, L.P.; Umasankar, A.; Vijendra Babu, D.; Anselin Nisha, A.S.; Prabhu, M.R.; Tilwani, S.A. Coordinating electric vehicle charging with multiagent deep Q-networks for smart grid load balancing. *Sustain. Comput. Inform. Syst.* **2024**, *43*, 100993. [CrossRef]
42. Dahiwale, P.V.; Rather, Z.H.; Mitra, I. A Comprehensive Review of Smart Charging Strategies for Electric Vehicles and Way Forward. *IEEE Trans. Intell. Transp. Syst.* **2024**, *25*, 10462–10482. [CrossRef]
43. Mazhar, T.; Asif, R.N.; Malik, M.A.; Nadeem, M.A.; Haq, I.; Iqbal, M.; Kamran, M.; Ashraf, S. Electric Vehicle Charging System in the Smart Grid Using Different Machine Learning Methods. *Sustainability* **2023**, *15*, 2603. [CrossRef]
44. Mololoth, V.K.; Saguna, S.; Åhlund, C. Blockchain and Machine Learning for Future Smart Grids: A Review. *Energies* **2023**, *16*, 528. [CrossRef]
45. Khan, M.A.; Saleh, A.M.; Waseem, M.; Sajjad, I.A. Artificial Intelligence Enabled Demand Response: Prospects and Challenges in Smart Grid Environment. *IEEE Access* **2023**, *11*, 1477–1505. [CrossRef]
46. Fescioglu-Unver, N.; Yıldız Aktaş, M. Electric vehicle charging service operations: A review of machine learning applications for infrastructure planning, control, pricing and routing. *Renew. Sustain. Energy Rev.* **2023**, *188*, 113873. [CrossRef]
47. Dharavat, N.; Golla, N.K.; Sudabattula, S.K.; Velamuri, S.; Kantipudi, M.V.V.P.; Kotb, H.; AboRas, K.M. Impact of plug-in electric vehicles on grid integration with distributed energy resources: A review. *Front. Energy Res.* **2023**, *10*, 1099890. [CrossRef]

48. Goodfellow, I.; Bengio, Y.; Courville, A. *Deep Learning*; MIT Press: Cambridge, MA, USA, 2016.
49. Hastie, T.; Tibshirani, R.; Friedman, J. *The Elements of Statistical Learning: Data Mining, Inference, and Prediction*; Springer: Berlin/Heidelberg, Germany, 2009.
50. Zhang, H.; Moura, S.; Hu, Z.; Song, Y. PEV Fast-Charging Station Siting and Sizing on Coupled Transportation and Power Networks. *IEEE Trans. Smart Grid* **2018**, *9*, 2595–2605. [CrossRef]
51. Jia, L.; Hu, Z.; Song, Y.; Zhan, K.; Ding, H. Planning of electric vehicle charging stations in highway network. *Dianli Xitong Zidonghua/Automation Electr. Power Syst.* **2015**, *39*, 82–89+102. [CrossRef]
52. Jia, L.; Hu, Z.; Liang, W.; Lang, W.; Song, Y. A novel approach for urban electric vehicle charging facility planning considering combination of slow and fast charging. In Proceedings of the POWERCON 2014–2014 International Conference on Power System Technology: Towards Green, Efficient and Smart Power System, Proceedings, Chengdu, China, 20–22 October 2014; pp. 3354–3360. [CrossRef]
53. Jia, L.; Hu, Z.; Song, Y.; Luo, Z. Optimal siting and sizing of electric vehicle charging stations. In Proceedings of the 2012 IEEE International Electric Vehicle Conference, IEVC, Greenville, SC, USA, 4–8 March 2012. [CrossRef]
54. Li, K.; Shao, C.; Hu, Z.; Shahidehpour, M. An MILP Method for Optimal Planning of Electric Vehicle Charging Stations in Coordinated Urban Power and Transportation Networks. *IEEE Trans. Power Syst.* **2023**, *38*, 5406–5419. [CrossRef]
55. Yi, Z.; Bauer, P. Optimization models for placement of an energy-aware electric vehicle charging infrastructure. *Transp. Res. Part E Logist. Transp. Rev.* **2016**, *91*, 173–191. [CrossRef]
56. Bilal, M.; Zeeshan, M. Electric vehicles in a smart grid: A comprehensive survey on optimal location of charging station. *IET Smart Grid* **2020**, *3*, 202–213. [CrossRef]
57. Bilal, M.; Ahmed, S.; Ullah, I. Optimal placement and sizing of electric vehicle charging stations in smart distribution systems. *Int. J. Electr. Power Energy Syst.* **2020**, *123*, 106253.
58. Gao, Y.; Zhang, Y.; Chen, S. Ant colony optimization-based hybrid algorithm for optimal charging station placement and sizing. *Energy* **2017**, *140*, 1186–1198.
59. Alhussan, A.A.; Khafaga, D.S.; El-Kenawy, E.S.; Eid, M.M.; Ibrahim, A. Urban Electric Vehicle Charging Station Placement Optimization with Graylag Goose Optimization Voting Classifier. *Comput. Mater. Contin.* **2024**, *80*, 1163–1177. [CrossRef]
60. Shukla, M.; Singh, D.K.; Yadav, A.; Singh, A.K.; Jumaa, N.A.; Mohammed, A. Optimizing Electric Vehicle Charging Station Placement in Urban Areas: A Data-Driven Approach. In Proceedings of the 2023 3rd International Conference on Advance Computing and Innovative Technologies in Engineering, ICACITE, Greater Noida, India, 12–13 May 2023; pp. 2835–2838. [CrossRef]
61. Kalakanti, A.K.; Rao, S. Charging Station Planning for Electric Vehicles. *Systems* **2022**, *10*, 6. [CrossRef]
62. Alsarhan, A.; Alnatsheh, A.; Aljaidi, M.; Makkawi, T.A.; Aljamal, M.; Alsarhan, T. Optimizing Electric Vehicle Charging Infrastructure through Machine Learning: A Study of Charging Patterns and Energy Consumption. *Int. J. Interact. Mob. Technol.* **2024**, *18*, 149–170. [CrossRef]
63. Alanazi, F.; Alshammari, T.O.; Azam, A. Optimal Charging Station Placement and Scheduling for Electric Vehicles in Smart Cities. *Sustainability* **2023**, *15*, 6030. [CrossRef]
64. Deb, S. Machine Learning for Solving Charging Infrastructure Planning: A Comprehensive Review. In Proceedings of the 5th International Conference on Smart Grid and Smart Cities, ICSGSC, Tokyo, Japan, 18–20 June 2021; pp. 16–22. [CrossRef]
65. Brahmachary, R.; Ahmed, I. Optimal Distribution Network Expansion and EV Charging Station Allocation based on Load Forecasted using Machine Learning. *IEEE Trans. Ind. Appl.* **2025**, 1–13. [CrossRef]
66. Xi, Z.; Guo, G.; Yang, X.; Sun, J.; Shi, X.; Xiao, X. Forecasting and managing urban EV charging demand with spatiotemporal graph models. *J. Renew. Sustain. Energy* **2025**, *17*, 0244546. [CrossRef]
67. Heo, J.; Chang, S. Optimal planning for electric vehicle fast charging stations placements in a city scale using an advantage actor-critic deep reinforcement learning and geospatial analysis. *Sustain. Cities Soc.* **2024**, *113*, 105567. [CrossRef]
68. Elomiya, A.; Křupka, J.; Jovčić, S.; Simic, V.; Švadlenka, L.; Pamucar, D. A hybrid suitability mapping model integrating GIS, machine learning, and multi-criteria decision analytics for optimizing service quality of electric vehicle charging stations. *Sustain. Cities Soc.* **2024**, *106*, 105397. [CrossRef]
69. Zhang, V.W. Optimal Deployment of Electric Vehicles Charging Network Using Point Clustering and Ray Casting. In Proceedings of the 32nd ACM SIGSPATIAL International Conference on Advances in Geographic Information Systems, ACM SIGSPATIAL, Atlanta, GA, USA, 1 November 2024; pp. 737–740. [CrossRef]
70. Roy, A.; Law, M. Examining spatial disparities in electric vehicle charging station placements using machine learning. *Sustain. Cities Soc.* **2022**, *83*, 103978. [CrossRef]

71. Elattar, H.; von Tüllenburg, F.; Karas, S.; Valdes, J. An Open-Source Model for Estimating the Need to Expansion in Local Charging Infrastructures. In Proceedings of the Communications in Computer and Information Science, Velizy, France, 28–30 September 2024; Volume 2107, pp. 69–91. [CrossRef]
72. Richard, R.; Cao, H.; Wachowicz, M. Discovering EV Recharging Patterns through an Automated Analytical Workflow. In Proceedings of the 2020 IEEE International Smart Cities Conference, ISC2, Piscataway, NJ, USA, 28 September–1 October 2020. [CrossRef]
73. Mohammad, F.; Kang, D.K.; Ahmed, M.A.; Kim, Y.C. Energy Demand Load Forecasting for Electric Vehicle Charging Stations Network Based on ConvLSTM and BiConvLSTM Architectures. *IEEE Access* **2023**, *11*, 67350–67369. [CrossRef]
74. Rashid, M.; Elfouly, T.; Chen, N. A Comprehensive Survey of Electric Vehicle Charging Demand Forecasting Techniques. *IEEE Open J. Veh. Technol.* **2024**, *5*, 1348–1373. [CrossRef]
75. Rao, P.U.; Sodhi, B. Hybrid quantum-classical solution for electric vehicle charger placement problem. *Soft Comput.* **2023**, *27*, 13347–13363. [CrossRef]
76. Pourvaziri, H.; Taghavi, M.; Sarhadi, H.; Afshari, H.; Azad, N. Multi-objective planning of electric vehicles charging stations by integrating drivers' preferences and fairness considerations: A case study in Halifax, Canada. *Comput. Ind. Eng.* **2025**, *201*, 110886. [CrossRef]
77. Sommer, C.; Hossain, M.J. Artificial Intelligence-driven optimization of V2G and charging point selection en-route: A systematic literature review. *Energy Convers. Manag. X* **2025**, *26*, 100978. [CrossRef]
78. Mousaei, A.; Naderi, Y. Predicting Optimal Placement of Electric Vehicle Charge Stations Using Machine Learning: A Case Study in Glasgow, UK. In Proceedings of the 2025 12th Iranian Conference on Renewable Energies and Distributed Generation, ICREDG, Toronto, ON, Canada, 27–28 December 2025. [CrossRef]
79. Sabzi, S.; Vajta, L. Machine Learning Based Electric Vehicle Drivers Charging Satisfaction Analysis and Prediction. In Proceedings of the 2024 IEEE Conference on Technologies for Sustainability, SusTech, Portland, OR, USA, 14–17 April 2024; pp. 383–389. [CrossRef]
80. von Wahl, L.; Tempelmeier, N.; São, A.; Demidova, E. Reinforcement Learning-based Placement of Charging Stations in Urban Road Networks. In *Proceedings of the 28th ACM SIGKDD Conference on Knowledge Discovery and Data Mining (KDD '22)*, Washington, DC, USA, 14–18 August 2022; Association for Computing Machinery: New York, NY, USA, 2022; pp. 3992–4000. [CrossRef]
81. Koley, C. (Ed). Proceedings of the 2022 IEEE 6th International Conference on Condition Assessment Techniques in Electrical Systems (CATCON 2022); Department of Electrical Engineering, National Institute of Technology, Durgapur, India, 17–19 December 2022; IEEE: Piscataway, NJ, USA, 2022. ISBN 978-1-6654-7381-1; CFP2220V-POD; 459p. Available online: https://www.proceedings.com/68473.html (accessed on 29 May 2025).
82. Zhao, Z.; Lee, C.; Ren, J.; Tsang, Y. Optimal EV Fast Charging Station Deployment Based on a Reinforcement Learning Framework. *IEEE Trans. Intell. Transp. Syst.* **2023**, *24*, 8053–8065. [CrossRef]
83. Pourvaziri, H.; Sarhadi, H.; Azad, N.; Afshari, H.; Taghavi, M. Planning of Electric Vehicle Charging Stations: An Integrated Deep Learning and Queueing Theory Approach. *Transp. Res. Part E Logist. Transp. Rev.* **2024**, *186*, 103568. [CrossRef]
84. Petratos, A.; Ting, A.; Padmanabhan, S.; Zhou, K.; Hageman, D.; Pisel, J.; Pyrcz, M. Optimal Placement of Public Electric Vehicle Charging Stations Using Deep Reinforcement Learning. *arXiv* **2021**, arXiv:2101.00000.
85. Deb, S.; Tammi, K.; Kalita, K.; Mahanta, P.K. Review of recent trends in charging infrastructure planning for electric vehicles. *Wiley Interdiscip. Rev. Energy Environ.* **2018**, *7*, e295. [CrossRef]

MDPI AG
Grosspeteranlage 5
4052 Basel
Switzerland
Tel.: +41 61 683 77 34

World Electric Vehicle Journal Editorial Office
E-mail: wevj@mdpi.com
www.mdpi.com/journal/wevj

www.ingramcontent.com/pod-product-compliance
Lightning Source LLC
Chambersburg PA
CBHW061844210926
51850CB00006B/964
* 9 7 8 3 7 2 5 8 7 6 8 3 9 *